Seismogenesis and **Earthquake Forecasting:** The Frank Evison Volume I

Edited by
Martha K. Savage
David A. Rhoades
Euan G. C. Smith
Matthew C. Gerstenberger
David Vere-Jones

Previously published in *Pure and Applied Geophysics*
(PAGEOPH), Volume 167, Nos. 6–7, 2010

Birkhäuser

Editors
Martha K. Savage
Institute of Geophysics
Victoria University of Wellington
Wellington
New Zealand
martha.savage@vuw.ac.nz

David A. Rhoades
GNS Science
Lower Hutt
New Zealand
d.rhoades@gns.cri.nz

Euan G.C. Smith
Institute of Geophysics
Victoria University of Wellington
Wellington
New Zealand
Euan.Smith@vuw.ac.nz

Matthew C. Gerstenberger
GNS Science
Lower Hutt
New Zealand
M.gerstenberger@gns.cri.nz

David Vere-Jones
Statistics Research Associates
PO Box 12649
Wellington
New Zealand
dvj@paradise.net.nz

ISBN 978-3-0346-0497-0 e-ISBN 978-3-0346-0498-7
DOI 10.1007/978-3-0346-0498-7

Library of Congress Control Number: 2010929575

Cover illustration: Composed of photographs of an experimental wall across the Alpine Fault, constructed in 1964 at Frank Evison's instigation, to detect any movement on the fault. Photo sources: Geophysics Division, Department of Scientific and Industrial Research (1964) and David Evison (2008).

Printed on acid-free paper

Springer Basel AG is part of Springer Science+Business Media

www.birkhauser-science.com

Contents

Pure Appl. Geophys. 167 (2010), 619–621

DOI 10.1007/s00024-010-0140-y

Pure and Applied Geophysics

Introduction

DAVID RHOADES,[1] MARTHA SAVAGE,[2] EUAN SMITH,[2] MATT GERSTENBERGER,[1] and DAVID VERE-JONES[2]

Professor Frank Evison, OBE, FRSNZ, (20.3.1922–25.01.2005). Inaugural Professor of Geophysics, Victoria University of Wellington, 1967–1988. Inaugural Director, Institute of Geophysics, Victoria University of Wellington, 1971–1988. Professor Emeritus, Victoria University of Wellington, 1988–2005. Photo credit: Robert Cross, VUW Image Services

This special issue is an augmented collection of papers originating from the Evison Symposium on Seismogenesis and Earthquake Forecasting held in Wellington, New Zealand, in February 2008. There are two volumes in the issue. The second volume is published in Pure and Applied Geophysics, vol. 167, Nos. 8/9, 2010. Here we set the scene, describe the papers within Volume I, and list the reviewers who helped with papers in either volume. There is a biography of Frank Evison and a list of his publications at the end of the introduction to Volume 1.

Seismogenesis and earthquake forecasting were the main focus of Frank Evison's research in the latter part of his life. Frank was persuaded that the Earth signals its preparation for large earthquakes well in advance of their occurrence, and that if one can read these signals, then warnings can be given and lives and property saved. He was also adamant that the search for earthquake precursors should be carried out with scientific rigour. He was one of the earliest advocates of putting earthquake forecasts through prospective statistical tests. The papers in this issue review progress made in observing and understanding the earthquake preparation process and in converting that understanding into forecasts of earthquake occurrence.

A singular achievement of Frank Evison's research was the discovery of the predictive scaling relations of long-term seismogenesis. These were derived initially from data on precursory earthquake swarms as long-term precursors of major earthquakes, and later confirmed from the more general precursory scale increase phenomenon. The first paper in this volume is an account of attempts by EVISON and RHOADES over several decades to conduct rigorous tests of forecasting models incorporating the precursory swarm hypothesis. In this paper, RHOADES describes lessons learned in the course of the tests, and how the failure to predict certain major earthquakes using swarms led to the discovery of the more general precursory phenomenon, and in turn to a weaker but more informative long-range forecasting model—the EEPAS model.

Statistical methods significantly contribute to the formulation of earthquake forecasting models. The paper by VERE-JONES traces the development of the emerging field of "Statistical Seismology", with particular attention to stochastic models for seismicity, enabled by the increasing availability of

[1] GNS Science, P.O. Box 30-368, Lower Hutt, New Zealand.
[2] Institute of Geophysics, Victoria University of Wellington, Box 600, Wellington, New Zealand. E-mail: Martha.savage@vuw.ac.nz

extensive, high-quality earthquake catalogues and other relevant data. Stochastic models are seen as arising in a hierarchy of roles in seismology, as in other scientific disciplines. As well as providing a convenient descriptive tool for summarising data patterns, they bridge the gap between the detailed modelling of a complex system and the need to fit models to limited data, and at the most fundamental level they arise as a basic component in the modelling of earthquake phenomena.

Next to the subduction regions of New Zealand, Japan and Greece, the Indian subcontinent is the region which has revealed most examples of earthquake swarms as long-term precursors of major events. The papers by SHANKER *et al.* and SINGH *et al.* review the data on precursory swarms in the northeast India region and the western Nepal Himalaya and its adjoining India region, respectively. SINGH *et al.* identify two examples of recent swarms followed by quiescence, which, they suggest, may be precursors of forthcoming major events.

A recent development in earthquake forecasting research is the initiative by the Collaboratory for the Study of Earthquake Predictability (CSEP) to undertake comparative tests of competing forecasting models in several different test regions and on a variety of time-scales. Several papers in this volume pertain to the formulation of forecasting models in a form suitable for submission for CSEP testing. For the California and Nevada region, KAGAN and JACKSON present both a long-term forecast, suitable for testing over a 5-year period, and a short-term forecasting model, incorporating clustering following the Omori-law, to be updated daily and tested against similar models by CSEP. For the New Zealand region, CONSOLE *et al.* present and retrospectively test a short-term clustering model for forecasting magnitude 4 and greater earthquakes.

In another paper, KAGAN *et al.* partition the Earth's surface into five tectonic regimes and extend existing branching models by incorporating potentially important estimates of tectonic deformation and by allowing the parameters in the models to vary between different tectonic regimes. The differences in branching parameters between the various tectonic regimes were found to be more modest than expected, but robust under changes in earthquake catalogue and magnitude threshold. They found that subduction zones have the highest earthquake rates, the largest upper magnitude limit, and the highest proportion of triggered events, and fast-spreading ridges have the smallest upper magnitude limit and the lowest proportion of triggered events. In a related paper, KAGAN and JACKSON present a simple method for long- and short-term earthquake forecasting using a global catalogue, modifying and extending the forecast methods used previously for the northwest and southwest Pacific to smaller magnitudes and wider regions. Their ultimate goal is to develop forecasting and testing methods to validate or falsify common assumptions regarding earthquake potential.

The paper by SHCHERBAKOV *et al.* is another approach to global modelling of earthquake occurrence. They examine the extent to which past seismicity can be used to forecast future earthquakes by the Relative Intensity (RI) method. The analysis is based on the relative operating characteristic (ROC) and Error Diagrams.

The Wenchuan, China, M_s 8.0 earthquake of 12 May, 2008 is one of the most damaging earthquakes to have occurred in recent years. It is for such earthquakes that an advance warning would be most useful. The paper by JIANG and WU conducts a retrospective forecast test of the Pattern Informatics (PI) algorithm for the earthquakes in Sichuan-Yunnan region since 1992. An ROC test showed that, at a target magnitude of M_s 5.5, the PI algorithm outperforms random guessing and the RI method. However, when considering a higher target magnitude, the PI method provided no indication of the approach of the Wenchuan earthquake.

It is often assumed in seismic hazard analysis that spatial averaging can be reliably substituted for temporal averaging, however, currently the validity of this "ergodic assumption" is increasingly being questioned. TIAMPO *et al.* consider the effect of ergodicity on the PI forecasting algorithm in a study of actual and synthetic earthquake catalogs. They find that the PI algorithm is more reliable in regions which satisfy the ergodic assumption.

ZHAO *et al.* also consider the Sichuan-Yunnan region of China. They apply the Reverse Tracing of Precursors (RTP) method to analyse the long-term seismic activity in the alarm regions identified by the

Annual Consultation on the Likelihood of Earthquakes in China. They report that the RTP analysis correctly removes some of the false-alarm regions and thereby reduces the false-alarm rate of the Annual Consultation, without reducing the hit rate.

In the paper by BEBBINGTON *et al.*, a stochastic model based on elastic rebound, which is capable of producing cycles of accelerating moment release, is fitted to the earthquake catalogue of the San Francisco Bay region. Repeated simulation from the stochastic model produces 30-year hazard estimates which are compared with the estimates from the 2003 Working Group on California Earthquake Probabilities.

Separating earthquake catalogues into background and induced events is standard practice in probabilistic seismic hazard analysis, nonetheless an ongoing matter of contention because of the perceived subjectivity of declustering algorithms. A paper by CONSOLE *et al.* presents an approach to estimating the background seismicity by applying the ETAS model. The degree of subjectivity is then limited by applying rigorous statistical procedures to adjust the free parameters of the model.

One promising strategy for improving earthquake forecasts is to combine the information from several different predictive parameters into a single model. IMOTO and RHOADES consider the a and b values of the Gutenberg-Richter relation and the EEPAS model rate as separate predictive parameters and use them to construct a combined hazard function. They demonstrate a significant information gain in forecasting moderate earthquakes in the Kanto region from 2000–2005.

The regularity of recurrence times of large earthquakes in a given fault zone is a matter of great interest for time varying hazard estimation, however limited data on past occurrences often hampers assessment of the degree of regularity. ZOELLER *et al.* describe a Bayesian method that can be used to continuously update the aperiodicity of the recurrence-time distribution, based on a catalogue which is complete for magnitudes above some threshold. They assert that, by using the smaller earthquakes, a stable estimate of the aperiodicity can be obtained even if only a small number of large earthquakes have been recorded. For the case of Parkfield, they find a higher value of aperiodicity than would be estimated from the large earthquakes alone.

We thank the following colleagues who have reviewed papers for this special issue: M. Bebbington, A. Christophersen, R. Console, J. Cousins, R. Davies, K. Felzer, C. Frohlich, B. Fry, J. Hardebeck, D. Harte, A. Helmstetter, M. Imoto, T. Iwata, Y. Kagan, A.M. Lombardi, B. Lund, W. Marzocchi, G. Molchan, R. C. Nicholson, P. M. Paradisopolo, R. Robinson, J. Rundle, D. Schorlemmer, D. Shanker, R. Shcherbakov, W. Smith, B. Stephenson, K. Tiampo, S. Toda, J. Townend, T. Van Stiphout, F. Wenzel, M. Werner, J. Woessner, M. Wyss, I. Zaliapin, A. Zavyalov, J. Zechar, J. Zhuang.

Pure Appl. Geophys. 167 (2010), 623–627

DOI 10.1007/s00024-010-0138-5

Pure and Applied Geophysics

Frank Evison Biography

DAVID RHOADES[1] and EUAN SMITH[2]

Frank Foster Evison, OBE, FRSNZ,
Professor Emeritus of Geophysics,
20 March 1922–25 January 2005

Frank Evison was born in Christchurch, New Zealand, in 1922. He graduated from Victoria University of Wellington with a BSc in physics in 1944 and an MA with Honours in mathematics in 1946. During this time he also served in the Royal New Zealand Air Force, where he gained a commission and became the commanding officer of a radar station. After the war, he traveled to Britain where he gained a Diploma from the Imperial College of Science and Technology and a PhD in geophysics from the University of London.

On his return to New Zealand, he joined the newly-formed Geophysics Division of the Department of Scientific and Industrial Research, where his early work was in exploration geophysics. He developed an electromagnetic transducer as a seismic source and then discovered coal-seam-guided S waves (Nature *176*, 1224–1225), which in 1985 EDWARDS *et al.* named 'Evison waves' (Geophysics *50*, 214–223).

The International Geophysical Year of 1957 included a major programme of exploration of Antarctica. Frank's contribution was a demonstration, using surface wave dispersion, that Antarctica was a continent with a crustal thickness of 30–40 km (Nature *83*, 306–308).

His research was becoming recognised and rewarded. He gained a Nuffield Fellowship in 1957 and a Fulbright Award in 1963.

In 1960 he became superintendent of the Seismological Observatory, a section of Geophysics Division, and in 1964 he became the Division's director. During this time he suggested an alternative explanation for observed palaeomagetic pole rotations, namely that they resulted from the slow flow of continental rocks under gravity (Nature *211*, 273–275). As a reason for poles' rotation it may have been wrong, nonetheless the idea that continental rocks could behave as fluids anticipated such thinking by decades.

Frank was an able administrator as well. During his directorship, the New Zealand seismograph network had a major upgrade. Many new stations were installed, including the first sets of three component seismographs at stations other than the Observatory's home in Wellington.

An example of Frank's direct approach to science is given by Evison's Wall. In the 1960s, there were many scientific questions about the then newly important Alpine fault, including whether, like parts of the San Andreas fault, it was creeping? Someone proposed that periodic surveys be conducted to test this. Repeated surveys would have been very expensive. Instead, Frank arranged to have a wall built across the fault at a location near Maruia. The wall still stands there, unbroken, as illustrated on the cover of this volume. The Alpine fault did not creep!

In 1967 he was appointed inaugural Professor of Geophysics at Victoria University of Wellington. Four years later he established the Institute of Geophysics. His vision was for an interdepartmental Institute with associate members outside the university as well. The

[1] GNS Science, P.O. Box 30-368, Lower Hutt, New Zealand. E-mail: d.rhoades@gns.cri.nz
[2] Institute of Geophysics, Victoria University of Wellington, Box 600, Wellington, New Zealand.

Institute would thus provide a linkage for all those in Wellington who were active or interested in geophysics. So it remains today, notwithstanding many institutional changes over the years. The Institute of Geophysics continues to be an area of research strength at Victoria, and many graduate students who gained knowledge of the physics of the Earth at Victoria owe thanks to Frank for their education and their careers.

It was during the early 1970s that Frank began his research into earthquake forecasting that would continue until his death. His interest in earthquake occurrence was firmly rooted in his views about what science should be about and what a scientist's duty was. Earthquakes were a social threat that caused loss of life and suffering. It was therefore the job of science to do something to mitigate this threat. And it was Frank's self-imposed duty to do what he could, as a scientist, to enable that mitigation.

Frank was persuaded that the Earth signaled its preparation for large earthquakes. If one could read those signals, then warnings could be given and lives and property saved. The fact that prediction was difficult, that others tried and failed, and that yet other people were skeptical that it was possible at all, did not deter him. The easy problems could be left to others. An idea must be persisted with until it was proved to be wrong.

Frank's persistence with an idea led him into many heated arguments. His suggestion that earthquake faults may be 'but a gross form of earthquake damage' (Bull. Seism. Soc. Am. *53*, 873–891) enraged his geologist colleagues. Later he admitted privately: "I was wrong, of course, but I had a lot of fun."

His first interest in earthquake precursors was in the then newly proposed idea of dilatancy: that at a critical state of stress cracks would open in rocks, thereby altering their mechanical properties. Frank had acquired the first portable seismographs in New Zealand. In an early experiment, the focal mechanisms of aftershocks of the 1968 M 7.4 Inangahua earthquake, recorded in 1972, showed a major difference from the mechanism of the main shock, which Frank and co-workers attributed to dilatancy (Nature *246*, 471–473).

However, Frank's first attempt at a forecasting model was the precursory swarm hypothesis. Fluctuations in seismicity prior to a large earthquake could, Frank believed, be used quantitatively to forecast the subsequent event (Nature *266*, 710–712). At about this time he started his collaboration on this work with a statistician, David Rhoades. The partnership continued until the day of his death.

Prediction of consequences from a theory is intrinsic in science. Therefore Frank argued that an earthquake forecaster's duty was to make predictions and test them against what would happen. He firmly believed that an assessment of reliability had to be part of any forecasting method. This required substantial testing, using data obtained subsequent to that from which the model had been constructed. He was so firm on this point that he successfully persuaded a New Zealand Prime Minister that a public announcement of a forecast was inappropriate because the method was inadequately tested. In discarding what he rightly called the anecdotal approach to earthquake prediction and insisting upon what is now known as prospective testing, Frank was decades ahead of the rest of the world.

In the late 1970s, scientific enthusiasm for earthquake prediction burgeoned. Frank helped to formulate an International Code of Practice for Earthquake Prediction. In 1979, he headed up a UNESCO conference on earthquake prediction in Paris, and was secretary and later chairman of the Commission on Earthquake Prediction of the International Association of Seismology and Physics of the Earth's Interior.

In time, scientific enthusiasm waned. The view that earthquake prediction is impossible began to prevail. The arguments for this did not persuade Frank. As time went by and new data came to hand, the model for precursory seismicity evolved, reaching its culmination with the Precursory Scale Increase—or Ψ—phenomenon (Pure and Applied Geophysics 2004, *161*, 47–72) in which Frank and David provided 47 examples of an increase in seismicity prior to large earthquakes in California, Greece-Turkey, Japan and New Zealand.

In 1988 Frank retired from the chair of geophysics at Victoria. This meant that he had more time for research! Until a few days before his sudden illness and death he was not just a regular, but indeed a daily attendee at the Institute of Geophysics. He was often there at weekends and after hours. Prior to his retirement he made a full contribution to the academic life of the university through teaching, research and administration. He was an active member of the Professorial Board. After retirement his participation in the research activities of the university continued. He also continued to make a collegial contribution through social intercourse with staff and students, not only about science but also in debates about the events of the day, politics, music, and any of Frank's other interests that happened to be under discussion. As in his science, Frank often questioned the prevailing view of the issue, always for good reasons which he enjoyed arguing. If Frank approached you and said, "Now, you're a geophysicist..." or "You're a seismologist..." it was the preamble to a sharp question that would require you to agree with him or argue convincingly against a new idea. Most of this activity took place in the School of Earth Sciences, but Frank did not forget his roots as a physicist, and continued to give seminars to his colleagues in the School of Chemical and Physical Sciences.

Frank passed away at the age of 82 on 25 January, 2005 in his home in Wellington after a relatively precipitant struggle with cancer.

Frank Evison Bibliography

Evison, F.F. (1951), *A New Approach to the Study of Elastic Propagation in Rocks*, Monthly Notices Roy. Astr. Soc. *6*, 209–221.

Evison, F.F. (1952), *The Inadequacy of the Standard Seismic Techniques for Shallow Surveying*, Geophysics *17*, 867–875.

Evison, F.F. (1953), *An Improved Electromechanical Seismic Source Tested in Shattered Rock*, New Zealand J. Sci. Tech.. B, General section *35*, 4–13.

Evison, F.F. (1954), *Early Arrivals in Seismic Prospecting*, Nature *173*, 1047.

Evison, F.F. (1955), *A Coal Seam as a Guide for Seismic Energy*, Nature *176*, 1224–1225. Reprinted in *Coal Geophysics*, Geophysics Reprint Series No. 6, (eds. Buchanan, D.J. and Jackson, L.J.) (Soc. Exploration Geophysicists 1986), pp. 278–280.

Evison, F.F. (1956a), *Seismic Waves from a Transducer at the Surface of Stratified Ground*, Geophysics *21*, 939–959.

Evison, F.F. (1956b), *The Seismic Determination of Young's Modulus and Poisson's Ratio for Rocks* In Situ, Geotechnique *6*, 118–123.

Evison, F.F. and Reilly, W.I. (1956a), *Deformation of Strata on Bord-and-Pillar Coal Mines I. Convergence Observations*, New Zealand J. Sci. Tech. B, General section *37*, 716–730.

Evison, F.F. and Reilly, W.I. (1956b), *Deformation of Strata in Bord-and Pillar Coal Mines II. Model Analysis*, New Zealand J. Sci. Tech. B, General section *38*, 129–138.

Evison, F.F. (1957), *The Pulsed Vibrator as a Seismic Source*, Geophys. Prospecting *5*, 381–391.

Evison, F.F., Orr, R.H., and Ingham, C.E. (1959), *Thickness of the Earth's Crust in Antarctica*, Nature *183*, 306–308.

Evison, F.F. (1960), *On the Growth of Continents by Plastic Flow under Gravity*, Geophys. J. Roy. Astr. Soc. *3*, 155–190.

Evison, F.F., Ingham, C.E., Orr, R.H., and Le Fort, J.H. (1960), *Thickness of the Earth's Crust in Antarctica and the Surrounding Oceans*, Geophys. J. Roy. Astr. Soc. *3*, 289–306.

Evison, F.F. (1961), *Rock Magnetism in Western Europe as an Indication of Continental Growth*, Geophys. J. Roy. Astr. Soc. *4*, 320–335.

Evison, F.F. and Whittle, P. (1961), *The Antipodal Location of Continents and Oceans*, Geological Magazine *98*, 377–379.

Evison, F.F. (1962), *Rock Magnetism and Low Angle Faulting*. Nature *194*, 644–646.

Hatherton, T. and Evison, F.F. (1962), *A Special Mechanism for Some Antarctic Earthquakes*, New Zealand J. Geol. Geophys. *5*, 864–873.

Thomson, A.A. and Evison, F.F. (1962), *Thickness of the Earth's Crust in New Zealand*, New Zealand J. Geol. Geophys. *5*, 29–45.

Evison, F.F. (1963a), *Earthquakes and Faults*, Bull. Seismol. Soc. Amer. *53*, 873–891.

Evison, F.F. (1963b), *Lessons from Agadir*, New Zealand Engineering *18*, 369–371.

Evison, F.F. (1963c), *Thickness of the Earth's Crust in Antarctica and the Surrounding Oceans: A Reply*, Geophys. J. Roy. Astr. Soc. *7*, 469–476.

Evison, F.F., *Earthquake Wave Anomalies in New Zealand*, In *Vesiac Report on Seismic Signal Anomalies* (University of Michigan, Ann Arbor 1966a), pp. 19–24.

Evison, F.F., *Earthquakes*, In *An Encyclopaedia of New Zealand*, Vol. 1 (ed. McLintock, A.H.) (Government Printer, Wellington, 1966b), pp. 509–516.

Evison, F.F. (1966c), *Polarity of the Earthquake Source*, Nature *211*, 273–275.

Evison, F.F. (1967a), *Note on the Aseismicity of Antarctica*, New Zealand J. Geol. Geophys. *10*, 479–483.

Evison, F.F. (1967b), *Nuclear Explosions and Natural Earthquakes*, New Zealand Engineering *22*, 33–34.

Evison, F.F. (1967c), *On the Occurrence of Volume Change at the Earthquake Source*, Bull. Seismol. Soc. Amer. *57*, 9–25.

Evison, F.F., *The Polymorphic Transition as a Possible Earthquake Source*. In *Vesiac Report on Source Mechanism of Shallow Events* (University of Michigan, Ann Arbor, 1967d), pp. 173–179.

Hamilton, R.M. and Evison, F.F. (1967), *Earthquakes at Intermediate Depths in South-west New Zealand*, New Zealand J. Geol. Geophys. *10*, 1319–1329.

Evison, F.F. (1968), *Active Regions of the Southwest Pacific*, Canadian J. Earth Sci. *5*, 1045–1049.

Hamilton, R.M. and Evison, F.F. (1968), *Reply to R.P. Suggate's Letter: Seismicity and Structure in Southwest New Zealand and the Macquarie Ridge*, New Zealand J. Geol. Geophys. *11*, 1277.

Evison, F.F. (1970), *Seismogenesis*, Tectonophysics *9*, 113–128.

Evison, F.F. (1971), *Seismicity of the Alpine Fault, New Zealand*. In *Recent Crustal Movements*, Royal Society of New Zealand

Bulletin 9 (eds. Collins, B.W. and Fraser, R.) (Royal Society of New Zealand, Wellington, 1971), pp. 161–165.

Evison, F.F., Robinson, R., and Arabasz, W.J. (1973), *Late Aftershocks, Tectonic Stress and Dilatancy*, Nature *246*, 471–473.

Evison, F.F. (1975), *Determination of Precursory Velocity Anomalies in New Zealand from Observatory Data on Local Earthquakes*, Geophys. J. Roy. Astr. Soc. *43*, 957–972.

Robinson, R., Arabasz, W.J., and Evison, F.F. (1975), *Long-term Behaviour of an Aftershock Sequence: The Inangahua, New Zealand, Earthquake of 1968*, Geophys. J. Roy. Astr. Soc. *41*, 37–49.

Evison, F.F., Robinson, R., and Arabasz, W.J. (1976), *Microearthquakes, Geothermal Activity, and Structure, Central North Island, New Zealand*, New Zealand J. Geol. Geophys. *19*, 625–637.

Evison, F.F. (1977a), *Fluctuations of Seismicity Before Major Earthquakes*. Nature *266*, 710–712.

Evison, F.F. (1977b), *Precursory Seismic Sequences in New Zealand*, New Zealand J. Geol. Geophys. *20*, 129–141.

Evison, F.F. (1977c), *The Precursory Earthquake Swarm*, Phys. Earth Planet. Inter. *15*, 19–23.

Evison, F.F. (1978), *Long-term Seismic Precursor to the 1968 Inangahua Earthquake, New Zealand*, New Zealand J. Geol. Geophys. *21*, 531–534.

Rhoades, D.A. and Evison, F.F. (1979), *Long-range Earthquake Forecasting Based on a Single Predictor*, Geophys. J. Roy. Astr. Soc. *59*, 43–56.

Evison, F.F. (1981), *Multiple Earthquake Events at Moderate-to-large Magnitudes in Japan*, J. Phys. Earth *29*, 327–339.

Evison, F.F. and Rhoades, D.A. *Synoptic Earthquake Forecasting as a Basis for Disaster Prevention*. In *Large Earthquakes in New Zealand: Anticipation, Precaution, Reconstruction*, Royal Society of New Zealand Misc. Series 5 (comp. Cresswell, M.M.) (Royal Society of New Zealand, Wellington, 1981), pp. 25–29.

Evison, F.F. (1982a), *Earthquake Forecasting and Countermeasures Planning*, Earthquake Prediction Res. *1*, 115–124.

Evison, F.F. (1982b), *Generalised Precursory Swarm Hypothesis*, J. Phys. Earth *30*, 155–170.

Evison, F.F., *Communications of Earthquake Predictions and Warnings*, In *Earthquake Prediction*, Proceedings of the International Symposium on Earthquake Prediction (Terra Scientific Publishing Company, Tokyo, 1984), pp. 937–946.

Evison, F.F., *Seismic Hazard Assessment in Continental Areas*. In *International Symposium on Continental Seismicity and Earthquake Prediction* (Seismological Press, Beijing, 1984b), pp. 751–762.

Rhoades, D.A. and Evison, F.F., *Method Assessment in Long-range Earthquake Forecasting*, In *Earthquake Prediction*, Proceedings of the International Symposium on Earthquake Prediction (Terrapub, Tokyo, 1984), pp. 497–504.

Evison, F.F. (1986), *Earthquake Generation and Prediction. In Earthquake Prediction and Mitigation of Earthquake Losses*, Proceedings of the UNDRO/USSR/UNESCO/UNDP Training Seminar (Dushanbe, Tadjikistan, 1986), pp. 209–234.

Evison, F.F. and Webber, S.J., *Seismogenic Stress Conditions in Central New Zealand*. In *Recent Crustal Movements of the Pacific Region*, Royal Society of New Zealand Bulletin 24 (Royal Society of New Zealand, Wellington, 1986), pp. 553–565.

Evison, F.F. (1989), *Time-variable Earthquake Hazard, Tectonophysics 169*, 155–157.

Rhoades, D.A. and Evison, F.F. (1989a), *On the Reliability of Precursors*. Phys. Earth Planet. Inter. *58*, 137–140.

Rhoades, D.A. and Evison, F.F. (1989b), *Time-variable Factors in Earthquake Hazard*, Tectonophysics *167*, 201–210.

Evison, F.F. and Rhoades, D.A. (1993), *The Precursory Earthquake Swarm in New Zealand: Hypothesis Tests*, New Zealand J. Geol. Geophys. *36*, 51–60.

Rhoades, D.A. and Evison, F.F. (1993), *Long-range Earthquake Forecasting Based on a Single Predictor with Clustering*, Geophys. J. Int. *113*, 371–381.

Evison, F.F and Rhoades, D.A., *On the Testing of Earthquake Precursors*. In *Electromagnetic Phenomena Related to Earthquake Prediction* (eds. Hayakawa, M. and Fujinawa, Y.) (Terrapub, Tokyo, 1994), pp. 1–11.

Rhoades, D.A. and Evison, F.F. (1996), *The VAN Earthquake Predictions*, Geophys. Res. Lett. *23*, 1371–1373.

Evison, F.F. and Rhoades, D.A. (1997), *The Precursory Earthquake Swarm in New Zealand: Hypothesis Tests II*, New Zealand J. Geol. Geophys. *40*, 537–547.

Evison, F.F. and Rhoades, D.A. (1998), *Long-term Seismogenic Process for Major Earthquakes in Subduction Zones*, Phys. Earth Planet. Inter. *108*, 185–199.

Rhoades, D.A., Evison, F.F., and Kozuch, M.J. (1998), *Earthquake Forecasting*, Tephra *17*, 21–23.

Evison, F.F. (1999), *On the Existence of Earthquake Precursors*, Ann. Geofis. *42*, 763–770.

Evison, F.F. and Rhoades, D.A. (1999a), *The Precursory Earthquake Swarm and the Inferred Precursory Quarm*, New Zealand J. Geol. Geophys. *42*, 229–236.

Evison, F.F. and Rhoades, D.A. (1999b), *The Precursory Earthquake Swarm in Japan: Hypothesis Test*, Earth Planets Space *51*, 1267–1277.

Evison, F.F. and Rhoades, D.A. (2000), *The Precursory Earthquake Swarm in Greece*, Ann. Geofis. *43*, 991–1009.

Evison, F.F. (2001), *Long-range Synoptic Forecasting: An Aim for the Millennium*, Tectonophysics *338*, 207–215.

Evison, F.F. and Rhoades, D.A. (2001), *Model of Long-term Seismogenesis*, Ann. Geofis. *44*, 81–93.

Evison, F.F. and Rhoades, D.A. (2002), *Precursory Scale Increase and Long-term Seismogenesis in California and Northern Mexico*, Ann. Geophys. *45*, 479–495.

Evison, F.F. and Rhoades, D.A. (2004a), *Demarcation and Scaling of Long-term Seismogenesis*, Pure Appl. Geophys. *161*, 21–45.

Evison, F.F. and Rhoades, D.A. (2004b), *Long-term Seismogenesis and Self-Organized Criticality*, Earth Planets Space *56*, 749–760.

Rhoades, D.A. and Evison, F.F. (2004), *Long-range Earthquake Forecasting with Every Earthquake A Precursor According to Scale*, Pure Appl. Geophys. *161*, 47–72.

Evison, F.F. and Rhoades, D.A. (2005), *Multiple-mainshock Events and Long-term Seismogenesis in Italy and New Zealand*, New Zealand J. Geol. Geophys. *48*, 523–536.

Rhoades, D.A. and Evison, F.F. (2005), *Test of the EEPAS Forecasting Model on the Japan Earthquake Catalogue*, Pure Appl. Geophys. *162*, 1271–1290.

Rhoades, D.A. and Evison, F.F. (2006), *The EEPAS Forecasting Model and the Probability of Moderate-to-large Earthquakes in Central Japan*, Tectonophysics *417*, 119–130.

Papadimitriou, E.E., Evison, F.F., Rhoades, D.A., Karakostas, V.G., Console, R., and Murru, M. (2006), *Long-term Seismogenesis in Greece: Comparison of the Evolving Stress Field and*

Precursory Scale Increase Approaches, J. Geophys. Res. *111*, B05318, doi:10.1029/2005JB003805.

Console, R., Rhoades, D.A., Murru, M., Evison, F.F., Papadimitriou, E.E., and Karakostas, V.G. (2006), *Comparative Performance of Time-invariant, Long-range and Short-range Forecasting Models on the Earthquake Catalogue of Greece*, J. Geophys. Res. *111*, B09304, doi:10.1029/2005JB004113.

(Published online April 17, 2010)

Pure Appl. Geophys. 167 (2010), 629–644

DOI 10.1007/s00024-010-0071-7

Pure and Applied Geophysics

Lessons and Questions from Thirty Years of Testing the Precursory Swarm Hypothesis

David A. Rhoades[1]

Abstract—In 1976 Frank Evison identified the first examples of earthquake swarms as long-term precursors of main-shock events, and thereby discovered the predictive scaling relations of long-term seismogenesis. From this time on, forecasting became the main focus of his research. After learning from an early attempt to communicate forecasts confidentially to government, he recognised the importance of hypothesis testing, and the precursory swarm hypothesis was cast in a form similar to a regional likelihood model. Tests of its performance relative to a stationary Poisson model at $M \geq 5.8$ in New Zealand were begun in 1977. The initial hypothesis was that of a 1–1 relation between swarms and main-shock events. Following a study of the Japan catalogue, the generalised swarm hypothesis, in which multiple swarms were precursory to multiple main-shock events, was formulated. Tests of this form of the hypothesis at $M \geq 6.8$ were initiated in a region of surveillance east of Japan in 1983. Eventually the generalised hypothesis was adopted in New Zealand also. In 1999, tests were begun in a region of Greece. In 1994–1995, several main-shock events favourable to the swarm hypothesis occurred, however four main-shock events near Arthur's Pass, New Zealand, occurred without precursory swarms. Subsequent analysis showed that events called "quarms", which were similar to swarms but more protracted in time, had preceded these events. This led to the proposal of a qualitative physical process to account for swarms, quarms and the predictive relations: A three-stage faulting process, in which a major crack induces aftercracks in its neighbourhood, just as a main shock induces aftershocks. An inference from this process was that the most general long-term precursor should be an increase of seismicity at similar magnitudes to the eventual aftershocks. It turned out that such a precursory scale increase nearly always occurs before major earthquakes and conforms to the predictive scaling relations. Setting aside the problem of identifying the scale increase before the major earthquake, the EEPAS (Every Earthquake a Precursor According to Scale) forecasting model was formulated. The success of this relatively weak model in forecasting major events in New Zealand, California, Japan and Greece shows that the predictive scaling relations are ubiquitous in earthquake catalogues. Although none of the formal tests of the swarm hypothesis were successful in their own terms, they were beneficial in identifying shortcomings in its formulation, thereby leading to improved understanding of long-term seismogenesis and a better forecasting model. Some puzzling aspects of the scaling relations are whether they vary regionally, and why the precursor area and aftershock area scale differently with magnitude. A more practical question is whether the EEPAS model can be strengthened, by making use of the clustering of some precursors in swarms and quarms, to bring us nearer to the original goal of forecasting individual major earthquakes.

Key words: Seismogenesis, seismicity patterns, earthquake prediction.

1. *Introduction*

In December of 1976, shortly after joining the Applied Mathematics Division of the Department of Scientific and Industrial Research located at Victoria University of Wellington as a statistician, I was visited by the University's Professor of Geophysics—Frank Evison. He had identified the first examples of earthquake swarms as long-term precursors of main-shock events (Evison, 1977a, b, c, 1978), and thereby discovered the predictive scaling relations of long-term seismogenesis. He perceived the potential of these empirical relations for earthquake forecasting, and wanted some help with the statistics.

That visit was the beginning of a long-term collaboration which continued until his death in 2005, in which we strove to turn his discovery into an informative forecasting model and to demonstrate the value of the model through formal monitored performance tests. This paper is an account of our efforts in that direction, and of lessons learned along the way.

At that time, earthquake prediction research was buoyant in the wake of the successful evacuation of the Chinese city of Haicheng prior to an M 7.3 earthquake on 1975 February 4, based largely on

[1] GNS Science, P O Box 30-368, Lower Hutt, New Zealand. E-mail: d.rhoades@gns.cri.nz

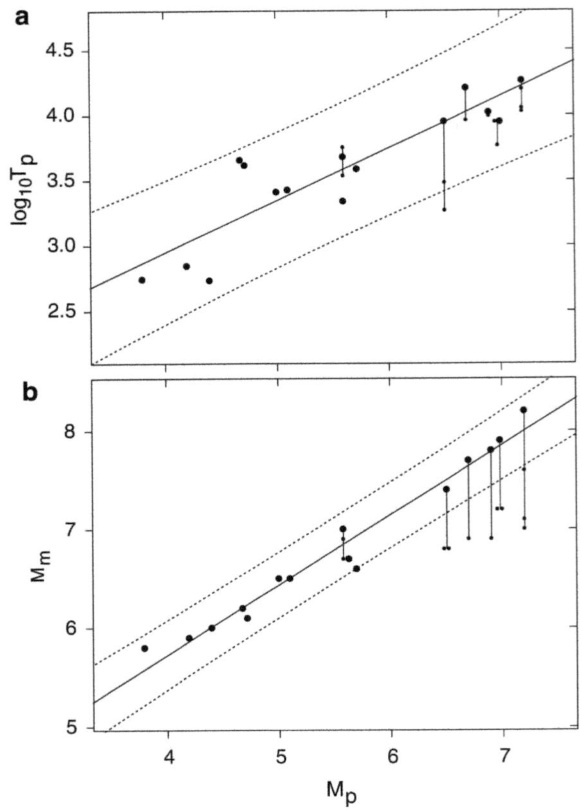

Figure 1
Predictive scaling relations for the generalised precursory swarm hypothesis, using New Zealand and Japan data. **a** Log precursor time T_P (days) versus swarm magnitude M_P, defined as the average of the three largest swarm earthquake magnitudes. **b** Mainshock magnitude M_m versus M_P. The regression lines do not apply to the secondary main-shocks, which are shown as *small symbols* connected to the relevant principal main shocks. *Dashed lines* are 95% tolerance limits. After EVISON and RHOADES (1997)

studies of seismicity, deformation and foreshocks (ADAMS, 1976). Precursory swarms became one of a number of phenomena known as Rikitake precursors of the first kind (RIKITAKE 1975, 1979, 1982), for which the logarithm of the precursor time is linearly related to the main-shock magnitude. The predictive scaling relations (Fig. 1)—regressions of main-shock magnitude, and log precursor time against precursor magnitude—distinguished swarms from all other such long-term precursors, in that they provided a way of estimating the time of occurrence and magnitude of future main-shock events from the precursor itself, namely from the magnitude of the largest swarm earthquakes. The location of the swarms was the same as that of the main-shock events. The problem was to reliably identify the precursory swarms before the main shock occurred.

Before his discovery of precursory swarms, Frank Evison's research had ranged widely over a number of important topics in seismology. But from then on, he focussed almost exclusively on the earthquake forecasting problem because he perceived that the predictive scaling relations were both empirically strong and physically reasonable, although at that stage he could not identify a particular physical process to account for them.

2. *Reliability Testing*

At about the same time as our collaboration began, Frank Evison had confidentially communicated four provisional earthquake predictions, based on four recent swarms, to the Minister of Science, stressing that they were not for public release. Once he became aware of the Prime Minister's intention to announce these predictions to the public, he withdrew them, citing the fact that Professor Clarence Allen, then President of the Seismological Society of America and a leading US expert in this field, had recently clarified the need for verification and reliability testing of earthquake predictions (ALLEN, 1976). He informed the government that he planned to undertake reliability tests, and that these would take "several months" to complete. A clear lesson from this episode was that the interaction between scientists and government on the matter of earthquake prediction is fraught with potential difficulty (EVISON, 1984). I am sure it was a useful background for Frank when he later became rapporteur of a UNESCO/IUGG committee which drafted a Code of Conduct for scientists engaged in earthquake prediction (INTERNATIONAL UNION OF GEODESY AND GEOPHYSICS, 1984). In any case it helped to commit him to reliability testing, in which I supported him wholeheartedly. As for the timeframe of "several months", we soon realised that real-time performance tests were the only sort of tests that would carry any scientific weight. For a long-term precursor, that meant a much longer time-period was needed.

The initial hypothesis was that of a 1–1 relation between two types of multiple earthquake event, called swarms and main-shock events. A third type of multiple event, called multiplets, was deemed irrelevant, being neither predictable nor precursory. For the purpose of testing, the swarm hypothesis was formulated with most of the elements of what is now called a regional earthquake likelihood model (e.g., FIELD, 2007). Mathematically, the rate density λ of the model could be represented at time t, main-shock magnitude m_M, and location (x, y) by

$$\lambda(t, m_M, x, y) = \sum_{i=1}^{n} v(m_{Pi}) f_i(t, m_M, x, y) + \mu(m_M)\lambda_0(m_M, x, y), \quad (1)$$

where there are n current unrelated swarms with magnitudes m_{Pi}, $i = 1,\ldots,n$, $v(m_{Pi})$ is the valid alarm rate, f_i is the probability density for time of occurrence, magnitude and location of a main shock related to the ith swarm, $\mu(m_M)$ is the failure rate and λ_0 the rate density of the stationary Poisson model.

Distributions for time, magnitude and location of a predicted main shock, given a swarm, were based on the predictive regressions and swarm earthquake locations. Estimates of the false-alarm rate (the proportion of swarms not followed by a main-shock event) and failure rate (the proportion of main-shock events not preceded by swarms), in different magnitude classes, were made from the score-sheet of past related and unrelated swarms and main-shock events, and updated as the tests proceeded using a simple Bayesian analysis (RHOADES and EVISON, 1979). Using the likelihood method, tests of the model's performance relative to a stationary Poisson model in forecasting shallow (hypocentral depth $h < 33$) main-shock events with $M \geq 5.8$ were begun in October 1977, in a large region which included most of the seismically active region of New Zealand except the volcanic zone (Fig. 2a), using the earthquakes with $M \geq 3.3$ as the data base. The tests were carried out with the help of a confidential monitor, whose role was to maintain a file of all forecasts.

A key element of the model was the identification of multiple earthquake events, and their classification as main-shock events, swarms or multiplets. Multiple events were recognised by simple quantitative criteria that could be applied by eyeballing an earthquake catalogue, which was always Frank's preferred way. The first criterion was that a multiple event must include at least three earthquakes above the minimum magnitude threshold in a 24-h period, and more than a certain number of earthquakes overall. Once these criteria were satisfied, the multiple event was extended backward and forward in time, to include all earthquakes within specified latitude and longitude limits, until a certain number of days elapsed without further earthquakes. Within the class of multiple events, main-shock events were distinguished from swarms and multiplets by having a relatively large value of $M_1 - M_3$, where M_1 is the largest magnitude and M_3 the third largest. Swarms and multiplets were distinguished by the area they occupied. Swarms occupied a relatively large area, and multiplets a relatively small area; the idea being that precursory swarms should occupy a region as large as the source area of the related main-shock event, which was typically 1–2 units higher in magnitude than the largest swarm earthquake.

The minimum number of earthquakes in a multiple event, the threshold value of $M_1 - M_3$, and the threshold area (which was in principle a function of magnitude), were naturally adjusted to optimize the score-sheet for past related and unrelated swarms and main-shock events, i.e., to make the number of related swarm main-shock pairs as large as possible. Once set for a test region, they were then applied rigidly as new multiple events were recognised during the tests. Because the learning data sets were never large, and because the proportions of related events estimated at the outset were not necessarily indicative of the proportions to be expected under forward testing, the updating procedure was put in place to improve the estimates of the false alarm and failure rates as the tests proceeded. In practice, good estimates of these rates were never obtained during the tests, because the number of relevant events in the test period was too small and the updating occurred too slowly.

One important question to consider was how much could be published about the tests as they proceeded, to satisfy both the need for scientific openness and the wider responsibility not to turn test

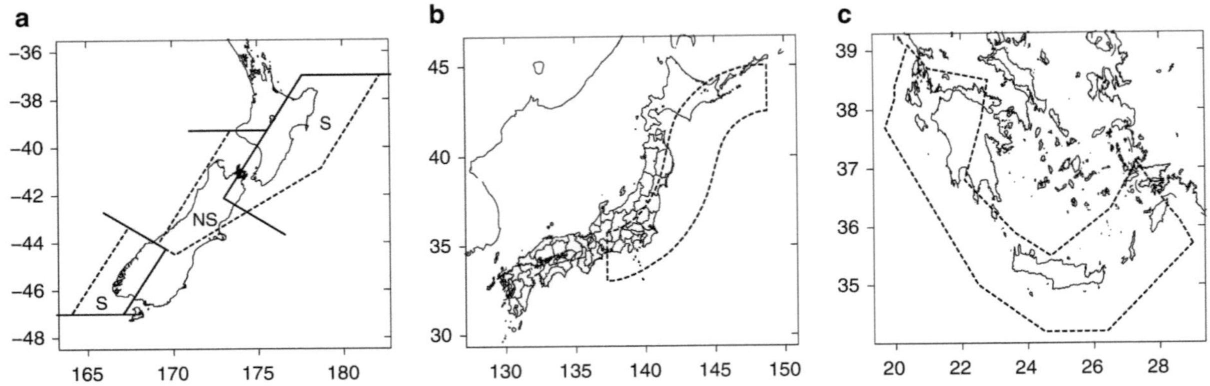

Figure 2
Regions of surveillance for tests of the precursory scale increase. **a** New Zealand, showing subduction (*S*) and non-subduction (*NS*) regions; **b** Japan; **c** Greece

predictions into de facto public warnings (Evison, 1984). The practice actually adopted was to publish the forecasting methods and the lists of multiple events and their classification as swarms, main-shocks events and multiplets, and eventually a summary of the results, but never the current forecasts, which were communicated only to the monitor of the tests. This seemed to be the right balance; although others could in theory duplicate test forecasts from published information, nobody ever showed any interest in doing so.

The test forecasts were presented using what was called initially the risk refinement factor and later the hazard refinement factor, but is known more commonly as the probability gain (Aki, 1981). This is the ratio of the rate density of earthquake occurrence under the swarm hypothesis to that under the stationary Poisson model. In the early days, the location distribution for the main shock was taken as uniform inside an elliptical envelope drawn around the swarm earthquake locations, but later, recognising that any abrupt change of probability of earthquake occurrence with location was physically unreasonable, a bivariate normal probability distribution was adopted (Rhoades and Evison, 1993). The forecasts were synoptic: At any time the rate of earthquake occurrence was estimated for all locations and magnitudes within the region of surveillance. The hazard was enhanced in precursory areas for a time after the occurrence of a swarm, but reduced at all other times and locations, relative to a stationary Poisson model, which was spatially uniform within the region of surveillance. This crude reference model was considered adequate at the time because the swarm hypothesis itself was similarly crude and the probability gains aimed for were large.

3. Generalised Swarm Hypothesis

During a visit to Kyoto in 1980, Frank conducted a detailed analysis of the Japan Meteorological Agency catalogue and formulated the generalised precursory swarm hypothesis, in which clusters of swarms were followed in the long term by clusters of main-shock events (Evison, 1981, 1982). The predictive scaling relations still applied to the amalgamated cluster of swarms and the largest of the cluster of main shocks, called the principal main shock. Tests of this form of the hypothesis at $M \geq 6.8$ were initiated in a mostly offshore region of surveillance east of Japan in 1983 (Fig. 2b). The detailed analysis had shown that for the rest of the Japan region, including most onshore areas, past earthquakes had not been preceded by swarms. This was not the only region to which the swarm hypothesis did not apply. The California catalogue yielded only a few isolated examples of swarm main-shock sequences, and many examples of main shocks occurring without precursory swarms.

The generalised hypothesis was in several ways more complex than the simple one. First, as new swarms were recognised during the tests, a decision had to be made as to whether they should be grouped into an existing cluster of swarms, or regarded as new precursory sequences. This seemed initially to be an easy subjective judgement, and so no formal rules for doing it were ever developed. On the other hand a logically consistent procedure, based on conditional probabilities, was developed for updating the forecasts when main shocks occurred. Unlike the 1–1 hypothesis, in which the occurrence of a related main shock fulfilled the forecast from an individual swarm and terminated it, in the generalised hypothesis the occurrence of a related main shock only partially fulfilled the forecast, so it continued in a modified form, affected by the relative probabilities that the main shock was the principal one, a secondary one, or an unrelated event (Rhoades and Evison, 1993).

For a time, tests of the 1–1 hypothesis in New Zealand continued in parallel with those of the generalised hypothesis in Japan. But in 1990, tests of the 1–1 hypothesis in New Zealand were abandoned (Evison and Rhoades, 1993). Figure 3a shows the evolution of the performance factor for this test. The performance factor is the ratio of the cumulative likelihood of the observed catalogue of main-shock events under the swarm hypothesis to that under the stationary Poisson model. The jumps represent earthquakes. In this test lasting 13 years, only three relevant main-shock events occurred, and the downward direction of the jumps shows that they were all less likely under the swarm hypothesis than under the stationary Poisson model. This result was not as disheartening as it may appear, because two of the main shocks occurred where swarms had occurred previously; in one case the swarm had not been recognised because of boundary effects, and in another, a cluster of swarms had occurred, although because they were treated separately, the main shock scored poorly. Fourteen swarms contributed to the forecasts over the period of the tests. The performance factor had dropped to near the proposed rejection level. Accordingly the 1–1 hypothesis test was terminated and replaced by a test of the generalised hypothesis.

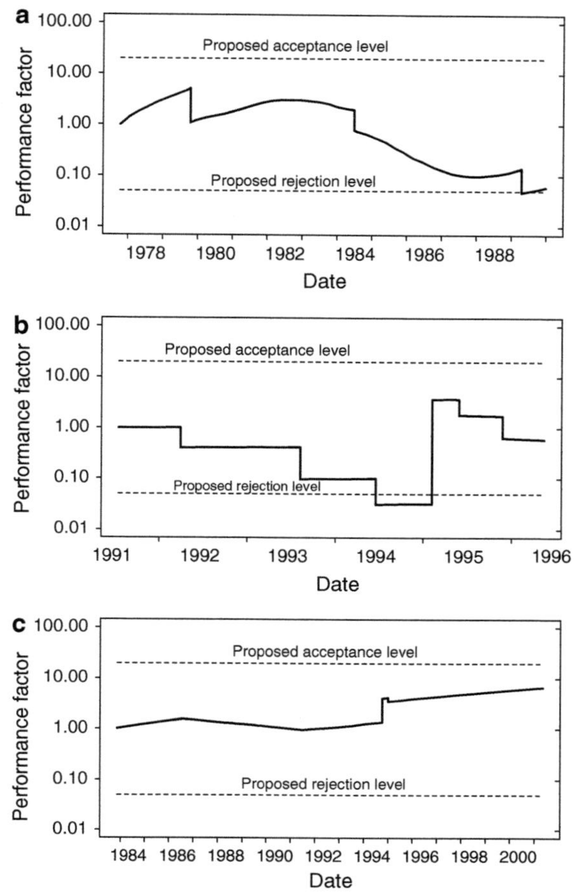

Figure 3
Evolution of the performance factor in real-time tests of the precursory swarm hypothesis: **a** Simple (1–1) hypothesis in New Zealand (Evison and Rhoades, 1993); **b** Generalised hypothesis in New Zealand (Evison and Rhoades, 1997); **c** Generalised hypothesis in Japan (Evison and Rhoades, 1999b, updated to 2001). The performance factor is the ratio of the likelihood under the swarm hypothesis to that under the stationary Poisson model

4. Successes and Failures

A sequence of earthquakes in 1994–1995 proved highly instructive. Several main-shock events favourable to the swarm hypothesis occurred (*M* 7.5 Sanriku-Haruka-Oki and *M* 8.1 Hokkaido-Toho-Oki, Japan, and *M* 7.0 East Cape, New Zealand), but four main-shock events near Arthur's Pass, New Zealand, occurred without precursory swarms. The successes served as an encouragement, however it was the contemplation of the failures that proved most valuable in the end.

By any measure, the *M* 7.0 East Cape earthquake of 1995 Feb. 5 was a resounding success for the swarm hypothesis (Fig. 4). The relevant forecast was supported by a cluster of six swarms occurring

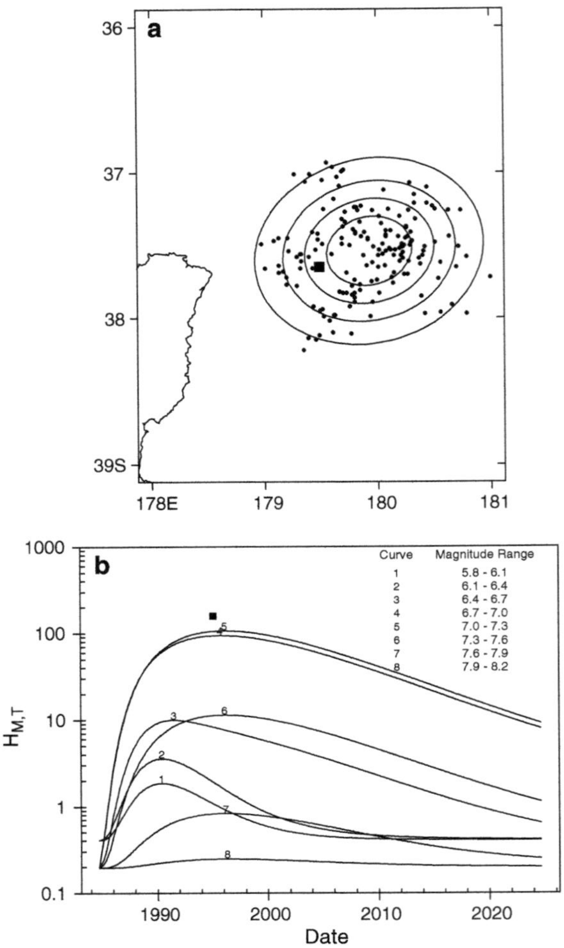

Figure 4
Test prediction generated by precursory swarms in tests of the generalised precursory swarm hypothesis in New Zealand. Data for the East Cape (M 7.0) earthquake of 5 February 1995 has been superimposed. **a** Location factor H_L. The value is unity at the centre of the target, and contours are shown for $H_L = 0.8, 0.6, 0.4$ and 0.2. The *filled circles* are the swarm epicentres and the *filled square* is the main-shock epicentre. **b** Magnitude and time-of-occurrence factor $H_{M,T}$. The *filled square* represents the main shock. The combined performance factor is 114. After Evison and Rhoades (1997)

between 1981 and 1991 (Evison and Rhoades, 1993, 1997). The magnitude and time of the East Cape main shock were at the maximum of the hazard refinement factor (Fig. 4) and the location was well within the envelope of swarm epicentres, giving a location factor H_f of about 0.7. Overall, this event was 114 times more likely under the swarm hypothesis than under the stationary Poisson model. Despite this success, because of the occurrence of four main-shock events near Arthur's Pass, the performance graph was not positive (Fig. 3b) and a simulation analysis showed that the performance factor overall was consistent with the stationary Poisson model, and significantly less than expected under the swarm hypothesis. Therefore this second New Zealand test was terminated (Evison and Rhoades, 1997). Retrospective analysis showed that events similar to swarms, but more protracted in time, had preceded the Arthur's Pass main-shock events. These were dubbed quasi-swarms (or "quarms"). The quarms fitted the predictive scaling relations (Evison and Rhoades, 1999a).

The quarm phenomenon was never completely defined, and no attempt was made to launch tests of a hypothesis based on it. Its usefulness lay in the new thinking about the seismogenic process that it stimulated. The central South Island region where the quarms occurred is the same region where a main-shock event without any related swarm occurred during the first New Zealand test. This is a continental transpressional zone, similar to most of California. On the other hand, nearly all of the examples of precursory swarms belonged to subduction regions. Frank inferred that high fluid pressure in the subduction regions was favourable to the occurrence of swarms, and that relatively low fluid pressure in the continental transpressional region was not. Therefore a revised hypothesis was launched: that precursory swarms occur in subduction regions and other regions of high fluid pressure, and precursory quarms in non-subduction regions. With this generalised (subduction) hypothesis in mind, a third New Zealand test was launched with a higher failure rate in the non-subduction region (Fig. 2a).

The M 7.5 Sanriku-Haruka-Oki and M 8.1 Hokkaido-Toho-Oki (Fig. 5) earthquakes occurred as main shocks related to previously identified swarms, and taken together with the absence of any events unrelated to swarms, were sufficiently positive for the swarm hypothesis that an overall assessment of the Japan tests in 1998 (Fig. 3c) showed that the performance factor, although short of the proposed acceptance level, was consistent with the swarm hypothesis and significantly greater than expected under the stationary Poisson model (Evison and Rhoades, 1999b).

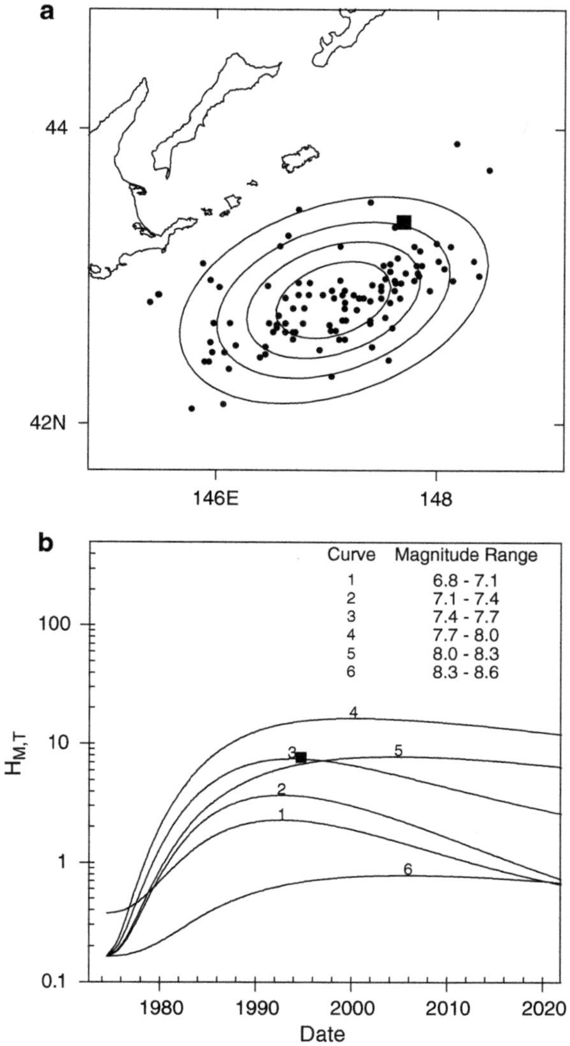

Figure 5
Test prediction generated by precursory swarms in tests of the generalised precursory swarm hypothesis in Japan. Data for the Hokkaido-Toho-Oki (M 8.1) earthquake of 4 October 1994 has been superimposed. **a** Location factor H_L. The value is unity at the centre of the target, and contours are shown for $H_L = 0.8$, 0.6, 0.4 and 0.2. The *filled circles* are the swarm epicentres and the *filled square* is the main-shock epicentre. **b** Magnitude and time-of-occurrence factor $H_{M,T}$. The *filled square* represents the main shock. The combined performance factor is 2.96. After Evison and Rhoades (1999b)

In 1998, Frank visited the University of Thessaloniki (AUTH) and undertook a detailed analysis of the AUTH earthquake catalogue, finding that the Hellenic subduction region displayed the same precursory swarm phenomenon as had been found in the New Zealand and Japan subduction regions (Evison and Rhoades, 2000). This was consistent with the generalised (subduction) hypothesis. Performance tests were subsequently initiated in the Hellenic subduction region (Fig. 2c). No detailed analysis of the outcome of these tests was ever completed, partly because of a lack of events favourable to the hypothesis, and partly because the research was developing rapidly and its scope broadening into new areas.

None of the precursory swarm hypothesis tests ever succeeded in its own terms. The proposed acceptance criterion was to reach a performance factor (likelihood ratio) of 20, which after a few relevant earthquakes should have been an easily achievable target for an effective model. In retrospect, it is clear that the swarm hypothesis was disadvantaged in the tests by the adoption at the outset of overly optimistic estimates of the false alarm and failure rates. This meant that the model being tested was too strong, in the sense that too high a proportion of the hazard was concentrated into too small a proportion of the total time–space volume. Such a model is inherently less likely to succeed than a weaker model, nonetheless the achievement is greater if it does succeed. Starting with more pessimistic estimates of the false alarm and failure rates would have given the tests a better chance to succeed, given a long enough test period.

As the tests proceeded, it became increasingly evident that they were prone to be affected by minor subjective elements in the identification of main-shock events, swarms, and multiplets, and by events close to the boundaries between classes. On one occasion, a main-shock event included a group of aftershocks located at some distance from the main shock. Taken by themselves, this group would have qualified as a swarm and could have been, but were not, interpreted as such. Subsequently a main-shock event occurred in the vicinity of this group, to the detriment of the model. On more than one occasion, a main-shock event would have scored better if some swarms had been regarded as separate events, rather than as a single cluster. An illustration of this was the 2008 July M 6.5 main-shock event in southwest Greece, in the vicinity of a swarm with M_P 5.1 which occurred in 1993 (swarm 19 in Table 2 of Evison and Rhoades, 2000). In the initial version of the tests in

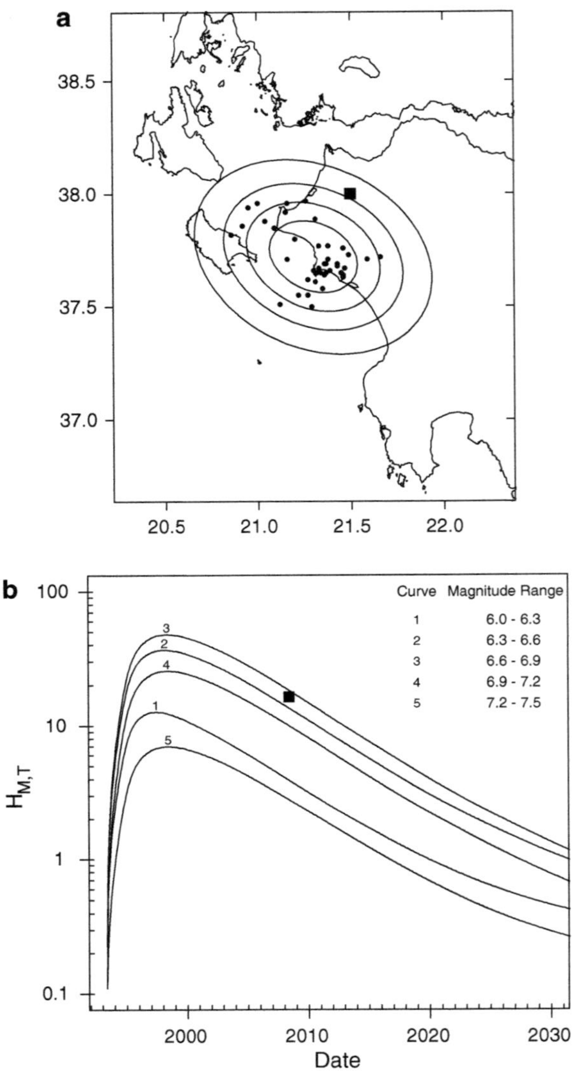

Figure 6
Test prediction launched in June 1999 in tests of the generalised precursory swarm hypothesis in Greece. Data for the M 6.5 earthquake of 8 June 2008 has been superimposed. **a** Location factor H_L. The value is unity at the centre of the target, and contours are shown for $H_L = 0.8, 0.6, 0.4$ and 0.2. The *filled circles* are the swarm epicentres and the *filled square* is the main-shock epicentre. **b** Magnitude and time-of-occurrence factor $H_{M,T}$. The *filled square* represents the main shock. The combined performance factor would have been 5.8. This prediction was later withdrawn and replaced by that in Fig. 7

Greece launched on 22 June 1999, a forecast based on swarm 19 by itself was launched (Fig. 6). In a reassessment following a revision of the AUTH catalogue, swarm 19 was combined into a single cluster with swarms 13 and 16 (from the same table) and became part of historical sequence Gg (Table 4 of Evison and Rhoades, 2000), which already had an M_m 6.6 main-shock event in 1997. The revised forecast, launched on 1 September 2000, is shown in Fig. 7. The performance factor for the 2008 main shock would have been 5.8 under the initial forecast, but was 0.36 under the revised forecast.

On other occasions, a decision to include or exclude one or two marginal earthquakes made the difference between the classification of a multiple event as a swarm or a multiplet. Some large earthquakes did not qualify as multiple events, because they did not have two aftershocks above the adopted magnitude threshold within 24 h, and so played no part in the tests. Also, for the M 7.0 East Cape main-shock event, which represented the greatest individual forecasting success in the tests, the aftershock sequence was so strong that if the magnitude of one aftershock had been 0.1 larger, the whole event would have qualified as a multiplet, rather than as a main-shock event, and would have been regarded as irrelevant to the tests. All of these examples point to the conclusion that the tests would have been more robust if a probabilistic, or perhaps fuzzy, identification scheme had been used, to reduce their sensitivity to marginal events and subjective elements of the identification scheme. Nevertheless, the tests were generally adequate for their purpose, and no simple refinement of the identification scheme would have substantially altered the overall outcome.

Around August of 2004, we discussed the possibility of terminating the swarm hypothesis performance tests, since it seemed that they had served their purpose. That decision was never taken, but Frank Evison's death in January 2005 brought an end to the identification of new precursory swarms in the catalogues, and meant that the formal evaluation of the performance could not extend much beyond that.

5. *Three-Stage Faulting Process*

A conceptual advance was made when an elegant qualitative physical model of the long-term seismogenic process, including swarms, quarms and the predictive relations, was devised. This was a

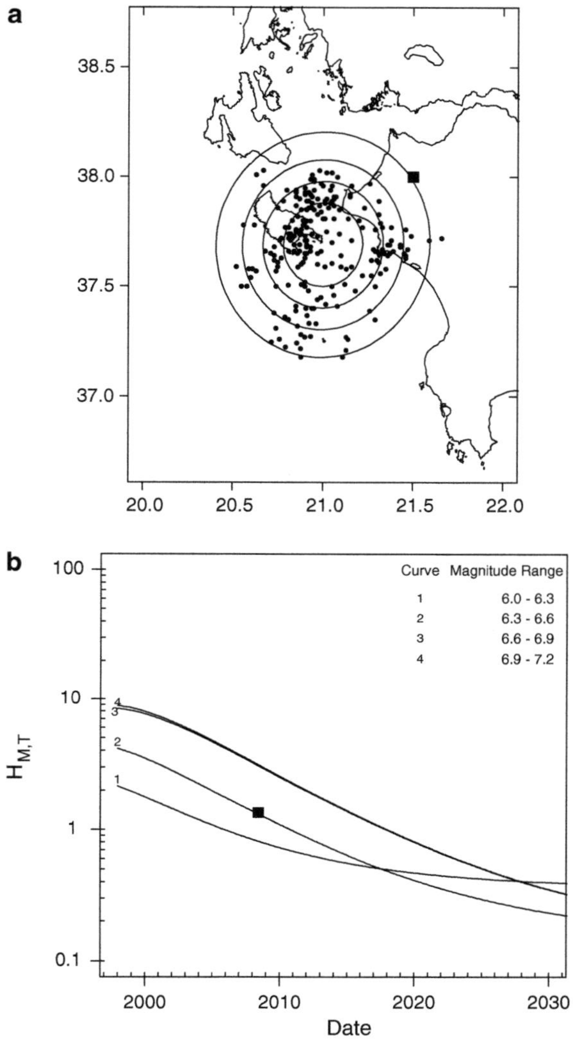

Figure 7
Test prediction launched in September 2000 in tests of the generalised precursory swarm hypothesis in Greece, superseding that of Fig. 6, after three swarms were amalgamated into a single cluster. Data for the *M* 6.5 earthquake of 8 June 2008 has been superimposed. **a** Location factor H_L. The value is unity at the centre of the target, and contours are shown for $H_L = 0.8, 0.6, 0.4$ and 0.2. The *filled circles* are the swarm epicentres and the *filled square* is the main-shock epicentre. **b** Magnitude and time-of-occurrence factor $H_{M,T}$. The *filled square* represents the main shock. The combined performance factor is 0.36

three-stage faulting process, involving crack formation, rupture, and healing (Evison and Rhoades, 1998, 2001, 2004b). It begins with the formation (or weakening) of a major crack, which induces the formation of aftercracks in its neighbourhood, just as a main shock induces aftershocks. The aftercracks render the medium inhomogeneous, and only after they have ruptured, as precursory earthquakes, and subsequently healed, does the medium become sufficiently uniform on the scale of the major crack for it to rupture, as the main shock. After rupture, the major crack itself heals (or strengthens) and becomes homogeneous with its surroundings. The size of the major crack thus determines the magnitude of the main shock, and indeed the scale of the whole process, including the area occupied by the whole process, the magnitude of the largest precursory earthquakes and the length of the precursor time. The scaling relation for precursor time results from the fact that each aftercrack is also, on its own scale, a major crack with its own aftercracks, which must rupture and heal before it does.

Frank consciously applied Occam's razor in formulating the qualitative model. He delighted in the fact that the only extension it involved, beyond what was already generally accepted, was the long separation in time between the formation of a crack and its rupture in an earthquake. In presenting the model, he would challenge the audience to suggest a simpler model, if one existed. To my knowledge, his challenge was never answered. This does not mean that there was wide acceptance of the qualitative model—far from it. Numerical/physical modellers of the earthquake process found in it few of the elements that they were accustomed to using in their modelling. And there was difficulty publishing the model in an environment which had become unreceptive to positive ideas concerning earthquake forecasting, following a well-publicized unpredictability campaign led by a small group (Geller, 1996; Main, 1996; Geller *et al.*, 1997).

In any event, the qualitative model passed its first test. Although it had been built to explain precursory clusters of earthquakes, i.e., swarm or quarms, Frank argued that if conditions favourable to such clustering did not exist, the precursory seismicity should occur anyway, in its most general form as a long-term precursory increase of seismicity at similar magnitudes to the eventual aftershocks. This inference from the model proved correct. A cumulative magnitude plot was developed to display this proposed "precursory scale increase",

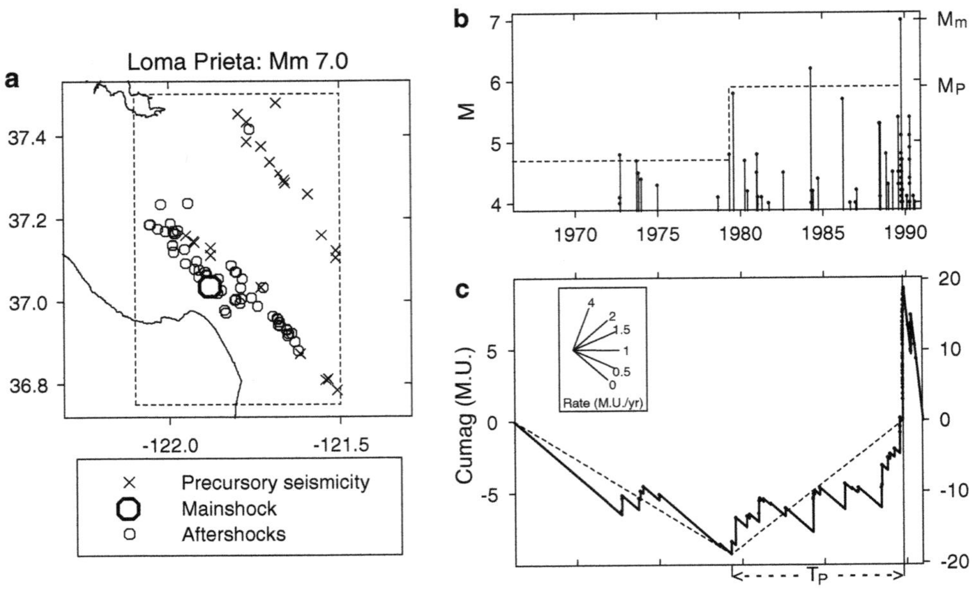

Figure 8
Example of the Ψ phenomenon. **a** Epicentres of precursory earthquakes, main-shock and aftershocks. *Dashed lines* enclose the precursory area A_P. **b** Magnitudes versus time of prior and precursory earthquakes, also main shock and aftershocks. *Dashed lines* show precursory increase in magnitude level. M_m is main-shock magnitude; M_P is precursor magnitude. **c** Cumulative magnitude anomaly (Cumag) versus time. *Dashed lines* show precursory increase in seismicity rate. Protractor translates cumag slope into seismicity rate in magnitude units per year (MU/year), for times before the main shock. Cumag values at the right hand ordinate refer to times beginning with the main shock

or "Ψ-phenomenon" as it became known. It was then easy to identify the Ψ-phenomenon before nearly all of the major earthquakes in four well-catalogued regions (EVISON and RHOADES, 2002, 2004a), and later in other regions. An example is shown in Fig. 8. This phenomenon was common to all tectonic settings studied, including California and the onshore regions of Japan. Each example of the Ψ-phenomenon gives values of the major earthquake magnitude (M_{m}), precursor magnitude (M_{P}) defined as the average of the three largest precursory earthquakes, precursor time (T_{P}) and precursor area (A_{P}). These quantities are all well correlated with each other, as for the special cases of precursory swarms and quarms, as shown in Figs. 9(a–c), and the predictive regressions of M_{m} and T_P on M_P are similar to those for swarms (EVISON and RHOADES, 2004a). Further, Fig. 9(d) shows the relation of M_P to M_a (the average of the three largest aftershocks). The closeness of this relation to the identity line shows that, indeed, the precursors and the aftershocks have similar magnitudes.

6. EEPAS Model

In the more general context of the precursory scale increase, the difficulty of distinguishing precursory earthquakes from others was clearly apparent. Consequently this problem was set aside and a testable model (EEPAS—"Every Earthquake a Precursor According to Scale") was built in which every earthquake makes a transient contribution to the future earthquake occurrence rate density, with the scale of its contribution in time, magnitude and location being controlled by its magnitude, consistent with the Ψ predictive scaling relations. The rate density of the EEPAS model can be described by an equation very similar to (1):

$$\lambda(t,m,x,y) = \sum_{i=1}^{N} \eta(m_i) w_i f_i(t,m,x,y) + \mu\lambda_0(t,m,x,y), \tag{2}$$

with the main differences being that the summation is now over all previous earthquakes in the catalogue above a magnitude threshold, instead of over a small number of identified swarms, and the valid-alarm probability is replaced by a weight w_i for each earthquake and a normalization function $\eta(m_i)$. Also the role of the stationary uniform Poisson model is now filled by a quasi-static model (PPE—"Proximity to Past Earthquakes") similar to a model proposed by JACKSON and KAGAN, (1999). The EEPAS model is simpler than the swarm model in that there is no estimation of false-alarm or failure rates, no conditional updating, and no subjective judgements of any kind to be made when applying the model. The advantage is that the performance of the model can be measured by testing it in on any catalogue independent from that used to estimate the parameters, with real-time tests serving mainly to demonstrate the performance to others.

The EEPAS model was fitted to the New Zealand catalogue (Fig. 10a), and tested with essentially unchanged parameters on the catalogue of California (Fig. 10b). The performance graph for the California region (Fig. 11a) clearly shows the efficacy of the model, and in this respect contrasts strongly with those for the swarm hypothesis tests. The success of this relatively weak model in subsequent independent tests (RHOADES and EVISON, 2005, 2006; CONSOLE *et al.*, 2006) across a variety of tectonic settings (Figs. 10b–d) confirms that the predictive scaling relations are ubiquitous in earthquake catalogues. This was progress towards improved forecasting of earthquake likelihoods, but it was not a fulfillment of Frank's goal, which was to forecast individual major earthquakes.

The EEPAS model is a weak forecasting model built around the strongest feature of the swarm hypothesis—the precursory scaling relations. Numerical/physical modellers are now beginning to take an interest in these scaling relations, and to recognize the potential importance of the physical activation process revealed by precursory swarms and the Ψ-phenomenon (BEN-ZION, 2008). Continued interest on their part should eventually result in a more detailed explanation of the Ψ-phenomenon than the three-stage faulting process provides, incorporating accepted physical elements such as: Stress transfer between faults by means of elastic, viscoelastic and other interactions; frictional or other material instabilities acting on fault surfaces; increasing stress accumulation due to tectonic forces; and the detailed geometry of the fault surfaces (RUNDLE *et al.*, 2002).

7. Further Lessons

The experience of formulating and testing the precursory swarm hypothesis has confirmed the essential role of formal hypothesis testing in this branch of science. In this case the most valuable function of testing was not in proving the swarm hypothesis to be correct, but rather in clarifying its shortcomings. Perhaps it is always so. Once it has been formalised, tested and found wanting, an hypothesis can be improved upon; if it is never put in a testable form, an hypothesis can never be improved upon, and its value is forever unknowable. Some proposals that seem to have promise for earthquake forecasting linger for a long time at the anecdotal stage. Progress can occur only when they are put into a testable form. The present effort of the Collaboratory for the Study of Earthquake Predictability (JORDAN, 2006) to promote testing of forecasting models is a forward step. The fruits of this effort will lie not only in determining which of the submitted models performs best, but in the impetus that the testing will give to the development of new models with improved forecasting skill.

An important lesson learned from Frank himself is the necessity of adopting a sound empirical approach in order to make progress in this field. Frank had an immense respect for data. He came from a generation that grew up before computers, and perhaps as a result he developed an unusual ability to carry out data analysis by hand. This meant he actually studied the data in detail, and not just standard analyses produced by computer programs. He regarded our earthquake catalogues, in spite of their well-known deficiencies, to be immensely valuable resources for hypothesis formulation and testing. While recognising that hypotheses could be tested against theory or observation, he held that the theory of seismogenesis is not yet sufficiently well developed to provide a basis for either formulating or testing precursory hypotheses. Accordingly an empirical methodology is the appropriate one.

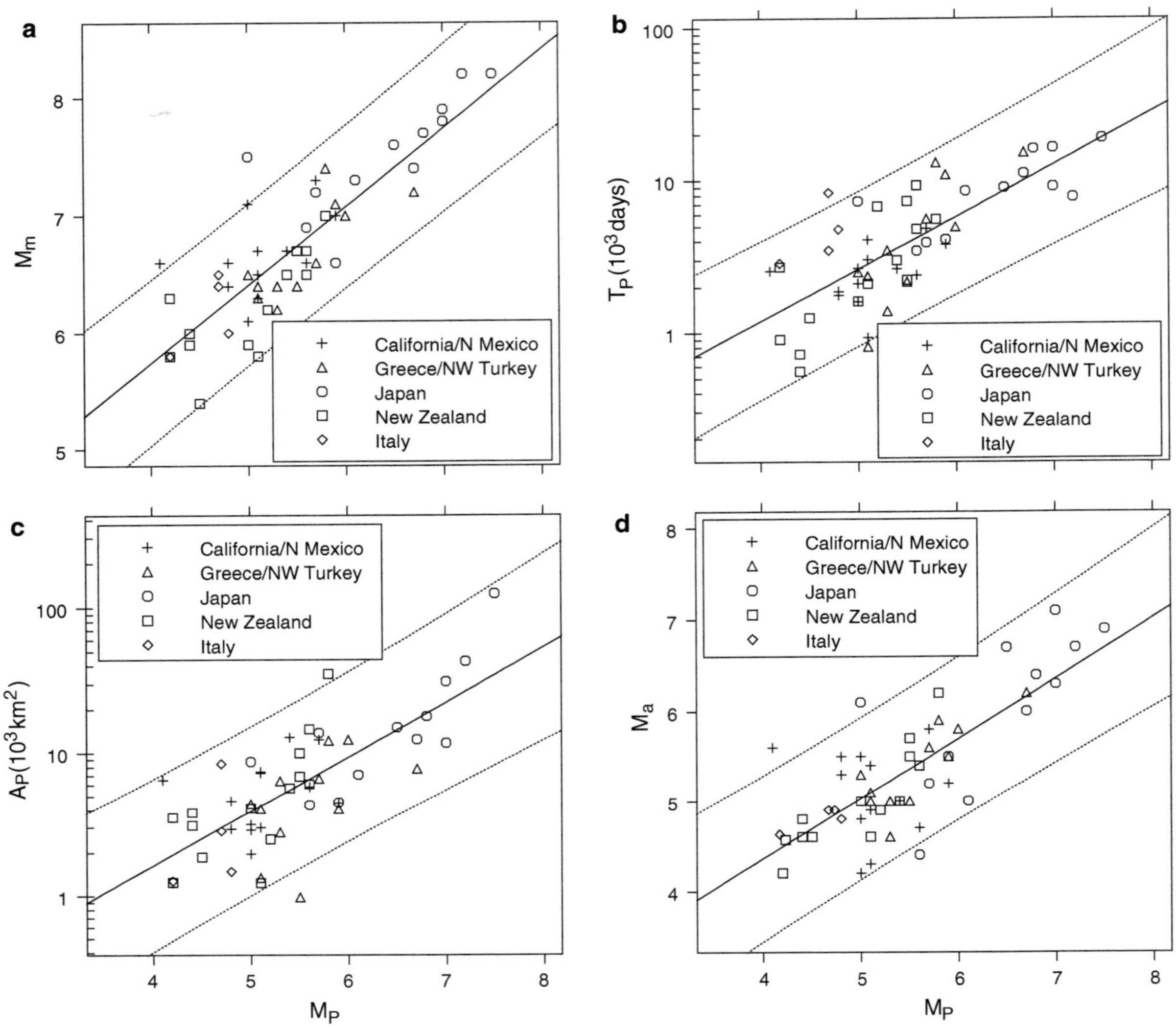

Figure 9
Ψ-phenomenon scaling relations: **a** Main shock magnitude M_m on precursor magnitude M_P; **b** Precursor time T_P on M_P; **c** Precursor area A_P on M_P; **d** Aftershock magnitude M_a on M_P. Data from Evison and Rhoades (2004a, 2005)

8. Questions for Future Studies

The scaling relations between M_P, M_m, T_P and A_P raise some interesting questions which should be considered in future studies.

The first question is whether the same relations apply to all regions, regardless of tectonic setting and regional seismicity rate. The linearity of the relations of log T_P and log A_P with M_m seem to be consistent with, or perhaps a consequence of, the Gutenberg-Richter relation, because for a time–space region of uniformly distributed seismicity, the number of earthquakes N is proportional to the product of time and area, and, if the earthquakes constitute a Gutenberg-Richter set, then the expected maximum magnitude (represented in the relations by M_m) increases linearly with log N. The relations of log T_P and log A_P with M_P thus seem to be a consequence of the Gutenberg-Richter law and the linearity of the relation between M_m and M_P. However, the data used to construct the Ψ scaling relations (Fig. 9) do not come from a single region of uniformly distributed seismicity. They come from diverse regions, and the data from all regions appear to fit the combined

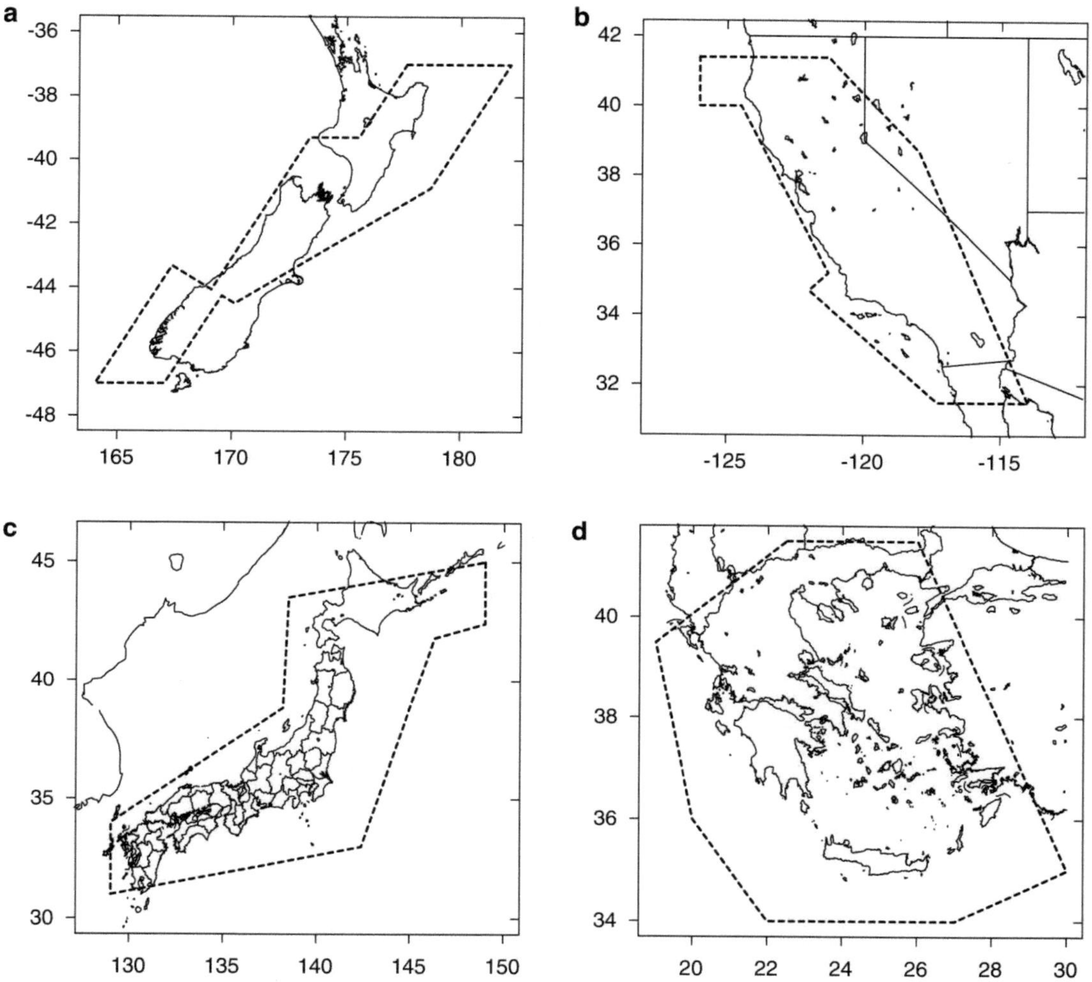

Figure 10
Regions of surveillance for independent tests of the EEPAS model **a** New Zealand; **b** California; **c** Japan; **d** Greece

relations well, in spite of the fact that Japan is on the whole several times more seismically active than Greece, which is in turn somewhat more active than California, New Zealand and Italy. Fitting separate parallel scaling relations to each region does not significantly improve the fit to the data. However, if all regions do conform to the same relations, the question then arises as to what becomes of the Ψ-phenomenon in regions where the seismicity rate is considerably lower, by an order of magnitude or more, than that in any of these regions? Considering the precursor time for instance, if the Ψ phenomenon does occur in low-seismicity regions, then either very few precursory earthquakes must be involved, or the precursor-time must be much longer than for regions of high seismicity. Either of these possibilities would be consistent with the fact that the large earthquakes recorded in low-seismicity areas do not display strong precursory scale increases on the time-scale indicated by the existing scaling relations. Fitting the EEPAS model to high quality catalogues with a variety of seismicity rates may assist in resolving this question.

A second question arising from the scaling relations is why the precursor area and the aftershock area scale differently with magnitude. In the Ψ scaling relation between log A_P and M_m, the magnitude coefficient is 0.48 (EVISON and RHOADES, 2004a), and a similar coefficient of about 0.5 was also found in the line dividing the areas of precursory swarms

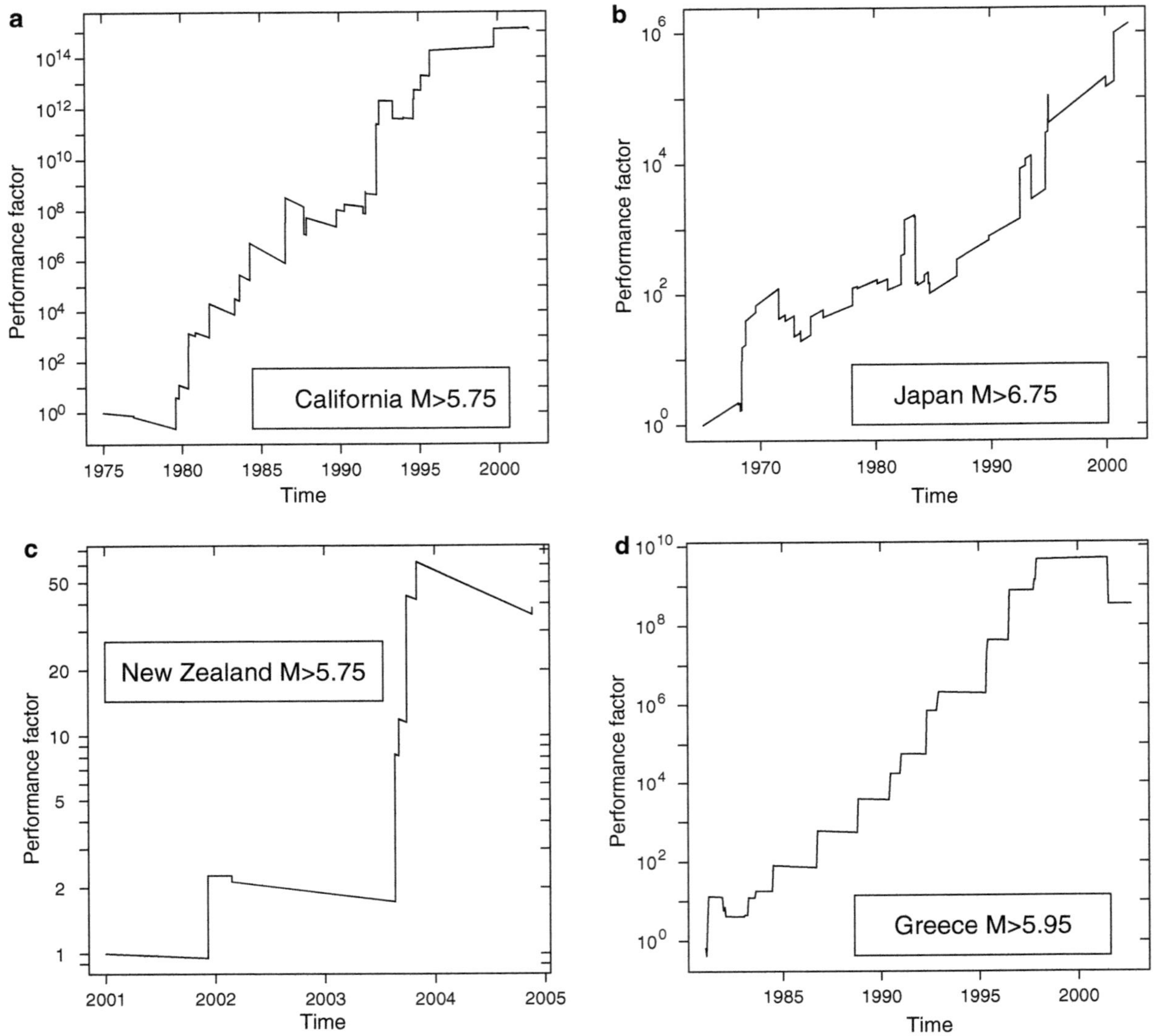

Figure 11
Evolution of the performance factor in tests of the EEPAS model on independent data sets **a** California (Rhoades and Evison, 2004); **b** Japan (Rhoades and Evison, 2005); **c** New Zealand; **d** Greece (Console *et al.*, 2006). The performance factor is the ratio of the likelihood under the EEPAS model to that under a static or quasi-static reference model. In (**a–c**) the reference model is PPE (Rhoades and Evison, 2004). In **d** the reference model is stationary Poisson with Frankel smoothing of locations (Console *et al.*, 2006)

and multiplets (Fig. 12). However in the Utsu (1961) aftershock area relation, the coefficient of magnitude is 1. Thus the precursory area is proportional to $10^{0.5M}$, whereas the aftershock area is proportional to 10^{M}. The relations are compared quantitatively in Fig. 13. They show that at about magnitude 9, the precursor area and aftershock area are similar, but that the ratio of precursor area to aftershock area becomes progressively larger as the magnitude is reduced. Considering the qualitative model mentioned above, the major crack and the earthquake fault are the same. Should not the aftercracks (precursors) and the aftershocks therefore occupy similar areas? The answer to this question may perhaps be found by detailed physical modelling of the cracking and aftershock processes.

An obvious question arising from this whole body of work is whether it is possible to move closer to the

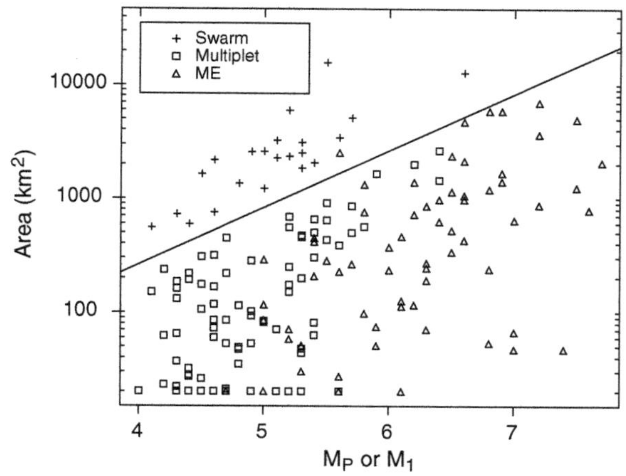

Figure 12
Areas occupied by precursory swarms, multiplets and main-shock events (ME) as a function of magnitude M_P for swarms and multiplets, or M_1 for main-shock events. The dividing line between swarm and multiplet areas is $\log A = 0.42 + 0.5\ M_P$. Data from Evison and Rhoades (2000)

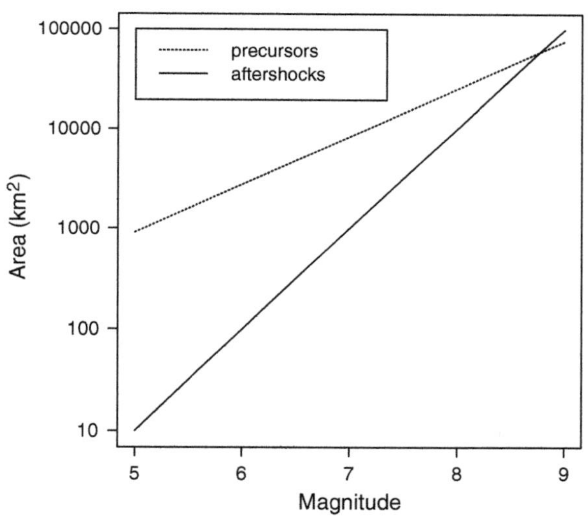

Figure 13
Comparison of Ψ-scaling relation for precursor area (*dashed line*) and the Utsu (1961) relation for area of aftershocks against main-shock magnitude (*solid line*)

original goal of forecasting individual major earthquakes by adding strength to the EEPAS model. Can the model be strengthened by giving a relatively high weight to earthquakes which are part of swarm- and quarm-like clusters, while avoiding re-introducing any subjective elements for recognition of such clustering? When one considers that the probability gain, relative to a time-invariant model, is on average no more than about two for each major earthquake occurrence at present, and that the probability gains for individual successes in the swarm tests reached 114, the scope for improvement seems obvious. The only circumstance that would seem to preclude any improvement would be if every earthquake actually is a long-term precursor, as the name of the model suggests. This is a challenge for statistical seismology.

9. Conclusion

The discovery of precursory swarms and the predictive relations of long-term seismogenesis was an achievement made possible by Frank Evison's unusual ability in exploratory data analysis. The subsequent long-running tests of the precursory swarm hypothesis in several regions reflected his conviction that the precursory scaling relations were useful for forecasting and his commitment to a systematic empirical approach to research on earthquake precursors. Although the formal tests of the precursory swarm hypothesis never succeeded in their own terms, they played an important role in the process of scientific discovery that eventually led to an improved understanding of some aspects of the long-term seismogenic process, and thereby to a model with unequivocal long-range forecasting skill, demonstrating that the precursory scaling relations are ubiquitous in earthquake catalogues. Further research is needed to resolve unanswered questions pertaining to the scaling relations, and to increase the strength of long-range forecasting models by exploiting the clustering of some precursory earthquakes in swarms and quarms.

Acknowledgments

This work was supported by the Foundation for Research, Science and Technology under Programme Number CO5X0402 and by GNS Science Capability Funding. Thanks are extended to R. Robinson, M. Reyners and R. Console for helpful reviews of the manuscript.

References

ADAMS, R.D. (1976), *The Haicheng, China, Earthquake of 4 February 1975; the first successfully predicted major earthquake*, Earthq. Eng. Struct. Dyn. *4*, 423–437.

AKI, K. *A probabilistic Synthesis of Precursory Phenomena*, In: *Earthquake Prediction—An International Review*, Maurice Ewing series, 4, (Simpson and Richards, eds.) (AGU, Washington D.C. (1981)) pp. 566–575.

ALLEN, C.R. (1976), *Responsibilities in earthquake prediction*, Bull. Seismol. Soc. Am. *66*, 2069–2074.

BEN-ZION, Y. (2008), *Collective behavior of earthquakes and faults: Continuum-discrete transitions, progressive evolutionary changes and different dynamic regimes*, Rev. Geophys. *46*(4), RG4006.

CONSOLE, R., RHOADES, D.A., MURRU, M., EVISON, F.F., PAPADIMITRIOU, E.E., KARAKOSTAS, V.G. (2006), *Comparative performance of time-invariant, long-range and short-range forecasting models on the earthquake catalogue of Greece*, J. Geophys. Res. *111*, B09304, doi:10.1029/2005JB004113.

EVISON, F.F. (1977a), *Fluctuations of seismicity before major earthquakes*, *Nature 266*, 710–712.

EVISON, F.F. (1977b), *Precursory seismic sequences in New Zealand*, New Zealand J. Geol. Geophys. *20*, 129–141.

EVISON, F.F. (1977c), *The precursory earthquake swarm*, Phys. Earth Planet. Inter. *15*, 19–23.

EVISON, F.F. (1978), *Long-term seismic precursor to the 1968 Inangahua Earthquake, New Zealand*, New Zealand J. Geol. Geophys. *21*, 531–534.

EVISON, F.F. (1981), *Multiple earthquake events at moderate-to-large magnitudes in Japan*, J. Phys. Earth *29*, 327–339.

EVISON, F.F. (1982), *Generalised precursory swarm hypothesis*, J. Phys. Earth *30*, 155–170.

EVISON, F.F. *Communication of earthquake predictions and warnings*. In *Earthquake Prediction: Proceedings of the International Symposium on Earthquake Prediction*, (Terra Scientific Publishing Company, Tokyo (1984)), pp. 937–946.

EVISON, F.F. and Rhoades, D.A. (1993), *The precursory earthquake swarm in New Zealand: Hypothesis tests*, New Zealand J. Geol. Geophys. *36*, 51–60.

EVISON, F.F. and RHOADES, D.A. (1997), *The precursory earthquake swarm in New Zealand: Hypothesis tests II*, New Zealand J. Geol. Geophys. *40*, 537–547.

EVISON, F.F. and RHOADES, D.A. (1998), *Long-term seismogenic process for major earthquakes in subduction zones*, Phys. Earth Planet. Inter. *108*, 185–199.

EVISON, F.F. and RHOADES, D.A. (1999a), *The precursory earthquake swarm and the inferred precursory quarm*, New Zealand J. Geol. Geophys. *42*, 229–236.

EVISON, F.F. and RHOADES, D.A. (1999b), *The precursory earthquake swarm in Japan: Hypothesis test*, Earth, Planets and Space *51*, 1267–1277.

EVISON, F.F. and RHOADES, D.A. (2000), *The precursory earthquake swarm in Greece*, Annali di Geofisica *43*, 991–1009.

EVISON, F.F. and RHOADES, D.A. (2001), *Model of long-term seismogenesis*, Annali di Geofisica *44*, 81–93.

EVISON, F.F. and RHOADES, D.A. (2002), *Precursory scale increase and long-term seismogenesis in California and Northern Mexico*, Ann. Geophys. *45*, 479–495.

EVISON, F.F. and RHOADES, D.A. (2004a), *Demarcation and scaling of long-term seismogenesis*, Pure Appl. Geophys. *161*, 21–45.

EVISON, F.F. and RHOADES, D.A. (2004b), *Long-term seismogenesis and self-organized criticality*, Earth, Planets and Space *56*, 749–760.

EVISON, F.F. and RHOADES, D.A. (2005), *Multiple-mainshock events and long-term seismogenesis in Italy and New Zealand*, New Zealand J. Geol. Geophys. *48*, 523–536.

FIELD, E.H. (2007), *Overview of the Working Group for the Development of Regional Earthquake Likelihood Models (RELM)*, Seismol. Res. Lett. *78*(1), 7–16.

GELLER, R.J., *VAN: A critical evaluation.* In A Critical Review of VAN, ed. J. Lighthill (World Scientific, Singapore (1996)), pp. 155–238.

GELLER, R.J., JACKSON, D.D.,KAGAN, Y.Y., and MULARGIA, F. (1997), *Earthquakes cannot be predicted*, Science *275*, 1616–1217.

INTERNATIONAL UNION OF GEODESY AND GEOPHYSICS (1984), *Code of Practice for Earthquake Prediction*, IUGG Chronicle 165, 115–124 (reprinted in Eos, 65, Feb. 14, 1984).

JACKSON, D.D. and KAGAN, Y.Y. (1999), *Testable earthquake forecasts for 1999*, Seismol. Res. Lett. *70*(4), 393–403.

JORDAN, T.H. (2006). *Earthquake predictability, brick by brick*, Seismol. Res. Lett. *77*(10), 3–6.

MAIN, I.G. (1996), *Statistical physics, Seismogenesis, and seismic hazard*, Rev. Geophys. *34*, 433–462.

RHOADES, D.A. and EVISON, F.F. (1979), *Long-range earthquake forecasting based on a single predictor*, Geophys. J. Royal Astronom. Soc. *59*, 43–56.

RHOADES, D.A. and EVISON, F.F. (1993), *Long-range earthquake forecasting based on a single predictor with clustering*, Geophys. J. Internat. *113*, 371–381.

RHOADES, D.A. and EVISON, F.F. (2004), *Long-range earthquake forecasting with every earthquake a precursor according to scale*, Pure Appl. Geophys. *161*, 47–72.

RHOADES, D.A. and EVISON, F.F. (2005), *Test of the EEPAS forecasting model on the Japan Earthquake Catalogue*, Pure Appl. Geophys., *162*, 1271–1290.

RHOADES, D.A. and EVISON, F.F. (2006), *The EEPAS forecasting model and the probability of moderate-to-large earthquakes in Central Japan*, Tectonophysics *417*, 119–130.

RIKITAKE, T. (1975), *Earthquake Precursors*, Bull. Seismol. Soc. Am. 65, 1133–1162.

RIKITAKE, T. (1979), *Classification of Earthquake Precursors*, Tectonophysics 54, 293–309.

RIKITAKE, T., *Earthquake Forecasting and Warning*, Center for Academic Publications Japan, (Kluwer (1982)), 402 pp.

RUNDLE, J.B., RUNDLE, P.B., KLEIN, W., DE SA MARTINS, J., TIAMPO, K. F., DONNELLAN, A. and KELLOGG, L.H. (2002), *GEM plate boundary simulations for the plate boundary observatory: A program for understanding the physics of earthquakes on complex fault networks via observations, theory and numerical simulation*, Pure Appl. Geophys. *159*, 2357–2381.

UTSU, T. (1961), *A statistical study on the occurrence of aftershocks*, Geophys. Mag. *30*, 521–605.

(Received September 5, 2008, revised April 23, 2009, accepted June 9, 2009, Published online February 25, 2010)

Pure Appl. Geophys. 167 (2010), 645–653

DOI 10.1007/s00024-010-0079-z

Pure and Applied Geophysics

Foundations of Statistical Seismology

David Vere-Jones[1]

Abstract—A brief account is given of the principles of stochastic modelling in seismology, with special regard to the role and development of stochastic models for seismicity. Stochastic models are seen as arising in a hierarchy of roles in seismology, as in other scientific disciplines. At their simplest, they provide a convenient descriptive tool for summarizing data patterns; in engineering and other applications, they provide a practical way of bridging the gap between the detailed modelling of a complex system, and the need to fit models to limited data; at the most fundamental level they arise as a basic component in the modelling of earthquake phenomena, analogous to that of stochastic models in statistical mechanics or turbulence theory. As an emerging subdiscipline, statistical seismology includes elements of all of these. The scope for the development of stochastic models depends crucially on the quantity and quality of the available data. The availability of extensive, high-quality catalogues and other relevant data lies behind the recent explosion of interest in statistical seismology. At just such a stage, it seems important to review the underlying principles on which statistical modelling is based, and that is the main purpose of the present paper.

1. Introduction

Nearly three decades ago, I was asked to write an account of statistical modelling of earthquake occurrence in time and space. To that end I endeavored to clarify, for myself as much as anyone else, what statistical models were doing in an observational science such as seismology (Vere-Jones, 1979).

Those 30 years have seen great changes. Earthquake prediction flowered, withered, and has begun to flower again, albeit with a much more cautious tone. We now have access to data of a scale and quality that would have been hard to imagine 30 years ago, just as we have computing devices of a power and speed that would have been equally hard to anticipate. In particular, the explosion of extensive, high-quality seismic data is a major reason behind the current increased interest in statistical seismology.

At just such a stage, it seems worthwhile to review again the purpose of stochastic modelling in what has been traditionally viewed as an observational science, and to ask what it has achieved. In particular, what is the benefit in treating the spatial (or temporal) locations of earthquakes as a random process?

This question seems to me a challenge to clarify the principles and achievements of statistical seismology itself. In these notes I have tried to set down thoughts in this direction.

The broad outline of the paper follows the same general pattern as my earlier (1979) paper. However, the models and their applications have filled out greatly since that time, and I have tried to choose illustrations that help to indicate these changes. The paper makes no attempt to provide a systematic review, however; the focus is on the different roles that stochastic models can play, and the considerations that have to be borne in mind if the modelling is to be useful. A more extended review is in Vere-Jones (2009).

2. Physical and Stochastic Models

To start with terminology, by a 'stochastic model' I mean a model in which some stochastic or random elements appear. 'Statistical model' is used almost synonymously, but I prefer the term 'stochastic'

[1] Statistical Research Associates, Victoria University, Wellington, New Zealand. E-mail: David.Vere-Jones@ecs.vuw.ac.nz

because it draws attention to the key element—the appearance of some features whose behaviour cannot be predicted exactly, but can be summarized in probability terms.

The difference between a physical and stochastic model is commonly taken to mean that the physical model embodies no random components, while the stochastic model embodies no physical components. Both are exaggerated views, and neither touches on the real requirements of useful models in a field such as seismology. Of course, there is a difference between deterministic and stochastic models, but neither seems to me more physical than the other. Both are physical to the extent that they succeed in describing and predicting the physical processes they purport to represent.

A physical model which contains no stochastic component deprives itself of the possibility of making any statements concerning the uncertainties of predictions based on the model. In a famous statement, Sir JEFFREYS (1939), who is also regarded as a pioneer in inferential statistics, argued that, to be worthy of its name, every physical theory should contain within itself the means not only of predicting the relevant observed quantities, but also of predicting their uncertainties. In our terminology, he was arguing that every physical theory should include some stochastic component.

From a philosophical point of view, the basic feature of a stochastic model is that it accepts that some aspects of the physical process are out of range, at least for practical purposes, and must be replaced in the model by some unknowable and hence random process. The model should incorporate in physical terms those aspects of the physical phenomenon which are stably observed and repeatable, but may relegate the rest to dice-tossing or one of its more contemporary avatars such as Brownian motion or the Poisson process.

In classical physics, the uncertainties in the model are traditionally attributed to observational errors. In quantum physics, by contrast, the uncertainties reflect a fundamental, irrefutable characteristic of nature. Earthquake seismology, at the present time, occupies an uncomfortable middle ground. General patterns of behaviour may be predicted qualitatively from physical theories, but the theories do not extend to the prediction of local earthquakes. Our uncertainties include, but are not restricted to, observational errors. A more fundamental difficulty is that the available observational data, wholly or largely based on instruments located on the Earth's surface, are only indirectly related to the physical processes within the Earth's crust. Stochastic models of earthquake occurrence must somehow marry the existing physical theory to the restricted data that bears directly on questions such as the initiation of a rupture and its development to a large-scale earthquake.

Under such circumstances, quantifying the uncertainties in the model predictions represents a major and formidable challenge. I believe it is fundamentally for this reason that the stochastic models for earthquake occurrence often appear to reflect the physical picture in such a limited way. They reflect the gap in our ability to observe and hence to understand the crucial physical processes at the microscopic level.

But all such general statements are too glib. In actual practice, a wide variety of models, both stochastic and otherwise, are used for a wide range of different purposes. In the next few sections I shall try to classify some of the different roles played by stochastic models in general, and in seismology in particular.

3. *Different Types of Stochastic Models*

Across their diverse fields of application in the natural sciences, two broad roles for stochastic models may be distinguished.

The first is epitomized by statistical mechanics. Here the stochastic model plays an integral role in understanding the physical processes themselves. Large-scale, macroscopic behaviours are deduced from the statistical properties of the interactions of large numbers of microscopic entities. Many details governing the behaviour of the microscopic entities do not need to be known precisely, as they do not effect the macroscopic effects. Hence very simple ("toy") models may suffice at the microscopic level, provided they capture the key aspects which influence behaviour at the macroscopic level.

In the second, by far more common, type of application, the stochastic model is used as a basis for planning, prediction or decision-making. In this case, whether or not it adequately represents the underlying physical processes may not be the crucial aspect. What is important in such applications is to know the reliability of the forecast values: not just the forecast values themselves, but some measure of their variability, given the information on which they are based.

It is also vital, since the model may be largely empirical in character, that the parameters in the model can be fully determined from the available data. Even if better models exist in principle, they will be of little practical value if they rely on information that cannot be obtained from the available data.

In my earlier paper I distinguished not two but three broad classes of models, which I referred to then as conceptual, descriptive, and engineering models. The two last both fall into the second general class referred to above. I no longer like the terminology, but the broad distinctions seem to me still useful, so in the next three sections I examine each of them in more detail, with seismological illustrations.

4. Conceptual Models

Within seismology, a role of this kind is implicit when the fracture processes within the Earth's crust are described as 'frozen turbulence' (KAGAN, 1991). It is also present in models for the fracture strength of materials as functions of the density and size distribution of microcracks within the material. Indeed, statistical models of this character have long played a role in fracture mechanics, from the time of GRIFFITHS (1921) and WEIBULL (1939) forward.

Weibull, for example, attributed the variations in strength from otherwise similar laboratory specimens to the random distribution of microcrack lengths in the specimen. The Weibull distribution takes its name from his studies of these variations.

The branching process, percolation, and cellular automata interpretations of the earthquake process all start from the underlying idea that, instead of progressing smoothly, as would a fault or fracture in a homogeneous elastic medium, the movement of an earthquake rupture is irregular, controlled by its essentially random progress from one weakness to another.

My own interest in this area was stimulated by the possibility of applying branching process ideas to the development of an earthquake rupture. This leads to a stochastic model which predicts a G–R law with b value around 2/3 in the critical case, and to tapered Pareto distributions ('Kagan distributions') when the process is subcritical (VERE-JONES, 1978).

It is remarkable that very similar branching process concepts reappear in Ogata's ETAS model (OGATA, 1988), lending credibility to one of Yan Kagan's old theses (e.g., KAGAN and KNOPOFF, 1981), that the distinction between stages of the rupture itself, and the associated events which accompany or follow it, are due more to the characteristics of our recording instruments than they are to the physical processes causing the motions.

It is also of interest to compare the roles of stochastic models for earthquake occurrence, such as the ETAS model or the branching model for fracture, with models for complex systems, whether stochastic, such as cellular automata, or deterministic, as in block-and-slider and related mechanical models for fault systems.

Under a wide range of conditions, many such models show characteristic features of earthquake occurrence: a G–R law, long-range correlations, aftershock sequences, etc. In this sense there may be no overriding reason for preferring one type of model over another. Each provides a different insight into the circumstances under which these features can be produced.

The merit of a model such as the branching model for crack propagation lies in the extent to which it can explain a complex phenomenon from simple premises. I view as negligible as in modelling a complex physical phenomenon by a model whose complexity approaches that of the phenomenon itself, particularly when both may be adequately predicted by a simple stochastic model, and only the model is known in detail at the microscopic level.

5. Descriptive Models

The term 'descriptive model' seems clear enough to require little further explanation, although a geophysicist might prefer the term 'empirical model' to denote much the same concept. In any case, the aim of the model is to describe, in the sense of being able to reproduce, the broad features of the observed data.

Within seismology, the canonical example would have to be the Gutenberg–Richter frequency–magnitude law (GUTENBERG and RICHTER, 1944). From the outset its purpose was purely descriptive, and in this it succeeds very well. The tragedy is that Gutenberg and Richter, following bad habits common among the physicists, first described their data in terms of numbers rather than proportions, then used the most familiar logarithm tables to base 10, and finally compounded their errors by fitting a least-squares regression line to the cumulative data. This led them to the familiar form

$$\log_{10} N(M) = a + b(M - M_0) + E_M \quad (1)$$

or equivalently

$$N(M) = 10^{a+b(M-M_0)+E_M}, \quad (2)$$

where $N(M)$ is the number of events in the data set under study which have magnitudes above M, M_0 is a magnitude threshold for reliable statistics, and E_M is an error term which, to quote Harold Jeffreys once more, 'is quickly forgotten or altogether disregarded in physical theories.'

Had it been regarded more carefully, the unsuitability of the above formulation might have been more apparent. Just think how different elementary textbooks in seismology might appear if, instead of numbers, Gutenberg and Richter had phrased their discovery in terms of the proportions, $p(M)$ of events above magnitude M:

$$\log p(M) = 10^{-b(M-M_0)}. \quad (3)$$

The pseudo-parameter 10^a disappears, being revealed as nothing more than a normalization constant (here the total number of events above the threshold magnitude). Nor is it any great step to replace the decimal logarithms by natural logarithms, and so to recognize the exponential distribution. Finally the term E_M is revealed as nothing like the error in a regression problem, but as the discrepancy between the true and empirical distribution functions at a particular point M on the cumulative curve, a beast of a totally different character.

My personal view on this issue is that anyone pretending to the title of an up-to- date seismologist should be required on oath to forsake the use of the traditional form of the G–R law (other than in its historical context) and to persuade their colleagues to do likewise, to rid both text-books and current practice of a misleading, unhelpful and totally unnecessary anachronism.

Within this revised formulation, that is, as a distribution function which gives a good fit to the data, the Gutenberg–Richter law is a clear example of a purely descriptive stochastic model. It has no pretence at being a physical law: The reasons why the distribution should be exponential are nowhere mentioned, let alone tied back to some physical theory.

The second obvious example is the Omori Law, at least when described, as suggested by JEFFREYS (1938), as a Poisson process with time-dependent rate of the form

$$\lambda(t) = A(c + t)^{-p}, \quad (4)$$

where A, c and p are parameters and t is the elapsed time since the main shock. This formulation of Omori's law is perfectly adequate for simulating a set of aftershocks with the same broad characteristics as the data set to which the parameters were fitted, and permits estimates of the errors in the parameters as well as in any prediction made from the model. Once again, however, it has no physical content beyond the central fact that it describes a physical phenomenon.

6. *Engineering Models*

By this I mean models developed in order to resolve a particular practical question in some planning, decision-making, or forecasting context.

There is a broad overlap between such models and descriptive models. The main difference is in the purpose of the model. In a descriptive model the main purpose is simply to describe the data effectively. In an engineering model the main purpose is to meet the specified practical end.

Traditional uses of such models in seismology have been those relating to earthquake zoning, earthquake engineering design, and earthquake insurance. For example, for many decades the

Poisson model has played just such a role in the applications to earthquake zoning. Nobody now pretends that the model fits earthquake data well, but its applications to zoning schemes are easy to understand, and lead to recommendations which are reasonable and even conservative in many cases, although it is clear that they ignore certain key features of earthquake occurrence such as aftershocks.

However, it appears to me that the major class of models in this category may now comprise the models used to produce earthquake probability forecasts. The task of the model is clear. It cannot be undertaken without a stochastic model. The question is whether the models are effective for the purpose in view. Recent developments such as the establishment of forecast testing centers testify to the perceived relevance of testing the models against the purpose for which they were designed, rather than against more general purposes.

Before addressing this issue in detail, it may be helpful to formulate a few guiding principles for the development of engineering models in general, which then can be applied to models for earthquake forecasts in particular.

1. The level of detail of the model should match the purpose in view. There is no point in modelling detail that is not needed for the purpose in hand. Moreover a simple model is likely to be more helpful than a complex one in understanding and communicating the issues involved, and also in avoiding nonsense recommendations.
2. The model must be able to be estimated from the available data. There is no room in this context for an elegant model that relies on unavailable data.
3. The model may 'borrow strength' from data outside the immediate context, for example by using Bayesian methods to make partial use of parameter values found from previous studies. Bayesian methods are almost unavoidable in decisionmaking contexts, and are likely to be used in most real-life applications of earthquake probability forecasts.
4. Even though following the physics may not be the main aim, a model which is based on a good, if simplified, physical picture, is likely to be safer for prediction than a model which is purely descriptive or ad hoc. A descriptive or ad hoc model cannot be trusted outside the range of the data to which it has been fitted.
5. Models used for limited-term forecasts may lack the long-term stationarity properties commonly required of time-series models. Such models also should be used with caution in making longer term forecasts.

Let me illustrate these points by two broad situations in seismological studies.

6.1. Hazard Estimates Based on Data from an Individual Fault

Renewal, time-predictable, slip-predictable and stress-release models all fall into this general picture. Their practical purpose is to provide estimates of the hazard on a given fault. They have some physical plausibility, perhaps enough to satisfy (4) above, but points (2) and (3) are very relevant because the data from an individual fault are generally very meager.

In using models for long-term simulations or forecasts, care is needed to avoid internal inconsistencies such as those mentioned in (5). For example, consider the time-predictable model (Shimazaki and Nakata, 1980), where the time intervals between events are modelled by

$$T_i = \exp\{A + BM_i\}, \qquad (5)$$

where the $T_i = t_{i+1} - t_i$ are the times between events, the M_i are the event magnitudes, and A, B are constants. Here the errors in the magnitudes, which may be considerable, become exaggerated by exponentiation, leading to uncontrolled fluctuations and potential biases in the behaviour of the accumulated stress, or accumulated moment release, if the model is simulated over long periods.

Other models, such as the stress-release model, may avoid this problem, but raise other problems of their own. For example, the behaviour of the stress release model is critically dependent on the assumptions made about the magnitude distribution. This leads to clustering if the magnitude distribution has the Gutenberg–Richter form, but to nearly periodic behaviour if the magnitude distribution is sharply peaked as for a characteristic earthquake model.

Unfortunately, the choice of magnitude distribution for events on a given fault is inadequately determined by the data, while the theory is still in contention. The renewal model is also widely used in this context, but its properties vary substantially with the choice of inter-event distribution, and the data rarely have enough information to determine the form of this distribution at all confidently, or to confirm or otherwise the crucial assumption of independent intervals for this model.

6.2. *Models for General Seismicity*

Another group of models that play a somewhat similar role in a different context are the models for general or background seismicity, such as the ETAS and Kagan–Jackson models.

The ETAS model (Ogata, 1988) has an important branching process interpretation, and is widely used as a basis for short-term data-fitting, for the exploration of model properties (foreshocks, Båth's law), and as a diagnostic tool for revealing regions of anomalous seismic activity.

The Kagan and Jackson (1994) model was expressly designed for the purpose of providing a base-line model more realistic than the Poisson model for use in testing proposed forecasting models. It is essentially a space–time smoothing model for earthquake catalogue data.

The EEPAS model (Rhoades and Evison, 2004) adds to the Kagan–Jackson model some explicit prediction terms taken from logarithmic regression studies, and is currently one of the most successful intermediate term forecasting models.

All three models can be defined by conditional intensities of deceptively similar form. For the space–time ETAS model (Ogata, 1998),

$$\lambda(t,x,M) = f(M)\{\mu(\mathrm{x}) + A\,\Sigma\, i : t_i < t\,\Phi(M_i - M_0)g(t-t_i)h(x-x_i)\}. \tag{6}$$

For the Kagan–Jackson model,

$$\lambda(t,x,M) = f(M)H(t)\{\delta + A_t\Sigma\, i : t_i < t\, g(x-x_i)\}. \tag{7}$$

For the EEPAS model,

$$\lambda(t,x,M) = \mu\lambda_0(t,x,M) + \Sigma\, i : t_i < t f(M-M_i)h(t-t_i|M_i)g(x-x_i|M_i). \tag{8}$$

In these expressions, $f(M)$ is the G–R law or one of its variants, while $g(x|M)$ and $h(x|M)$ are probability densities for time and space intervals, respectively, about the initiating event, conditioned by the magnitude M of the initiating event. In the Kagan–Jackson model $H(t)$ is a fixed function used for down-weighting contributions to the current risk from past events, as those events recede into the past.

The basic characters of the models are very different, however. Φ in the ETAS model is an exponential productivity term. It has to be balanced against the G–R term to determine the conditions for criticality. μ governs the background (independent) events and sets the overall spatial pattern. There are simple conditions for the existence of a stationary version, and when simulated from given initial conditions the model converges to its stationary form (ergodicity).

In the Kagan–Jackson model, the constant A_t is adjusted each time a new earthquake is added to the sum, to ensure that the total contribution from the bracketed term is unity and hence that $H(t)$ continues to denote the overall rate. When simulated, the model behaviour is heavily dependent on the initial condition, and the role of the 'surprise events' controlled by δ. It is not clear whether it can be linked to a stationary point process model, even if H is constant, and if so whether that model would be ergodic.

In the EEPAS model, λ_0 is first obtained from a model similar to the Kagan–Jackson model. The terms f, g, h in the sum are taken from logarithmic regression studies of the ratios of the seismic moment, time and space coordinates of an initial event to those of the events it anticipates. Again the model involves sequential renormalization, and it is not clear whether it can be associated with a stationary point process model.

Despite their varied backgrounds, all three models are widely used in their own contexts. However, their varied characters raise many further questions about the nature of the seismic regime, the models by which it can be best represented, and the requirements of a good forecasting model.

7. *On the Statistical Education of a Geophysicist*

In another early paper (VERE-JONES, 1975), I bemoaned the lack of time given to statistics courses in a typical degree program in geophysics.

Traditional applied mathematics, physics, chemistry, geology—all these make up a really crowded program for a geophysics student. And there is no time for statistics.

Until, that is, the student embarks on a project or thesis, when he or she is faced with some harsh realities in the form of a stack of observational data requiring interpretation, display, and the drawing of some kind of statistically legitimized conclusion.

There may be worse ways of learning statistics than being pitched in at the deep end. But it seemed a pity 30 years ago, and even more of a pity now if it is still true, that no serious attempt is made to incorporate statistics, or better stochastic modelling, into the geophysics degree program.

The advent of new and improved data, the growing interest in probabilistic forecasting and time-dependent hazard estimation, the powerful computing facilities now available to handle simulation and optimization techniques; all these point to a need to reassess the priorities, and to open up some pathway to a more mature form of statistical thinking among geophysics graduates, and seismology graduates in particular. However I emphatically do not advocate compulsory attendance at a cookbook statistics course. Many such courses are an insult to a mathematically literate student, and many geophysics students are more than a little mathematically literate. Rather, the aim should be to acquaint students with the basic style of statistical thinking—probability models, their link to data, checking properties by simulation. Some familiarity with basic distributions and classical statistical tests will ultimately be needed, but is relatively easily learned. Familiarity with the basic style of thinking in terms of stochastic models is harder to teach and more important.

My suggestions for a half-year course at around third year or fourth-year level might be something like this:

1. Take advantage of the modern statistical software which includes excellent techniques for displaying data in many different forms. The importance of effective data display should be lesson one.
2. The concept of a statistical model is best taught through simulation, generating random numbers according to the model specifications, starting from independent random samples moving to samples showing simple forms of chaining or dependence.
3. The empirical laws of seismology, and comparisons between actual and simulated data, offer plenty of scope for instructive and even rewarding discussions of statistical inference including both estimation and model testing. The aim in the latter should not be unquestioned obedience to 5% t tests, but some understanding of the universal problem of trying to determine when a signal stands out above the noise.
4. Some introduction to simple stochastic processes, especially branching processes, simple Markov chains, autoregression models in time series.

At the MSc level and beyond, when it comes to training students to move into a field such as statistical seismology, there is a question as to whether one is looking to convert statistical graduates into seismology, or geophysics graduates into statisticians.

As a general rule, it is easier to do the mathematics (here I mean the statistics) first, and the more applied subject later, but both routes are possible. I have found it easier to interest statistics students in seismology than seismology students in statistics. Perhaps that is because in my own university I am recognized as a statistician. But the statistics students do not proceed far with seismology. Most of them are quickly lured away into careers in finance and the like. The prospect of a secure career in statistical seismology seems doubtful, at least at the present time.

The seismology students, on the other hand, fail to see in statistics a subject that warrants their attention when they could alternatively be going on enjoyable field trips in their own subject. It looks hard, and they may be subject to off-putting propaganda from physics teachers and students about the value of statistics. This is where I think it is important to counterattack, to capture the interest of students while they are still young and to impress them by the challenges and ideas in statistical seismology. It is surely an area where there is considerable important research to be done, and a need for self-promotion as such.

8. *Conclusions and Future Perspectives*

In this lecture I have enjoyed the opportunity to indulge my own prejudices and opinions. I have tried to make, yet again, the case that stochastic models should not be seen as alternatives to physical models, but as extended versions of the physical models in which an attempt is made to explain the variability, or uncertainties, in the observations, as well as their basic causes.

At the same time stochastic models come in a number of guises and serve a number of purposes. These are likely to be more closely linked to physical considerations in some cases than in others. I have distinguished between

- Descriptive models, which are no more or less than what they claim to be: a simplified description of the data and its main properties;
- The so-called 'engineering models', the majority in fact, which are there to answer a practical need and should be judged in the first instance on whether they succeed in their stated task.
- Conceptual models, in which the aim is not only to describe but also to explain and understand the processes underlying some physical phenomenon. The role of statistical modelling ideas in seismology has increased to the stage where serious attention should be given to the possibility of incorporating some statistical modelling ideas into the late undergraduate or early postgraduate teaching program in seismology. Better late than never, I think, and with the emphasis on statistical modelling and simulation, not on cook-book recipes.

In the meantime there is no shortage of new and important questions looming in statistical seismology. Let me just mention a few somewhat far-out interests of my own.

A few years ago I found a rigorously self-similar modification of the ETAS model (Vere-Jones, 2005), and suspect there may be a similar version of the EEPAS model. There may be some way of linking these self-similarity ideas with the discovery, a few years ago by Brémaud and Massoulié (2001), of versions of the Hawkes (ETAS type) processes, which run in the critical regime but without immigrants. It is possible that the Kagan–Jackson type models are linked to these.

I now believe that there may be some mathematical paradigm of the earthquake process which is exactly self-similar, and self-perpetuating. Five years ago I would have thought this ridiculous, but now I feel that nature may have beaten us to it in suggesting a remarkable new mathematical model.

Finally, the 'rate and state' friction ideas of Jim Dieterich (1979) seem to me to invite incorporation into a rigorous stochastic model, but the best way of setting up such a model is not yet clear, at least to me.

Acknowledgements

This paper has grown out of many years of thinking about stochastic models for earthquakes, starting from the encouragement initially given to a green young statistician by Frank Evison, and sustained since then by many colleagues: Yan Kagan, Cinna Lomnitz, Yosi Ogata, David Harte, Mark Bebbington and many others.A preliminary version of this paper was presented to the 5th Statsei Meeting in Erice, Sicily, in June 2007.

References

Brémaud, P. and Massoulié, L. (2001), *Hawkes branching point processes without ancestors*. Jl Appl. Prob. *38* 122–135.

Dieterich, J. H. (1979), *Modeling of rock friction: 1. Experimental results and constitutive equations*. J. Geophys. Res. *84*, 2161–2168.

Griffith, A. A. (1921), *The phenomena of rupture and flow in solids*, Philos. Trans. R. Soc. London A *221*, 163–198.

Gutenberg, B. and Richter, C. F. (1944), *Frequency of earthquakes in California*, Bull. Seism. Soc. Am. *34*, 185–188.

JEFFREYS, H. (1938), *Aftershocks and periodicity in earthquakes*, Beitr. Geophys. *53*, 111–139.

JEFFREYS, H. *Theory of Probability* (1st Ed. 1939, 3rd Ed. 1961) (Cambridge, University Press. 1939).

KAGAN, Y. (1991), *Fractal dimension of brittle fracture*, J. Nonlinear Sci. *1* 1–16.

KAGAN, Y. and JACKSON, D. D. (1994), *Probabilistic forecasting of earthquakes*, Geophys. J. Int. *143*, 438–453.

KAGAN, Y. and KNOPOFF, L. (1981), *Stochastic synthesis of earthquake catalogues*, J. Geophys. Res. *86*, 2853–2862.

OGATA, Y. (1988), *Statistical models for earthquake occurrence and residual analysis for point processes*, J. Amer. Stat. Soc. *83*, 9–27.

OGATA, Y. (1998), *Space-time point process models for earthquake occurrences*, Ann. Inst. Stat. Math. *50*, 379–402.

RHOADES, D. A. and EVISON, F. F. (2004), *Long-range earthquake forecasting with every event a precursor according to scale*, Pure Appl. Geophys. *161*, 147–171.

SHIMAZAKI, K. and NAKATA, T. (1980), *Time-predictable recurrence model for larger earthquakes*, Geophys Res. Lett. *7*, 179–282.

VERE-JONES, D. (1975), *Stochastic models for earthquake sequences*, Geophys. J. Roy. Astron. Soc. *42*, 811–826.

VERE-JONES, D. (1978), *Earthquake prediction: a statistician's view*, J. Phys. Earth *26*, 129–146.

VERE-JONES, D. (1979), *Distributions of earthquakes in space and time*, In *Chance and Nature*, ed. P.A.P. Moran (Austral. Acad. Sci., Australia) ACT.

VERE-JONES, D. (2005), *A class of self-similar random measure*, Adv. Appl. Prob. (SGSA) *37*, 908–914.

VERE-JONES, D. (2009), *Stochastic models for earthquake occurrence and mechanisms*, Encyclopedia of Complexity and Systems Science (to appear).

WEIBULL, W. (1939), A *statistical theory of the strength of materials*, Ingvetensk. Akad. Handl. no 151.

(Received August 1, 2008, revised July 20, 2009, accepted July 21, 2009, Published online May 18, 2010)

Pure Appl. Geophys. 167 (2010), 655–666
© 2010 Birkhäuser / Springer Basel AG
DOI 10.1007/s00024-010-0067-3

Pure and Applied Geophysics

Searching for an Earthquake Precursor—A Case Study of Precursory Swarm as a Real Seismic Pattern Before Major Shocks

D. SHANKER,[1] H. N. SINGH,[2] H. PAUDYAL,[3] A. KUMAR,[2] A. PANTHI,[2] and V. P. SINGH[2]

Abstract—A long-range correlation between earthquakes is indicated by some phenomena precursory to strong earthquakes. Most of the major earthquakes show prior seismic activity that in hindsight seems anomalous. The features include changes in regional activity rate and changes in the pattern of small earthquakes, including alignments on unmapped linear features near the (future) main shock. It has long been suggested that large earthquakes are preceded by observable variations in regional seismicity. Studies on seismic precursors preceding large to great earthquakes with M $\geq$ 7.5 were carried out in the northeast India region bounded by the area 20°–32°N and 88°–100°E using the earthquake database from 1853 to 1988. It is observed that all earthquakes of M $\geq$ 7.5, including the two great earthquakes of 1897 and 1950, were preceded by abnormally low anomalous seismicity phases some 11–27 years prior to their occurrence. On the other hand, precursory time periods ranged from 440 to 1,768 days for main shocks with M 5.6–6.5 for the period from 1963 to 1988. Furthermore, the 6 August, 1988 main shock of M 7.5 in the Arakan Yoma fold belt was preceded by well-defined patterns of anomalous seismicity that occurred during 1963–1964, about 25.2 years prior to its occurrence. The pattern of anomalous seismicity in the form of earthquake swarms preceding major earthquakes in the northeast India region can be regarded as one of the potential seismic precursors. Database constraints have been the main barrier to searching for this precursor preceding smaller earthquakes, which otherwise might have provided additional information on its existence. The entire exercise indicates that anomalous seismicity preceding major shocks is a common seismic pattern for the northeast India region, and can be employed for long-range earthquake prediction when better quality seismological data sets covering a wide range of magnitudes are available. Anomalous seismic activity is distinguished by a much higher annual frequency of earthquake occurrence than in the preceding normal and the following gap episodes.

Key words: Precursory phenomena, swarm episode, seismic precursors, long-range earthquake prediction, northeast India region.

1. Introduction

There is a worldwide effort to study spatial and temporal variations in seismic activity and use them to recognize patterns that precede large earthquakes. Gaps in seismic activity and seismicity fluctuations have been recognized as long-range earthquake precursors. Large earthquakes occur frequently at trenches along the plate boundaries, separated by seismic gaps. SYKES (1971), MOGI (1969), and KHATTRI and WYSS (1978) identified some of the gaps as the most probable sites for future large earthquakes. Some precursory phenomena such as low seismic activity (BRADY, 1974) and foreshocks (ISHIDA and KANAMORI, 1977; JONES and MOLNAR, 1979) have been reported before large earthquakes. EVISON (1977a,b) identified the swarm as a precursor to earthquakes and proposed that a causal relationship exists between the swarm and the main shock. He also asserted that the swarm magnitude can be used as a predictive parameter for estimating the time of occurrence and magnitude of the forthcoming large earthquake in the region occupied by the swarm sequence. SINGH *et al.* (1982) discussed the possibilities of premonitory swarms for three sequences of earthquakes of the Burma-Szechwan region. On the basis of quantitative empirical evidence that swarms, main shocks and aftershocks are closely related phenomena, EVISON and RHOADES (1998) proposed a qualitative physical process for the long-term seismogenesis of major earthquakes in subduction zones. SHANKER *et al.* (1995) discussed anomalous seismic

[1] Department of Earthquake Engineering, Indian Institute of Technology Roorkee, Roorkee 247667, India. E-mail: dayasfeq@iitr.ernet.in
[2] Department of Geophysics, Banaras Hindu University, Varanasi 221005, India. E-mail: hridaynarain.singh@gmail.com
[3] Department of Physics, Tribhuvan University, Kirtipur, Nepal. E-mail: hariharpaudyal@gmail.com

activity preceding major earthquakes in Himachal Pradesh, India.

Thus, accumulating evidence, from both numerical/physical modelling and empirical studies in certain environments, suggests that a long-term precursory relation exists between earthquake swarms and major earthquakes. A sufficiently strong relation of this kind would suggest that the precursory phenomenon is an early stage in a long-term seismogenic process for major earthquakes. Studies in several parts of the Himalayas have established that the pattern of anomalous seismic activity in the form of earthquake swarms preceding small to great earthquakes occurs in the region (Gupta and Singh, 1986; Shanker *et al.*, 1995; Singh *et al.*, 2005). In view of its high seismic activity, including frequent disastrous earthquakes, the northeast India region is potentially the best tectonic belt in which to search for seismic precursors using the distribution of seismic activity in the space and time domains. In the present work, we have identified patterns in the fluctuation of seismic activity prior to medium-sized to great earthquakes in the north India region. The study has two distinct phases, i.e., 1860–1962 and 1963–1988, due to appreciable changes in the detection level of earthquakes.

2. *Study Area and Geological Background*

Northeast India, bounded by 20°–32°N latitude and 87°–100°E longitude, is seismically one of the most active intra-continental regions in the world. It was the site of ten damaging earthquakes with M ≥ 7.5 in a short span of 55 years from 1897 to 1952 (Table 1). This episode of abnormally high

Table 1

List of large to great earthquakes (M ≥ 7½) from 1897 to 1962, and moderate to large earthquakes from 1963 to 1988 (m_b ≥ 6.1) and 30 December, 1984 earthquake of m_b 5.6 that have occurred in the northeast India region

Sl. Nos.	Source parameters of main shocks in the northeast India region					
	Date	Epicenters		Focal depth	Magnitude	Tectonic zones
	mm/dd/yyyy	°N	°E	km	M	
1	**06/12/1897**	**26.0**	**91.0**	**60**	**8.7**	**Shillong Plateau**
2	12/12/1908	26.5	97.5	–	7 1/2	Eastern Syntaxis
3	05/23/1912	21.0	97.0	60	8.0	Arakan Yoma fold belt
4	07/08/1918	24.5	91.0	–	7.6	Shillong Plateau
5	01/27/1931	25.6	96.8	60	7.6	Arakan Yoma fold belt
6	**09/12/1946**	**23.5**	**96.0**	**60**	**7 3/4**	**Arakan Yoma fold belt**
7	**07/29/1947**	**28.5**	**94.0**	**60**	**7 3/4**	**Eastern Syntaxis**
8	**08/15/1950**	**28.5**	**96.5**	**25**	**8.7**	**Eastern Syntaxis**
9	**11/18/1951**	**31.1**	**91.4**	**–**	**8.0**	**North of Frontal Arc**
10	**08/17/1952**	**30.5**	**91.5**	**60**	**7.5**	**North of Frontal Arc**
11	01/22/1964	22.4	93.6	88	6.1	Arakan Yoma fold belt
12	02/27/1964	21.7	94.4	102	6.4	Arakan Yoma fold belt
13	**03/27/1964**	**27.2**	**89.3**	**32**	**6.3**	**Frontal Arc**
14	07/12/1964	24.9	95.3	155	6.7	Arakan Yoma fold belt
15	07/13/1964	23.7	94.7	117	6.5	Arakan Yoma fold belt
16	**01/12/1965**	**27.6**	**88.0**	**23**	**6.1**	**Frontal Arc**
17	07/29/1970	26.0	95.4	59	6.5	Arakan Yoma fold belt
18	07/08/1975	21.5	94.7	157	6.5	Arakan Yoma fold belt
19	05/29/1976	24.6	98.9	08	6.1	Arakan Yoma fold belt
20	**08/12/1976**	**26.7**	**97.1**	**27**	**6.2**	**Eastern Syntaxis**
21	11/19/1980	27.4	88.8	27	6.1	Frontal Arc
22	**12/30/1984**	**24.6**	**92.9**	**22**	**5.6**	**Shillong Plateau**
23	**08/06/1988**	**25.12**	**95.17**	**115**	**7.5**	**Arakan Yoma fold belt**

The earthquakes listed with bold letters were found to be associated with anomalous seismicity/earthquake swarms (please refer to the text and Table 2 for further details)

seismic activity included two great earthquakes of M 8.7, on 12 June, 1897 and 15 August, 1950, in the vicinity of the Shillong Plateau and the India–China border, respectively. GUPTA and SINGH (1986) and Singh et al. (2005) systematically analyzed the time series of earthquakes associated with several large main shocks in the region, including both of the great earthquakes.

The region is complex in terms of geology and tectonics. Several authors have described the geology of the region (EVANS, 1964; DESIKACHAR, 1974; NANDY, 1980). Following DUTTA (1964), the entire northeast India region could be divided into the following tectonic zones: Zone I (The Eastern Syntaxis); Zone II (the Arakan Yoma and Naga Thrust Fold Belts); Zone III (the Shillong Plateau) and Zone IV (the Main Central Thrust and the Main Boundary Faults of the Himalayan Frontal Arc). These zones and the major tectonic elements of the region are depicted in Figure 1 along with the spatial distribution of events with $m_b \geq 5.1$ from 1963 to 1985.

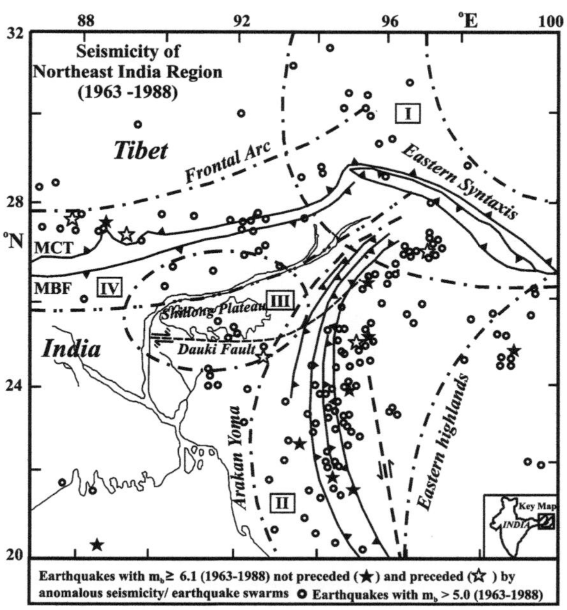

Figure 1
Seismicity of the northeast India region from 1963 to 1988 showing earthquakes with $m_b \geq 5.1$ over the structural features and the four major tectonic zones. MCT—Main Central Thrust; MBF—Main Boundary Fault

3. Seismicity Database

Since the earthquake of 12 June, 1897 appears to be the earliest instrumentally recorded earthquake in this region, we have arbitrarily assigned this to be the demarcation between the historical and recent earthquakes. Countries such as China, Japan and USSR have succeeded in preparing reliable catalogues of historical earthquakes; however similar efforts have not yet been made to compile a seismicity database for the northeast India region. Earlier, a large portion of the northeast India region was known as Assam. The recurring destruction by natural hazards (e.g., earthquakes, floods, etc.) and attacks by several invaders destroyed the monuments, buildings and scriptures, thereby erasing much of the data on the seismicity of the northeast India region from the pages of history. These points are well brought out by SIR EDWARD GAIT (1905) in his book entitled *A History of Assam*.

There are two important prerequisites for estimating abnormal fluctuations in seismicity prior to the occurrence of large earthquakes: (1) *A clear assessment of background (normal) seismicity; and* (2) *identification of anomalous changes*. For the present investigations on seismic precursor studies, we have used a database compiled by GUPTA and SINGH (1986) for the northeast India region from 1897 to 1962. This database is observed to be free from obvious errors since the available records have been critically examined and the detection capabilities of the seismological network taken into account. The earthquake catalogues published by the International Seismological Centre (ISC, 1963–1988) and PDE, 1963–1988 catalogue (preliminary determination of epicenters) of the United States Geological Survey (USGS) have been considered for the period from 1963 to 1988 and events with $m_b \geq 4.5$ are used in the present study. A review of a cumulative number of earthquakes in the four zones demonstrates that the data are more consistent for the events larger than magnitude 5 (Fig. 2). The annual frequency of events in each tectonic zone was extracted from this database using three different selections of events: all events, and events with $m_b \leq 5.0$ and $m_b > 5.0$. The temporal patterns thus obtained are represented in Figure 2 along with the time of occurrence of medium size earthquakes.

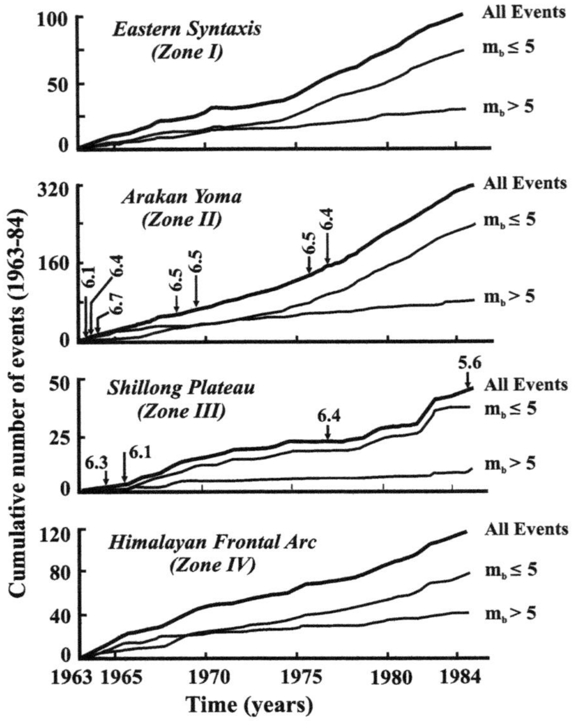

Figure 2
Temporal pattern of seismic events in four tectonic zones of the northeast India region from 1963–1984. Occurrence times of the medium-sized earthquakes during the period have been indicated with an *arrow* along with their respective magnitude

4. *Identification of Earthquake Swarms/Anomalous Seismicity Preceding Medium to Great Earthquakes in the Northeast India Region*

The procedures to identify anomalous seismicity in the form of earthquake swarms preceding medium size to large earthquakes are based on spatial, temporal and magnitude distributions of events, keeping in mind the following:

- An appropriate grid enclosing the epicenters of each of the main shocks and the period of the seismicity data was selected, taking into account the size of the main shocks and the major tectonic features with which they were possibly associated.
- The main database of the region was searched for seismicity data in each grid for a specified period, irrespective of magnitude and focal depths of the events.
- Preliminary studies were carried out on the spatial and temporal distribution of all earthquakes within the grid associated with each main shock, and approximate preparatory areas were delineated, which were found to be elliptical and in most cases elongated in the direction of tectonic features.
- If clustering of events in both space and time prior to main shocks was evident, then a second search was made using a variety of event selections: i.e., all events, events with $M \geq 6$; and ≥ 6.5 for the period 1860–1962; and events with $m_b \geq 4.5$; ≥ 5.0; ≥ 5.5 and ≥ 6 for the period 1962–1988.
- For shallow-focus main shocks, only precursory events with shallow focus ($h \leq 60$ km) were considered; whereas for intermediate focus main shocks, only intermediate precursory events ($h \geq 60$ km) were used for recognition of anomalous seismicity/earthquake swarm patterns.
- Using the above procedures, all medium to large-sized earthquakes that occurred during 1860–1988 in the northeast India region were investigated. For each, the space and time domains corresponding to four seismicity stages were identified: normal seismicity; anomalous seismicity gap (or quiescence); and foreshock-main shock-aftershock sequences. The resulting spatial, temporal and magnitude relationships are presented here.

5. *Anomalous Seismicity during 1860–1962*

The medium-sized to great earthquakes that occurred in the northeast India region from 1860 to 1988 are listed in Table 1. Two great earthquakes and eight others with $M \geq 7.5$ occurred in a short span of 55 years from 1897 to 1952, and the seismic activity in this period was much higher than previously (prior to 1897) or since (after 1950). In view of this, we investigated the seismicity pattern prior to the occurrence of these earthquakes, to determine whether these large to great earthquakes were preceded by any common seismicity pattern, which may be useful to formulate a theory that can aid our understanding of future seismicity trends in the region. The spatial distribution of these earthquakes during 1897–1952, along with their delineated preparatory areas and orientations, are shown in Figure 3. Evidently, the preparatory areas of 1897, 1946, 1947, 1950 and 1951–1952 main shocks have orientations mostly in the direction of major tectonic features.

The data showed that the main shocks of 1897, 1946, 1947, 1950 and main shock sequence of 1951–1952 in northeast India were preceded by four well-defined seismicity phases, clearly identifiable both in space and time, as shown in Figure 4. In each case, clustering of events in space and time was observed. This was evident in the plot of Cumulative Number of Events against Time (CNET) as shown in Figure 4 (right). Such clustering in space and time was found to be present even when only events with higher magnitudes ($M \geq 6$ and 6.5) were considered (Fig. 4) prior to all the five large events. On considering higher magnitude subsets of precursory events, it became possible to resize and reorient the preparatory areas (Fig. 4), especially for the 1946, 1947 and 1950 main shocks. Such deductions are generally helpful to minimize the spatial uncertainties for the location of epicenters of impending earthquakes. It is also evident that the size of the preparatory area increases with the size of the main shock (Table 2). When considering all events, the preparatory areas of the 1897 and 1947 main shocks were found to be oriented in a north-east–southwest direction; those of the 1946 and 1950 main shocks in a northwest-southeast direction; and those of the 1951–1952 main shocks in an east–west direction. There was no observable change in the orientation or size of the preparatory area of the 1951–1952 main shock when higher magnitude subsets were considered, however a change in orientation was noticeable in the cases of the 1946, 1947 and 1950 main shocks. Similar characteristics of clustering of events were observed when magnitudes of the events in all the identified episodes were considered, as shown in Figure 5.

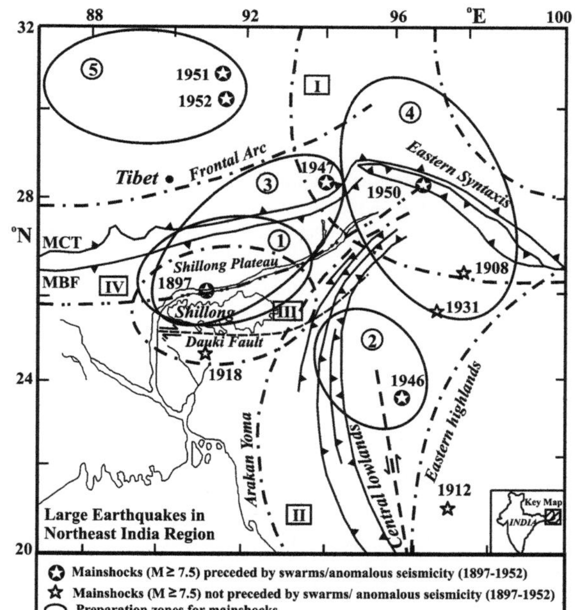

Figure 3
Large to great earthquakes of $M \geq 7.5$ from 1897 to 1952 in the northeast India region are shown on the major tectonic zones (I–IV) of the region. Orientations of preparatory areas associated with the main shocks of 1897, 1946, 1947, 1950 and 1951–1952, which were preceded by anomalous seismic activity/precursory swarms, are also depicted. MCT—Main Central Thrust; MBF—Main Boundary Fault

6. *Annual Frequency and the Main Shocks*

The observations described above suggest that anomalous seismic activity occurred prior to the 1897, 1946, 1947, 1950 and 1951–1952 main shocks in the northeast India region. In each of these cases, the anomalous seismicity episodes were characterized by an appreciably higher annual frequency (Fig. 6) than in the preceding normal and the following gap episodes. In some cases, e.g., the main shocks of 1897, 1976 and 1984, existence of an extremely high annual frequency is due to a high concentration of events in a short time-span. In the remaining cases, a prolonged anomalous episode was observed with a comparatively low, but still significantly elevated, annual frequency. The main shocks were preceded by long preparatory gaps associated with a significantly low annual frequency, and the main shocks occurred as an interruption of the gap episodes. The indication is that the smaller the duration of the anomalous episode, the smaller the associated main shock will be, and *viceversa.* Also, the logarithm of the preparatory time period is related to the magnitude of the main shock. Precursory parameters estimated for these main shocks are provided in Table 2.

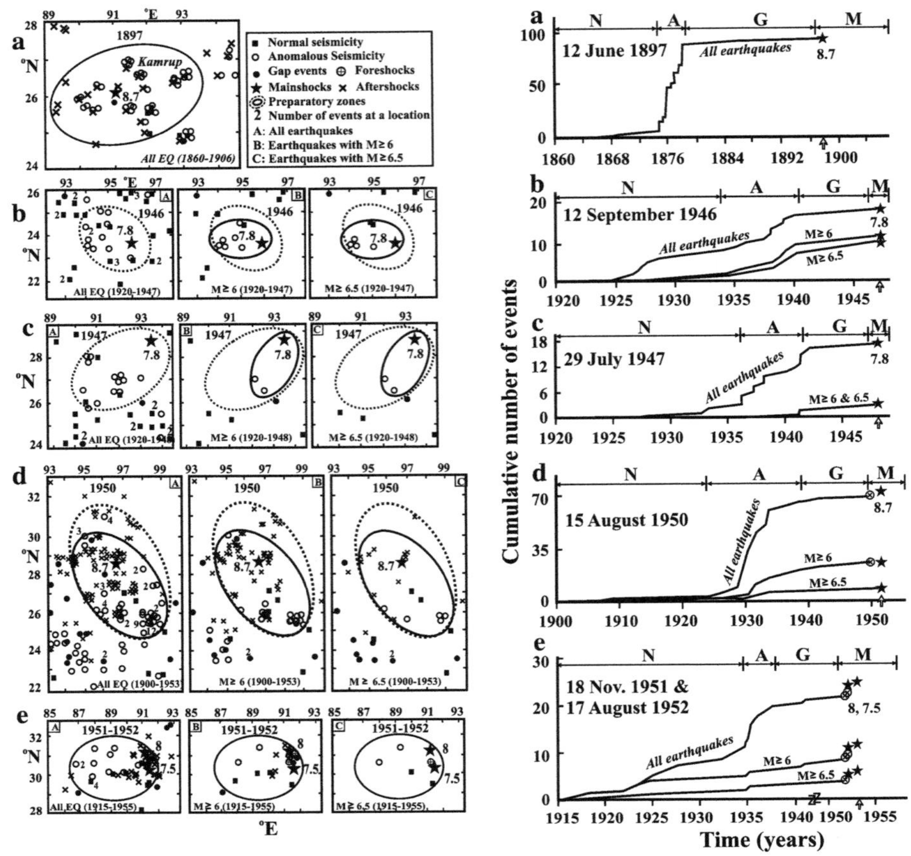

Figure 4
Patterns of spatial (*left*: **a–e**) and temporal (*right*: **a–e**) distribution of events during four identified anomalous episodes, i.e., background seismicity (N), precursory swarm/anomalous seismic activity (A), gap/quiescence seismicity (G) and foreshocks/main shocks/aftershocks (M), preceding five large to great earthquakes ($M \geq 7\frac{1}{2}$) of 1897, 1946, 1947, 1950 and 1951–1952 that occurred in the northeast India region prior to 1963

7. Predictive Regressions

The values of parameters M_m, M_p and T_p, furnished in Table 2 for the event Nos. 1–5, indicate interdependencies between them. These were used to establish predictive regressions for the northeast India region. After making minor modification to the regressions established by Gupta and Singh (1986), the following three predictive regressions were established in the present work:

$$M_m = 1.37M_p - 1.40, \quad (1)$$

$$\log T_p = 0.45M_p + 0.66, \quad (2)$$

$$\log T_p = 0.34M_m + 1.03. \quad (3)$$

The three sets of regression lines shown in Figure 7 are drawn with the help of the above three equations. All these best fit lines have three degrees of freedom. The equations are predictive; if M_p is known, then the precursor time T_p and the magnitude of principal main shock (M_m) can be estimated. The equations are applicable to the northeast India region only. It is necessary to recognize the anomalous seismic activity (or earthquake swarm) well before the main shock. These equations hold good provided no major outstanding earthquake with size greater

Table 2

List of main shocks from 1897 to 1988 that were preceded by anomalous seismicity/earthquake swarms and their related seismic precursory parameters in the northeast India region (refer Table 1 for details)

Sl. Nos.	Date	Study Grid		Period of data examined	Onset of swarm activity	Precursory time gap	Magnitude	Average magnitude of swarm	Orientation of preparatory zones and areas
	mm/dd/yyyy	°N	°E		mm/dd/yyyy	T_p (days)	M_m	M_p	Km^2
1 (1)	06/12/1897	24–28	89–95	1860–1906	05/02/1874	8,436	8.7	7.3*	1.18×10^5(E–W)
2 (6)	09/12/1946	21–26	92–98	1920–1947	12/05/1934	4,296	7 3/4	7.0	7.88×10^4(NW–SE)
3 (7)	07/29/1947	24–29.5	88.5–94.5	1920–1948	02/11/1936	4,183	7 3/4	6.6	1.33×10^5(NE–SW)
4 (8)	08/15/1950	22–33	93–100	1900–1953	01/30/1924	9,687	8.7	7.3	3.79×10^5(NW–SE)
5 (9, 10)	11/18/1951 08/17/1952	28–33	85–93	1915–1955	12/14/1934	6,210	8.0 7.5	6.8	1.67×10^5(E–W)
6 (13, 16)	03/27/1964 01/12/1965	23–30	86–92	1950–1965	05/24/1959	1,768	6.3 6.1	5.0*	12.0×10^4(NW–SE)
7 (20)	08/12/1976	24–29	94–100	1974–1976	05/30/1975	440	6.2	5.3	03.9×10^4(NE–SW)
8 (22)	12/30/1984	23–28	89–94	1980–09/30/1985	01/11/1982	1,083	5.6	5.0	08.4×10^4(NW–SE)
9 (23)	08/06/1988	20–26	92–98	1950–08/06/1988	06/01/1963	9,199	7.5	6.6	12.8×10^4(N–S)

$\mathbf{M_m}$—*main shock magnitude;* $\mathbf{M_p}$—*average magnitude in the anomalous/swarm sequence;* $\mathbf{T_p}$—*precursory time period from the onset of the anomalous activity until the occurrence of the main shock*

* Approximate body wave magnitude prior to 1963; *M_p 7.3 related to 1897 main shock is arbitrary (refer Gupta and Singh, 1986; Singh et al. 2005)

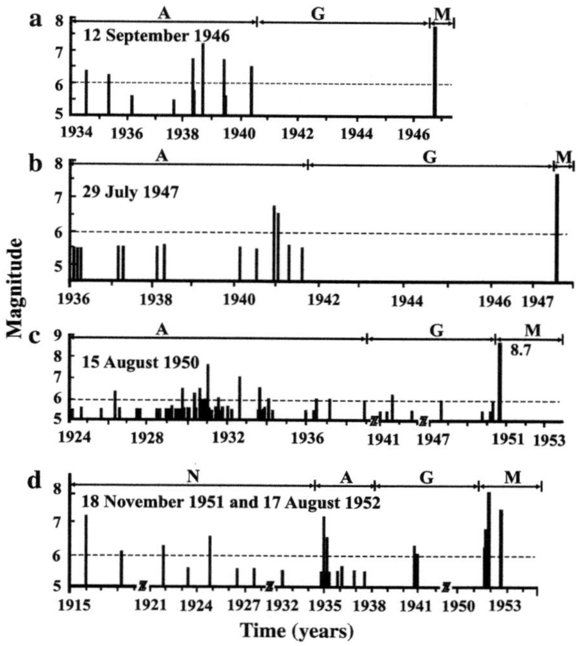

Figure 5
Magnitude versus time plot of the events during four anomalous episodes preceding four large earthquakes of 1946, 1947, 1950 and 1951–1952 in the northeast India region prior to 1962 (please refer to Figure 4 for details on spatial and temporal patterns). N, A, G and M have their usual meaning as in caption of Figure 4. Magnitude relationships between the events in three episodes identified prior to the 1897 great earthquake could not be established due to the lack of details on felt reports and magnitude

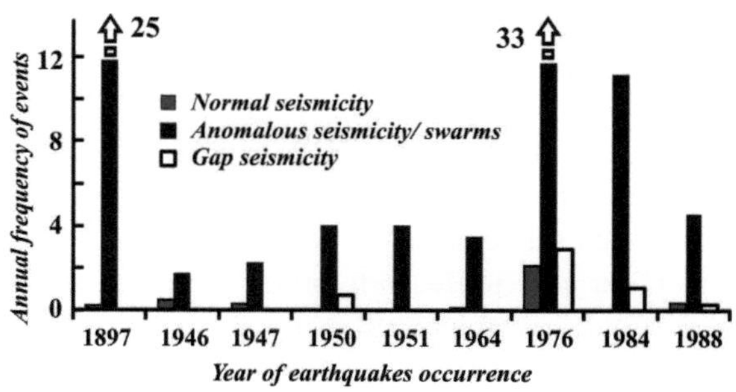

Figure 6
Annual frequency of events in the identified normal, anomalous (swarm) and gap episodes within the delineated preparatory areas for the main shocks that occurred during 1897–1988 in the northeast India region. It is evident that the anomalous episode is characterized by extremely high annual frequency as compared to the preceding normal episode and the following gap episode for all the earthquakes. The *numbers* above the *bars* of the main shocks of 1897 and 1976 show total annual frequency

than or equal to the size of the largest earthquake M_p occurs within the delineated preparation area during the gap period. The occurrence of a swarm sequence

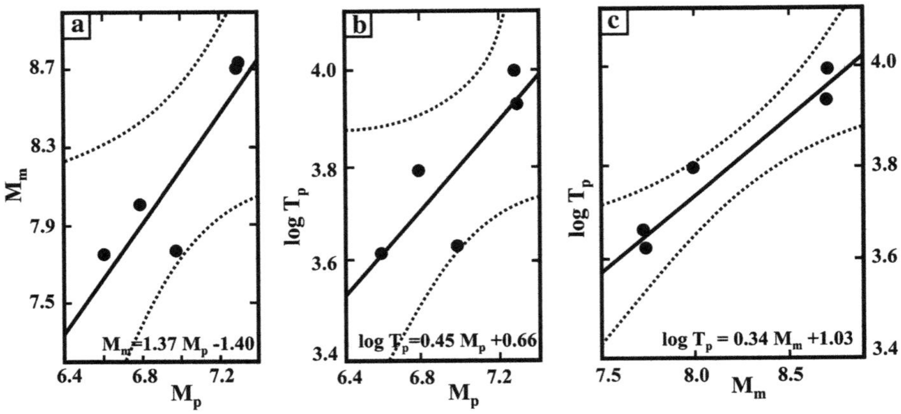

Figure 7
Relation among M_m, M_p and T_p for the main shocks in the northeast India region that occurred during 1897–1962. The *solid lines* are the best fit equations to the data: **a** $M_m = 1.37\ M_p - 1.40$ ($R_2 = 0.76$); **b** $\log T_p = 0.45\ M_p + 0.66$ ($R_2 = 0.67$); and **c** $\log T_p = 0.34\ M_m + 1.03$ ($R_2 = 0.95$); and the outer *dotted curves* are the 95% tolerance limits

is indicative of a converging trend of stress accumulation towards the pending focal region (Singh and Singh, 1984). The occurrence of any outstanding events in the preparatory area during the gap period may divert the converging trend of stress accumulation, which may result in extending the gap period, and also in increasing the magnitude of the impending main shock. If so, certain considerations are needed in these equations before estimating M_m and T_p for a pending earthquake.

8. *Anomalous Seismic Activity during 1963–1988*

Table 1 lists the earthquakes with $m_b \geq 6.0$ that occurred in the region from 1963 to 1988. There were twelve such earthquakes (Sl. Nos. 11–21, and 23), of which four were preceded by well-defined sequences of anomalous seismicity (Sl. Nos. 13, 16, 20 and 23). The 30 December, 1986 Cachar earthquake with $m_b = 5.6$ (Sl No. 22) was also preceded by a similar pattern. The precursory gap period for the first three main shocks (Sl Nos. 6–8 in Table 2) ranges from 440 to 1,768 days. The preparatory areas were observed to be elliptical and oriented in the direction of major tectonic features as shown in Figure 8. Detailed data analysis to search for the anomalous pattern prior to these main shocks was carried out by Gupta and Singh (1986) and Singh et al. (2005). The

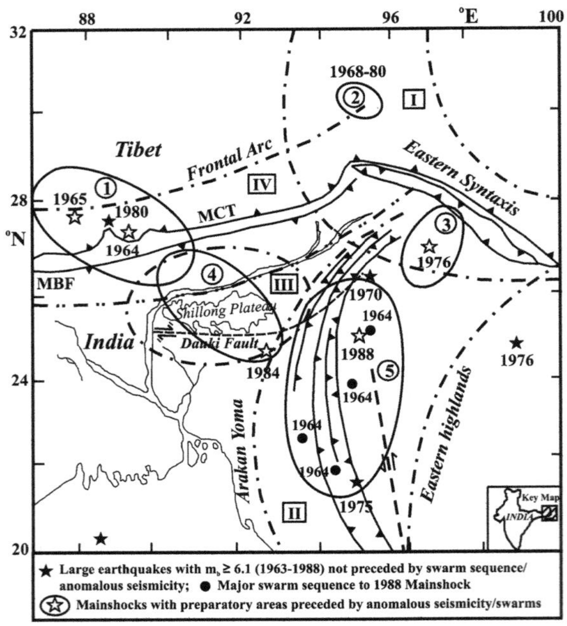

Figure 8
Spatial distribution of earthquakes with $M \geq 6.0$ from 1963 to 1988 in the northeast India region in four tectonic zones (I–IV). The preparatory areas of main shocks of 1964, 1976, 1984 and 1988 which were preceded by anomalous seismicity/earthquake swarms are shown in relation to the major tectonic features of the region. Northwest–southeast trending elliptical preparatory area (2) is delineated for three consecutive swarm sequences of 1968, 1977 and 1982 in which no main shock has occurred yet. Such concentrated activity is generally associated with the movement of magma in the crust in volcanic terranes (Hill *et al.*, 1984). MCT—Main Central Thrust; MBF—Main Boundary Fault

four earthquakes of $m_b \geq 6.0$ in the Arakan Yoma fold belt, which occurred in 1964 in quick succession, constituted a well-defined anomalous pattern on the basis of which Gupta and Singh (1986) predicted a medium-to-large-sized earthquake to occur by 1990. The 1988 main shock (Sl. No. 9 in Table 2) occurred within the delineated north–south oriented preparatory area and we think that this pattern was real and causally related to the main shock. The anomalous seismicity episodes, in all these cases, were characterized by an appreciably higher annual frequency than in the background and gap episodes, as depicted in Figure 6. Next we furnish the details of this anomalous pattern associated with 1988 main shock.

9. Anomalous Seismicity Pattern Prior to 06 August, 1988 Main Shock

Four earthquakes of $m_b \geq 6.1$ occurred in quick succession in the Arakan Yoma fold belt during 1964 (Tables 1, 2), constituting a well-defined swarm sequence. No major earthquake had occurred in this region since the 1946 main shock of M 7.7 which was preceded by an earthquake swarm (Gupta and Singh, 1986). The sequence of four earthquakes in 1964 was followed by quiescence and hence could be considered as an anomalous seismicity pattern. The medium-sized earthquakes which occurred after August 1964 within the aerial extent of the 1964 swarm sequence were those of 1970 and 1975, both with magnitude 6.5. Spatio-temporal and magnitude features of the normal, anomalous, and gap episodes and the main shock are depicted in Figure 9 for events with $m_b \geq 5.0$, 5.5 and 6.0, respectively.

The four swarm earthquakes occurred within a short span of 6 months, i.e., between January 22, 1964 and July 13, 1964 in the Arakan Yoma fold belt (Fig. 8). To examine this swarm sequence, an area bounded by 20°–26°N latitudes and 92°–98°E longitudes, which covered the aerial extent of the 1964 earthquake sequence, was selected and the data from 1950 to 1988 was scrutinized (Fig. 9).

The earthquakes that occurred from 1950 until the 1988 main shock were categorized into normal, anomalous-seismicity and gap episodes for three magnitude subsets, $m_b \geq 5.0$, 5.5 and 6.0, in order to better understand the characteristics of this sequence. The spatial distributions of these earthquakes were used to demarcate an area that encloses most of these events, and it is the preparatory area for the 1988 main shock. When earthquakes of $m_b \geq 5.0$ were analyzed, episodes of normal seismicity (1950–May, 1963), anomalous (June 1963–1967) and quiescence (1968 to 05 August, 1988) could be defined marginally (Fig. 9aA). But this pattern helped to demarcate the area occupied by these events in the sequence. Of course the spatial distribution of background seismicity along with the 1964 swarm sequence jointly helped delineate the north–south trending preparatory area for the location of future major seismic events, if any (Figs. 8, and 9). Resizing of these episodes is facilitated by considering earthquakes of $m_b \geq 5.5$ and $m_b \geq 6.0$ only (Figs. 9a,b,c). For example during the period from 1950 to 27 September, 1963, only five earthquakes of $m_b \geq 5.5$ occurred within the delineated area, with an average of about one event every 2 years, which we take as the background seismicity level (Fig. 9b) of the considered region.

During the period 28 September, 1963 to June 1965, however, eight earthquakes of $m_b \geq 5.5$ occurred within the same elliptical area, giving an annual frequency of four earthquakes. This represents an eightfold increase in seismic activity over the background seismicity level, which we infer to be the swarm sequence (Fig. 6). During July 1965 to 5 August, 1988, only six events of $m_b \geq 5.5$ occurred in the elliptical area, resulting in an extremely low annual frequency. This is regarded as the gap episode (or quiescence). The above-mentioned episodes of background seismicity, anomalous seismicity and precursory gap in the elliptical area, demarcated on the basis of 1964 swarm sequence, were equally obvious even when events of $m_b \geq 6.0$ only were analyzed. In all these cases, the anomalous seismicity episode was associated with a significantly higher annual frequency (Fig. 6) than in the preceding and following episodes. It is noteworthy that only one earthquake with $m_b = 6.5$ (in 1975) occurred during a long quiescence phase of about 25 years prior to the occurrence of the 1988 main shock.

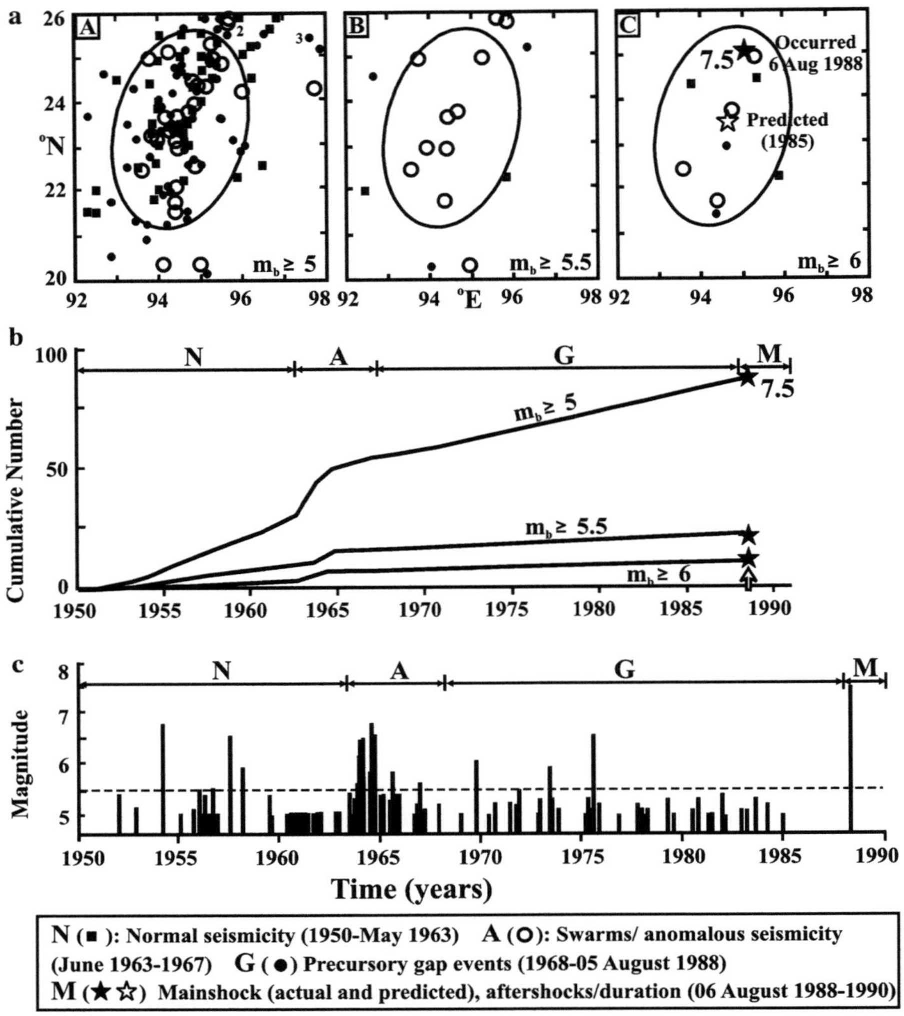

Figure 9

Patterns of spatial (**a**), temporal (**b**) and magnitude distribution (**c**) of premonitory seismicity during normal (N), anomalous seismicity/swarm (A) and gap (G) episodes in an approximately north–south trending preparatory zone delineated with the help of events for $m_b \geq 5$ from 1950–1988 which defines the 1963–1965 swarm sequence in Arakan Yoma fold belt. However, the main shock of 6 August, 1988 with M 7.5 occurred in the delineated preparatory area at its northern extremity (Fig. aC)

10. Discussion and Conclusions

It has already been established that medium to large earthquakes are preceded by epochs of abnormal seismicity patterns including background (normal) seismicity, anomalous seismicity (swarms), quiescence and foreshocks (KELLEHER and SAVING, 1977; EVISON, 1977a,b; MOGI, 1977; KANAMORI, 1981; SINGH *et al.*, 1982; SINGH and SINGH, 1984; OHNAKA, 1984–1985; Gupta and Singh, 1986, 1989; Singh et al. 2005; PAUDYAL and SINGH, 2007). We have also observed that moderate to great earthquakes in the northeast India region were preceded by well-defined anomalous seismicity phases. Our analyses show that over 70% of major earthquakes (M $\geq$ 7 ½) during 1860–1962 in the northeast India region were preceded by well-defined patterns of anomalous seismicity with long quiescence periods. The precursory period, thus estimated, ranges from 23 to 27 years prior to the M 8.7 great earthquakes of 1897 and 1950, and from 11 to 17 years for M 7 ¾–8 earthquakes. In comparison, EVISON (1982) reported longer precursory periods ranging from 11 to 40 years for the Japanese earthquakes of M 6.6–7.9.

Ohnaka (1984–1985) estimated a 27-year precursory period for the 1923 Kanto earthquake of M 7.9. Sekiya (1977) estimated a precursory duration of about 10.5 years for M 6.9; and about 19.3 years for M 7.3 earthquakes of the Fukui area of 28 June, 1948 based on anomalous seismic activity. These reports suggest that precursory time periods vary significantly from region to region, even for main shocks of similar magnitude; and hence it can be said that the precursor time is dependent on both the tectonics of a region and the magnitude of the impending main shock. Contrastingly, the precursory time periods for main shocks since 1963 in the northeast India region were observed to vary from as low as 440 to 1,768 days for main shocks with $m_b \geq 5.6$; and about 25 years for M 7.5 (Gupta and Singh, 1986; Singh et al. 2005). Recently, Paudyal and Singh (2007) reported preparatory time periods ranging from 617 to 1,043 days for main shocks of 18 May, 1984 (m_b 5.6) and 29 July, 1980 (m_b 6.1), respectively, that occurred in the Bajhang area of western Nepal Himalaya using the anomalous seismicity pattern.

In the case of the 1988 main shock in the Arakan-Yoma fold belt, we estimated $T_p = 9{,}200$ days ($\sim$25 years) and $M_p = 6.6$ (m_b) based on the swarm sequence observed during 1963–1964. On substituting $M_p = 6.6$ in the predictive equations derived earlier for the northeast India region; a magnitude (M_m) 7.5 is estimated for the impending main shock associated with the 1963–1964 swarm sequence. Considering the inherent level of error in the magnitude estimation, this sequence would indicate that an earthquake with M 8 ± 0.5 could occur at any time between the precursory time periods of $\sim$23 and 27 years estimated for the two great earthquakes of 1897 and 1950 respectively of the northeast India region. If this fact is applied for the 1963–1964 swarm sequence, a similarly large earthquake should occur in the region between 23 and 27 years from the onset of the associated anomalous seismicity. In the present case, a preparatory area enclosed by 21°–25.5°N and 93°–96°E is delineated for the location of the forthcoming such earthquake. Based on similar reasoning, Gupta and Singh (1986) inferred that the impending earthquake should occur somewhere within the delineated preparatory area as shown in Figure 9. The identified swarm and the gap events within the preparatory area are of intermediate focal depths ranging from 88 to 157 km; hence it was anticipated that the main shock should also be located in the focal depth range of 100 ± 40 km. Interestingly, the main shock that occurred on 6 August, 1988 was within the predicted times (after 25.2 years), magnitude (M 7.5), within the delineated area (epicenter 25.12°N, 95.17°E) and focal depth (at 115 km) (Fig. 9). This observation suggests a clear case of the existence of anomalous seismicity and its relation with the occurrence time, magnitude and location of the associated main shock.

The foregoing discussions suggest that the swarm sequence during 1963–1964 was preceded and followed by well-defined low seismicity epochs of background and quiescence. Further, the region was quiet for about 25 years throughout the precursory gap period (1964–1988) (Gupta and Singh, 1986). In view of this, the anomalous seismicity observed during 1963–1964 in the Arakan Yoma fold belt is considered to be a real seismic signal for the 1988 main shock. The lack of a seismicity database over a wide range of magnitude has hindered the search for similar patterns prior to smaller earthquakes, which otherwise might have provided additional examples.

In conclusion the following need to be stressed:

1. Anomalous seismic activity in the form of earthquake swarms commonly precedes medium to large size earthquakes in the northeast India region, and a causal relationship between this pattern and the source parameters of the impending main shocks is proposed.
2. The predictive regressions established in the present work facilitate estimation of the main shock magnitude and time of occurrence using the swarm magnitude. The main-shock magnitude lies well within the limit, and the precursor time slightly outside the 95% tolerance levels of the regressions. The main-shock magnitude and precursor time lay close to published regressions for precursory swarms in other countries (Evison, 1977b). In the field of earthquake precursor studies, there are still no regressions for most of the precursors; hence these regressions are purely subjective.

Acknowledgments

The first author is thankful to the Professor and Head, Department of Earthquake Engineering, IIT-R, Roorkee for providing necessary facilities for the preparation of this manuscript. Financial support provided by the organizing committee of the F. F. Evison Symposium is gratefully acknowledged; without which this presentation was not possible. The authors thank both the anonymous reviewers for their critical comments on the manuscript and David Rhoades for his efforts throughout publication.

REFERENCES

BRADY, B. T. (1974), *Seismic precursor before rock failures in mines*, Nature *252,* 549–552.

DESIKACHAR, S. K. (1974), *A review of tectonic and geological history of Eastern India in terms of plate-tectonic theory*, Jour. Geol. Soc. India *15,* 137–149.

DUTTA, T. K. (1964), *Seismicity of Assam-zone of tectonic activity*, Bull. Nat. Geophy. Res. Inst *2,* 152–163.

EVANS, P. (1964), *The tectonic framework of Assam*, J. Geol. Soc. India *5,* 80–96.

EVISON, F. F. and RHOADES, D. A. (1998), *Long-term seismogenic process for major earthquakes in subduction zones*, Phys. Earth and Planet. Inter. *108,* 185–199.

EVISON, F. F. (1977a), Fluctuations *of seismicity before major earthquake,* Nature (London) *226,* 710–712.

EVISON, F. F. (1977b), *The precursory earthquake swarm*, Phys. Earth Planet. Inter. *15,* 19–22.

EVISON, F. F. (1982), *Generalized precursory swarm hypothesis*, J. Phys. Earth *30,* 155–170.

GAIT, E., *A history of Assam*, LBS Publications, Assam, 1905, 390 pp.

GUPTA, H. K. and SINGH, H. N. (1989), *Earthquake swarms precursory to moderate to great earthquakes in northeast India region*, Tectonophysics *167,* 285–298.

GUPTA, H. K. and SINGH, H. N. (1986), *Seismicity of northeast India region: Part II: Earthquake swarms precursory to moderate magnitude to great earthquakes* J. Geol. Soc. India *28,* 367–406.

HILL, D. P., BAILEY, R. A., and RYALL, A. S. (eds.) (1984), *Proc. Workshop XIX, Active Tectonics and Magmatic Processes beneath Long Valley, Caldera, Eastern California*, USGS Open File Report, 942 pp. *89–939.*

ISC-INTERNATIONAL SEISMOLOGICAL CENTRE BULLETIN, 1963–1988.

ISHIDA, M. and KANAMORI, H. (1977), *The spatio-temporal variations of seismicity before the 1971 San Fernando earthquake, California,* Geophys. Res. Lett. *4,* 345–346.

JONES, L. and MOLNAR, P. (1979) Some characteristics of foreshocks and their possible relationship to earthquake prediction and premonitory slip on fault, J. Geophys. Res. *84, 3596–3608.*

KANAMORI, H., *The nature of seismicity patterns before large earthquakes. In Earthquake Prediction.* D. W. Simpson and P. G. Richards, (Eds.). Maurice Ewing Series, IV, AGU, Washington, D.C., (1981), pp. 1–19.

KELLEHER, J. and SAVING, J. (1977), *Distribution of seismicity before large strike-slip and thrust-type earthquakes,* J. Geophys. Res. *80,* 260–271.

KHATTRI, K.N. and WYSS, M. (1978), *Precursory variation of seismicity rate in the Assam area, India*, Geology *6, 685–688.*

MOGI, K. (1977), *An interpretation of the recent tectonic activity in the IZU-Tokai district*, Bull. Earthq. Res. *52,* 315–330.

MOGI, K. (1969), *Some features of recent seismic activity in and around Japan (2), activity before and after great earthquakes*, Bull. Earthq. Res. Inst., Univ. Tokyo *47,* 395–417.

NANDY, D. R. (1980), *Tectonic patterns in northeastern India*, Indian J. Earth Sc. *7,* 3–107.

OHNAKA, M. (1984–1985), *A sequence of seismic activity in the Kanto area precursory to the 1923 Kanto earthquakes,* Pure Appl. Geophys. *122,* 848–862.

PAUDYAL, H. and SINGH, H. N. (2007), *Anomalous seismic activity in farwestern Nepal and its adjoining regions*, Proc. National seminar on *Modern Trend in Geophysical Sciences and Techniques*, Indian School of Mines University, Dhanbad, 40–43.

PDE-PRELIMINARY DETERMINATION OF EPICENTERS Catalogue of USGS, from 1963 to 1988.

SEKIYA, H. (1977), *Anomalous seismic activity and earthquake prediction*, J. Phys. Earth *25,* S85–S93.

SHANKER, D., SINGH, H. N., and SINGH, V. P. (1995), *Anomalous seismic activity and long-range earthquake prediction in Himachal Pradesh, India,* Acta Geod. Geophy. Hung. *30(2–4),* 379–395.

SINGH, H. N., SHANKER, D., and SINGH, V. P. (2005), *Occurrence of anomalous seismic activity preceding large to great earthquakes in northeast India region with special reference to 06 August, 1988,* Physics Earth Planet. Interiors *148,*561–284.

SINGH, V. P. and SINGH, H. N. (1984), *Precursory swarm and medium-size earthquake occurrences in Pamirs and its adjoining regions*, Earthq. Predic. Res. *2,* 245–258.

SINGH, V. P., SINGH, H. N., and SINGH, J. (1982), *On the possibilities of premonitory swarms for three sequences of earthquakes of Burma-Szechwan region,* Tectonophysics *85,* T21–T29.

SYKES, L. R. (1971), *Aftershock zones of great earthquakes, seismicity gaps and earthquake prediction for Alaska and the Aleutians*, J. Geophs. Res. *76,* 8021–8041.

(Received August 1, 2008, revised February 6, 2009, accepted March 9, 2009, Published online February 27, 2010)

Pure Appl. Geophys. 167 (2010), 667–684

DOI 10.1007/s00024-010-0072-6

Pure and Applied Geophysics

Anomalous Seismicity and Earthquake Forecast in Western Nepal Himalaya and its Adjoining Indian Region

H. N. SINGH,[1] H. PAUDYAL,[2] D. SHANKER,[3] A. PANTHI,[1] A. KUMAR,[1] and V. P. SINGH[1]

Abstract—Precursory swarms associated with major earthquakes in the Western Nepal Himalaya and its adjoining region (bounded by 28.0°–31.0°N and 79.5°–82.2°E) have been studied using seismicity data from 1963 to 2006. The delineation of preparation zones for future seismic disturbances is carried out using the temporal and the spatial distribution of earthquakes, considering the events with cutoff magnitude $m_b \geq 4.3$ in four anomalous episodes: normal/background (N); anomalous/swarm (A); precursory gap (G) and main shock sequence (M), respectively. Five cases of anomalous seismicity have been identified, including two cases for which quiescence episodes still continue. Three moderate earthquakes of 1980 (m_b 6.1, Bajhang), 1984 (m_b 5.6, Bajura) and 1999 (m_b 6.6, Chamoli) in Western Nepal and its adjoining Indian region were preceded by well-defined patterns of anomalous seismicity/precursory swarm. Two additional cases of anomalous seismicity patterns were observed: (1) 1999–2006, and (2) 2003–2006. In these two cases no main shock has yet occurred. However, the seismicity from 1999 onwards has fluctuated from low to high to low, as in the precursory sequences for previous earthquakes. The occurrence of the swarm sequence followed by a quiescence phase, which is still continuing, is an indication of a precursory seismicity gap in the region. From the predictive equations developed for the Himalayan frontal arc, it is estimated that an earthquake of M 6.5 ± 0.5 may occur at any time up to 2011 in an area bounded by 29.3°–30.5°N and 81.2°–81.9°E, in the focal depth range 10–30 km.

Key words: Anomalous seismicity, Nepal Himalaya, precursory swarm, earthquake forecast.

1. *Introduction*

Various precursors such as the land deformation, changes in sea level, tilt, strain, and crustal stress, foreshocks, anomalous seismic activity, precursory swarm, *b* value, changes in seismic wave velocity, underground water, radon content, water and gas spouting, etc. are known to precede medium to large earthquakes (RIKITAKE, 1976, 1982). The precursory phenomena differ considerably between earthquakes and also from region to region, including total absence of any known precursor (LAY and WALLACE, 1995), presumably because of differences in the subsurface geology. Significant fluctuations in seismicity, mostly prior to large earthquakes, are the most commonly observed precursory indicators. In the pending focal region of a large earthquake, numerous ruptures or heterogeneities probably exist on the main fault that can produce earthquakes in response to the loading process. Anomalous seismicity is the first precursory phenomenon to take place, due to the formation of various ruptures where considerable strain energy is accumulated, and would be useful for predicting earthquake related hazards (SEKIYA, 1977). Such anomalous seismic activity has been observed prior to many earthquakes, including the San Fernando earthquake of 1971 (ISHIDA and KANAMORI, 1977) and the intermediate earthquake of 1977 in Vrancea, Romania (MARZA, 1979).

Precursory seismic quiescence, defined by a decreased level of seismicity activity, has been recognized prior to several earthquakes, and has been used authentically for earthquake predictions (HABERMANN and WYSS, 1987; HABERMANN, 1988). MOGI (1969, 1985) was the first to suggest that

[1] Department of Geophysics, Banaras Hindu University, Varanasi 221 005, India.
[2] Department of Physics, Birendra Multiple Campus, Tribhuvan University, Bharatpur, Chitwan, Nepal. E-mail: hariharpaudyal@gmail.com
[3] Department of Earthquake Engineering, Indian Institute of Technology Roorkee, Roorkee 247667, India. E-mail: dayasfeq@iitr.ernet.in

seismic quiescence may precede large main shocks, based on his observation of a doughnut pattern associated with earthquakes with $M > 6$, in which the seismicity activity was reduced in and near the epicentral area and enhanced at farther distances. The quiescence is often broken by stress buildup resulting in an increase in the number of earthquakes, known as pre-shocks. Often these take the form of foreshock activity immediately preceding the main shock. The foreshock activity represents rapid micro-cracking immediately before the final rupture and is consistent with the dilatancy model (Scholz *et al.*, 1973).

A swarm is a succession of earthquakes clustered in space and time with no outstanding principal main shock. This pattern has been identified in different seismogenic regions of the world. Swarm activity is often associated with volcanic activity; however, swarms also occur in many non-volcanic regions (Bullen and Bolt, 1985). A precursory earthquake swarm occurs in and around the focal region of a major earthquake several years before its occurrence (Evison, 1977a, b). Evison (1982) proposed a generalized precursory swarm hypothesis for the occurrence of multiple earthquake swarms, precursory gap and multiple main-shock events, based on the detailed studies of precursory swarms in Japan, following an earlier hypothesis of a one-to-one relationship between swarms and main shocks based on four and nine significant earthquakes in California and New Zealand, respectively (Evison, 1977a, b). Versions of the swarm hypothesis were tested in Japan and New Zealand (Evison and Rhoades, 1993, 1997, 1999). Following a similar methodology, several cases of the earthquake swarm pattern were reported in different parts of Himalaya and its adjoining region, including the Burma Szechwan region (Singh *et al.*, 1982); Pamir and its adjoining region (Singh and Singh, 1984, 1985, 1986); northeast India (Gupta and Singh, 1986, 1989; Singh *et al.*, 2005a); and Himachal Pradesh, India (Shankar et al. 1995). Singh and Singh (1986) proposed a hypothesis for the occurrence of an earthquake swarm sequence and related main-shock sequence based on identification of such a pattern in the Himalayan region.

Moderate to great earthquakes in the northeast and the western parts of the Himalayan thrust belt are found to be preceded by well-defined patterns of earthquake swarms. Khattri and Wyss (1978) observed that a period of significant quiescence precedes all earthquakes of $M \geq 6.6$ in the northeast India region. Using such a pattern, a precursory gap period of 1–5 years for the main shocks M 5.3–6.2 was reported in the Himalayan region by Singh and Singh (1984) and Singh *et al.* (1982). On the other hand, a precursory gap period ranging from 11 to 17 years for main shocks of M 7.5–8.0, and from 23 to 27 years for M 8.7 in northeast India region was estimated using the same pattern (Singh and Singh, 1984; Gupta and Singh 1986; Singh *et al.*, 2005a). They also established predictive regressions among the main-shock magnitudes, the largest magnitude in the earthquake swarm sequence and the quiescence periods, and reported that the logarithm of the gap period is proportional to the magnitude of the main shocks. Using the patterns of earthquake swarm and associated seismic characteristics observed in 1964–1965 in the Arakan Yoma fold belt, Gupta and Singh (1986) predicted a large earthquake (M 7.5 $\pm$ 0.5) to occur by the end of 1990 in a delineated preparatory zone, in which an earthquake of M 7.5 subsequently occurred on 6 August 1988. Although similar patterns observed in the Indian shield region (Singh *et al.*, 2005b, 2007) and in Himachal Pradesh, India (Shankar et al. 1995) have to date resulted in no moderate earthquakes.

Generally, the recurring tendencies in time and space of major earthquakes have not been widely accepted, for the reason that catalogues of historical earthquakes suffer from incomplete reporting and nonrandom selection due to the lack of recording stations (Bullen and Bolt, 1985). Heterogeneous reporting interferes with the correct identification of precursory seismic quiescence (Rikitake, 1982; Gupta and Singh 1986; Wyss, 1997a). A reliable seismicity database spread over a wide range of magnitude in a region is essential for the understanding of earthquake processes and precursory phenomena (Rikitake, 1982). The importance of seismicity data for earthquake prediction has been demonstrated in a number of studies (Habermann and Wyss, 1987). Further, Habermann and Wyss (1984) have stressed the importance of evaluating the background seismicity level and its changes for estimating abnormal fluctuation in the seismicity pattern prior to large earthquakes. Several earthquake predictions have been made on the basis of the analysis of seismicity patterns (OHTAKE *et al.* 1977a, b; VAN WORMER and RYALL, 1980; BAKUN and MCEVILLY, 1984; WYSS and BURFORD, 1985).

The Nepal Himalaya region displays all major tectonic features of the Himalayan mobile belt and represents intense seismic activity producing frequent strong earthquakes. Identification of anomalous seismicity and earthquake swarms prior to large earthquakes in the region has thus far not been attempted. Here we have attempted to examine the existence of anomalous seismicity patterns prior to the occurrence of large earthquakes. The earthquake database compiled for the period 1963–2006 for the Central Himalaya region by PAUDYAL (2008) considering all existing catalogues has been used for the identification of seismicity patterns. Such pattern recognition could not be attempted in the data prior to 1963, due to poor detection and location capabilities of the seismograph network in the study region.

2. *Identification of Seismicity Pattern*

A gradual increase in seismic activity in a region has been explained by a slow increase of tectonic stress through the dilatancy hypothesis; whereas a decrease in seismic activity was observed in the dilatancy hardening stage (SCHOLZ *et al.*, 1973). A burst of seismic activity reflects the onset of the precursory sequence that is followed by a period of abnormal quiescence which continues until the occurrence of the major event (EVISON, 1977a). The entire preparatory period may be classified into four episodes as: normal (or background) seismicity sequence (measured from the onset of swarm activity); anomalous seismicity (or precursory swarm) sequence (period from the onset to end of swarm sequence); precursory gap (or seismic quiescence) sequence (from the date of termination of swarm activity to the onset of the main-shock sequence); and the main-shock sequence (duration of main shock and its associated aftershocks) (EVISON, 1977a; SINGH and SINGH, 1984). Within the preparatory area, the episodes of normal (N), anomalous (A), gap (G) and main-shock (M) sequences represent anomalously low, high, low and high seismic activities, respectively.

2.1. *Western Nepal Himalaya and its Adjoining Region*

In the western and the eastern Nepal regions, the cutoff magnitude for the period 1963–2006 is estimated to be $m_b \geq 4.3$, using the *b* value method, and so only such earthquakes have been considered for identifying anomalies (PAUDYAL, 2008). The cutoff magnitude thus estimated was observed to be different for different sectors of the Central Himalayan region ranging from 4.1 to 4.4. A search has been conducted for anomalous seismicity changes prior to main shocks with $m_b \geq 5.6$. It is observed that there is a significant fluctuation in seismicity at different times but mostly prior to large earthquakes. Five cases of anomalous seismicity have been identified: prior to three earthquakes that have already occurred, and two cases for which quiescence episodes still continue.

During 1950–1962, 31 earthquakes occurred in this region. They were sparsely distributed, mostly to the north of 29°N. The earthquakes that occurred subsequently from 1963 to 1966 were clustered in a smaller area in which four large earthquakes of magnitude $m_b \geq 6$ occurred in succession, not preceded by anomalous seismic activity. This cluster of events lies almost in the center of the area occupied by the events during 1950–1962. Aftershocks associated with the 27 June 1966 earthquake continued until 11 March, 1967. From 12 March, 1967 to 2005 March, 1969, only five earthquakes occurred, mostly confined to the western portion. After about 9 months of quiescence, a pattern of increasing seismic activity occurred until 16 May, 1977 in a small area between 81–81.6°E and 29.3–29.82°N. Then, after about 4 months of quiescence, an alarming seismicity increase was observed to the west approximately 50 km from the previous active zone, although with comparatively lower magnitudes. In the present work, anomalous seismicity and the delineation of preparation zones are carried out using the temporal and the spatial distribution of events, considering all events with $m_b \geq 4.3$ in four anomalous episodes, respectively.

2.2. Bajhang Earthquake of 29 July, 1980 (m_b 6.1)

To study the anomalous seismicity preceding the Bajhang earthquake of 29 July, 1980, the data from 1967 to 1980 for the area bounded by 28.5°–31°N and 80°–82.2°E were considered. Four anomalous episodes of seismic activity associated with this main shock were delineated, using the cumulative number of events with time (CNET) and the spatial distribution of events for the individual episode (Fig. 1). An elliptical preparatory area trending approximately east–west was delineated (Fig. 1a) using the spatial distribution of events in the four identified episodes. This area is supposed to enclose all the events closely related to preparatory processes at different stages of the Bajhang earthquake. The four episodes with their seismic activities in the preparatory area are given in Table 1. During the normal seismicity period (1967–19 September 1977) 17 earthquakes occurred, sparsely distributed in a large area. Only four of these events fall within the delineated east–west trending preparatory area. During the anomalous seismicity period five earthquakes (m_b 4.4–5.0) occurred, including one intermediate event which is excluded in this analysis. During the gap period three events in the magnitude range from 4.3 to 5.8 occurred, however none of these was located in the identified preparatory area.

The seismicity fluctuates from low to high to low in the normal, anomalous and gap episodes, respectively (Table 1). There was about a 22-fold increase in seismic activity during the anomalous episode as compared to the normal seismicity episode. On the other hand, the gap episode is characterized by extremely low seismic activity, even below the normal level. The events in the preparatory zone were in the focal depth range of 15–55 km (Fig. 1b). A foreshock of 29 July, 1980 with magnitude 5.7 occurred about 2 h before the main shock, within the preparatory area. After a quiescence period of 881 days from the termination of the anomalous seismicity episode, the main shock of 29 July, 1980 occurred at a focal depth of 18 km, close to the northern boundary of the preparatory area (Fig. 1a). The main shock was followed by several aftershocks which were mostly

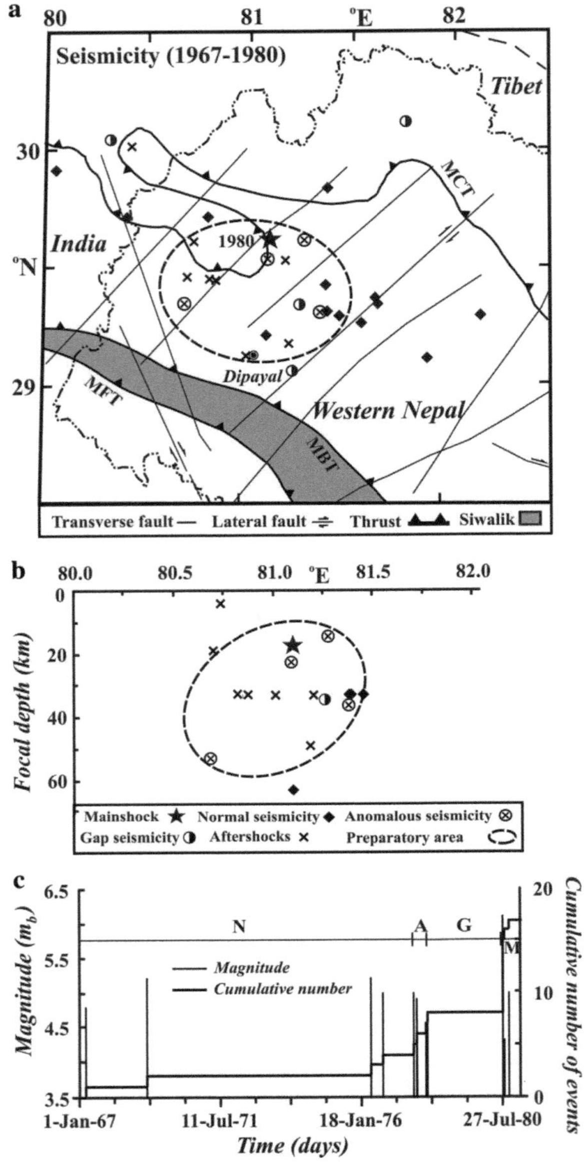

Figure 1
Spatial, focal depth and temporal distribution of events ($m_b \geq 4.3$) for the period 1967–1981 associated with the Bajhang earthquake of 29 July, 1980 (m_b 6.1). The *dotted elliptical area* in **a** and **b** is the preparatory area for the Bajhang earthquake on the surface and with focal depth, respectively. **b** Represents the focal depth and longitude distribution of events and **c** shows the magnitudes and cumulative number of events versus time. The four identified anomalous seismic phases are: normal/background seismicity (N); anomalous seismicity/swarm (A); gap or quiescence (G); and main shock and its associated aftershocks (M). *MCT* Main Central Thrust, *MBT* Main Boundary Thrust, and *MFT* Main Frontal Thrust

Table 1

Seismic characteristics in the identified seismic episodes in the preparatory area of Bajhang main shock of 29 July, 1980 (m_b 6.1) in the western Nepal Himalaya

Seismic episodes	Duration	Days	Total events	Level of activity
Normal/background (N)	11 March 1967–19 September 1977	3,846	4	Very low
Anomalous/swarm (A)	20 September 1977–28 February 1978	162	4	Extremely high
Precursory gap (G)	01 March 1978–28 July 1980	881	0	Extremely low
Main-shock sequence (M)	29 July 1980–10 October 1980	74	8	–

clustered to the west and the south of the main shock within the preparatory area. The aftershock sequence continued is shown until October, 1980. The analysis suggests that the Bajhang earthquake was preceded by anomalous seismic activity persisting for about 5 months followed by a gap of about 2 years and 10 months prior to the main shock (Table 1).

2.3. *Bajura Mainshock of 18 May, 1984 (m_b 5.6)*

This earthquake is located to the south of the Main Central Thrust (MCT), about 80 km east of the Bajhang earthquake, and is close to the Karnali transverse fault at its northern extremity. To study the precursory seismic activity for this earthquake, the seismic events from 1981 to 1984 were considered in an area bounded by 28.5°–30.3°N and 80°–82.3°E (Fig. 2a). Four identified anomalous episodes and their characteristics are given in Table 2. The delineated preparatory area is oriented approximately parallel to the transverse faults (Fig. 2a).

The normal seismicity episode is spread over about 20 months from 1981 (Table 2) with the occurrence of four events, of which two are within the delineated preparatory area showing very low seismic activity. An anomalous seismicity episode in the following four and a half months brought about a 12-fold increase in the seismic activity. Then the seismic activity decreased drastically in the following gap episode with an annual frequency of less than one event. The main shock occurred after a quiescence of 476 days on 18 May, 1984 and is located at the northeast corner of the preparatory area, and was followed by a few aftershocks. Most of the events in all four identified episodes were confined to a small area in the 30–45 km depth range between longitude 81.1°–82.0°E (Fig. 2b). The temporal pattern of events and their magnitude relationships within the elliptical preparation area, depicted in Fig. 2c, clearly show that the seismicity preceding the mainshock fluctuated from low to high to low. It is inferred here that the 18 May, 1984 earthquake of the western Nepal region was preceded by anomalously high seismic activity that commenced about 1.7 years before the main shock. The region was quiet for the next 2 years from the date of termination of aftershock activity on 23 November, 1984 (Table 2).

2.4. *Chamoli Earthquake of 28 March 1999 (m_b 6.6)*

This main shock occurred in the Chamoli region of India, in close relation with the surface expression of the MCT, and is the largest among the recent earthquakes in the Central Himalaya region. The distribution of the events in this region (79–80.2°E) prior to the Chamoli earthquake (1963–1990) followed clearly the surface trend of the MCT. The area considered is bounded by 29.9°–31.0°N and 79°–80.2°E (Fig. 3a) and the time period from 1981 to 1999 was considered to investigate the precursory seismic activity. Four identified anomalous episodes preceding the main shock are given in Table 3. The delineated preparatory area is oriented in the northwest–southeast direction (Fig. 3a).

In the normal episode, a total of 15 earthquakes occurred in two clusters, extending from the epicenter of the Chamoli earthquake to the west-southwest (within ~10–30 km); and towards the southeast (within ~50–80 km). Seven of these events were within the delineated preparatory area (Fig. 3a) with focal depths ranging 20–70 km (Fig. 3c). These represent very low seismic activity, with a frequency of one event every 2 years. The normal seismicity

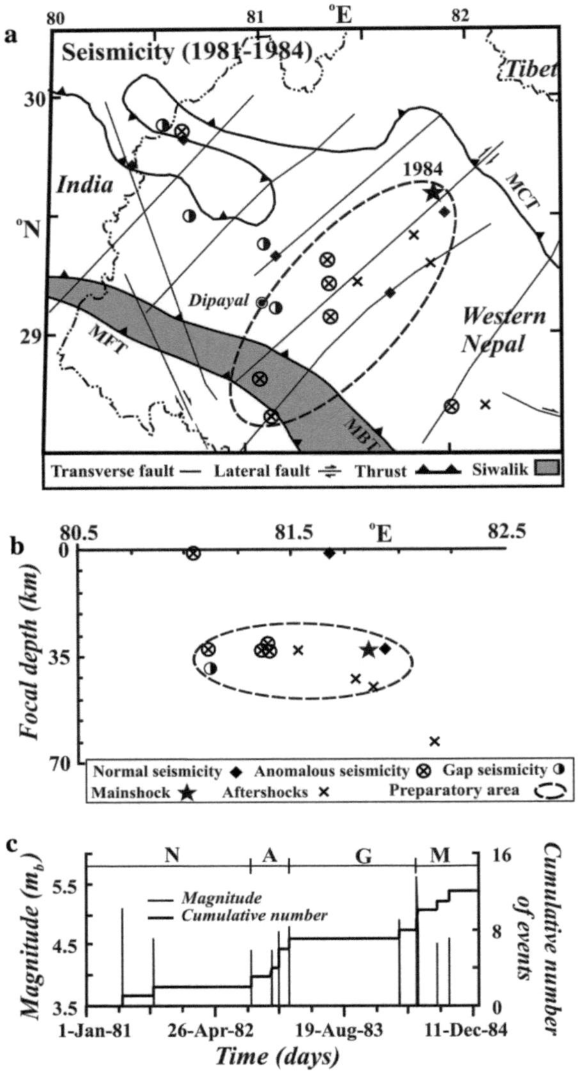

Figure 2
Spatial, focal depth and temporal distribution of events ($m_b \geq 4.3$) for the period 1981–1984 associated with the Bajura main shock of 18 May, 1984 (m_b 5.6). The *dotted elliptical area* in **a** and **b** is the preparatory area for this main shock on the surface and with focal depth, respectively. **b** Represents the focal depth and longitude distribution of events. **c** Shows the magnitudes and cumulative number of events versus time. *N*, *A*, *G* and *M* have their usual meaning as given in caption of Fig. 1

was terminated by a burst of seismic activity over a short period of about 205 days (Table 3), with four events in sequence in the magnitude range of 4.3–5.3. This produces a high annual frequency of ~8 events that represents a 16-fold increase in the seismicity rate. These four events are distributed nearly in a line about 30-km-long trending in a NW–SE direction out from the epicenter of the main shock. The events in the anomalous sequence are confined to a narrow depth segment of 15 km (27–42 km focal depth) and hence constitute a well-defined pattern of anomalous seismic activity (Fig. 3a, c).

The anomalous seismicity episode was followed by a quiescence of about 33 months until the occurrence of the main shock (Table 3; Fig. 3b). During this period two events occurred, of which only one event of 28 February, 1999 is located within the delineated preparatory area, providing an extremely low seismic activity compared to that of the anomalous episode. The quiescence was terminated by the occurrence of the Chamoli main shock on 28 March, 1999 (focal depth 23 km) to the north of the MCT, close to the second swarm event (Fig. 3). The main shock was followed by a series of aftershocks which continued until 2 June, 1999, with seven aftershocks in the magnitude range of 5.0–5.5. The main shock and most of the aftershocks are located in a narrow zone on either side of the MCT (Fig. 3a). The delineated preparatory area is oriented in the northwest–southeast direction almost perpendicular to the major trend of aftershocks activity. The time and magnitude distribution of events in the preparatory area is depicted in Fig. 3b. The majority of events in all four identified episodes are confined to a small area in the depth range 10–50 km and the longitude range 79.1°–79.6°E (Fig. 3c). The overall focal depth distribution shows that events became progressively deeper from east to west. In summary, the Chamoli earthquake was preceded by a well-defined swarm of earthquakes which continued for 205 days from November 1995. The main shock was preceded by a precursory time period of 1,217 days. It is noted that the region was quiet for 2 years following the termination of the aftershock sequence on 03 June, 1999, with the occurrence of only two small events since then.

3. *Anomalous Seismicity for Future Earthquakes in Western Nepal*

Upon analyzing the seismicity data from 1999 to 2006, two additional cases of characteristic seismicity

Table 2

Characteristics of the identified seismic episodes in the preparatory area of 18 May, 1984 Bajura main shock (m_b 5.6) in the western Nepal Himalaya

Seismic episodes	Duration	Days	Total events	Level of activity
Normal/background (N)	01 January 1981–08 September 1982	616	2	Very low
Anomalous/swarm (A)	09 September 1982–27 January 1983	141	5	Extremely high
Precursory gap (G)	28 January 1983–17 May 1984	476	1	Extremely low
Mainshock sequence (M)	18 May 1984–22 November 1984	179	3	–

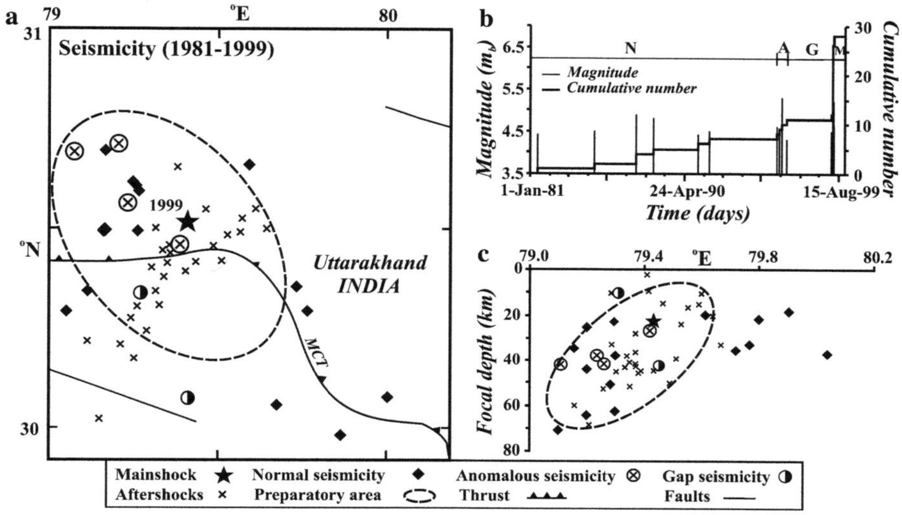

Figure 3
Spatial, focal depth and temporal distribution of events ($m_b \geq 4.3$) for the period 1981–1999 associated with Chamoli earthquake of March 28, 1999 (m_b 6.6). The *dotted elliptical regions* delineated in **a** and **c** are the preparatory areas on the surface and with focal depth, respectively. **b** Shows the magnitudes and cumulative number of events versus time. *N*, *A*, *G* and *M* have their usual meaning as given in caption of Fig. 1. **c** Shows the focal depth and longitude distribution of events

Table 3

Seismic characteristics in the identified seismic episodes in the preparatory area of the Chamoli main shock (m_b 6.6) of 28 March, 1999 in the Central Himalaya region

Seismic episodes	Duration	Days	Total events	Level of activity
Normal/background (N)	19 June 1981–26 November 1995	5,574	7	Extremely low
Anomalous/swarm (A)	27 November 1995–18 June 1996	205	4	Extremely high
Precursory gap (G)	19 June 1996–27 March 1999	1,012	1	Extremely low
Mainshock sequence (M)	28 March 1999–02 June 1999			–

patterns were observed: (1) 1999–2006, and (2) 2003–2006. In these two cases, though the anomalous seismicity exists, no main shock has occurred to date. There are three anomalous episodes in each of the sequence of 1999–2006 and 2003–2006, and the seismicity fluctuates from low to high to low as in the examples of precursory seismicity described above. The characteristics of these episodes are discussed here.

3.1. *Anomalous Seismicity of 1999–2006*

In western Nepal, a spurt in seismic activity occurred from 15 April, 2001 to 4 June, 2002, preceded

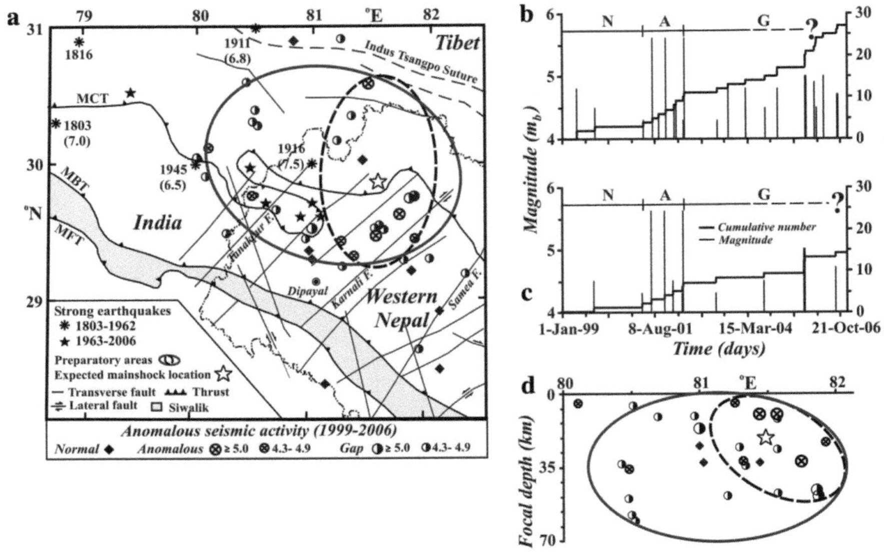

Figure 4

Spatial, focal depth and temporal distribution of events ($m_b \geq 4.3$) for the period 1999–2006 which constitute a well-defined anomalous seismicity/precursory swarm pattern for an expected future earthquake. The location of the expected main shock is shown by an *open star*. The elliptical areas delineated through *solid* (large) and *dotted* (small) curves in **a** are the preparatory areas for future seismic hazard using spatial and temporal distribution of normal (*N*), anomalous (*A*) and gap events (*G*). **b, c** Represent temporal distribution of anomalous events in large and small preparatory areas, respectively. **d** Represents the focal depth and longitude distribution of events

and followed by abnormally low seismic phases (Fig. 4a, b; Table 4). Based on the spatial and temporal patterns of the clustering of events, a large elliptical preparatory area ($\sim 3 \times 10^4$ km^2) trending approximately east–west, which encloses all the eight swarm events (magnitude 4.3–5.6), has been demarcated (Fig. 4a, b). Two preparatory episodes can be clearly identified, i.e. the normal seismicity and the anomalous seismicity episodes; whereas the gap episode continues, no main shock having yet occurred. The normal episode is considered to be from 1999 until 14 April, 2001, and is characterized by abnormally low activity, with only three events within the preparatory area. During the anomalous period, a fivefold increase in the seismic activity over the normal episode is observed (Table 4). The gap period, initiated from 05 June 2002, continues with a lower level of activity in comparison to that of the anomalous episode. This well-defined anomalous seismic pattern, both in space and time domains, may indicate that the delineated area is preparing for a future moderate to large earthquake.

Three swarm earthquakes of magnitude 5.6 occurred within 11 months during the anomalous episode and were confined to a north–south segment in the eastern portion of the large preparatory area. These large swarm events were not preceded by any type of seismic pattern although the later two events were followed by a few aftershocks. It is probable that these large swarm events re-orient the existing direction of tectonic stress to N–S in the preparatory area. Using these observations, a smaller preparatory area ($\sim 1.3 \times 10^4$ km^2) has been delineated within the larger preparatory area (Fig. 4a), oriented north–south and enclosing six out of eight swarm events. Seismic activity in the identified normal, anomalous and gap episodes is in the ratio 1:13:4, respectively. The temporal distribution of these events in three episodes within the smaller preparatory area shows the familiar low–high–low pattern of seismic activity (Fig. 4c; Table 4). The spatial and focal depth distribution shows the clustering of swarm events in a vertical column of 5–35 km and may be helpful in locating the probable site for an expected strong event as depicted by an open star in Fig. 4a and d. The occurrence of the precursory swarm sequence, followed by a significant decrease in seismicity which continues, is thus an indication of the existence of an anomalous seismicity gap in the region (Table 4).

Table 4

Characteristics of identified seismic phases in the proposed preparatory area in western Nepal and its adjoining region during 1999–2006, possibly related to a future large earthquake

Seismic episodes	Duration	Days	Number of events		Level of activity
			LPA	SPA	
Normal/background (N)	01 January 1999–14 April 2001	835	3	1	Extremely low
Anomalous/swarm (A)	15 April 2001–04 June 2002	416	8	6	Extremely high
Precursory gap (G)	05 June 2002 to (continuing)	1,671	14	7	Extremely low (continuing)
Main-shock sequence (M)	Not yet occurred				

Large preparatory area (LPA) and small preparatory area (SPA) represent the areas enclosed by solid and dotted curves in Fig. 4a, respectively

3.2. Anomalous Seismicity of 2003 to March 2007

While examining the recent earthquake database of western Nepal and its adjoining region from 1999 forward, two earthquake swarms during 2001–2002 and 2005 were observed almost in overlapping areas but without any main shock as yet. All events in the earthquake swarm of 2005 were in the precursory gap period of the earlier earthquake swarm of 2001–2002 described above. The 2005 swarm is now discussed.

Based on the spatial and temporal distribution of events from 2003 onwards (Fig. 5a, b), the normal seismicity, swarm events, and the events in the continuing gap period were identified both in space and time, and the preparatory area ($\sim 1.1 \times 10^4$ km^2) for the expected future seismic disturbances was delineated. Six events constitute the earthquake swarm sequence, which occurred over a short span of 47 days (Table 5) but was distributed widely in the northwest–southeast trending preparatory area: the four northernmost events were located within the preparatory area defined by the swarm sequence of 2001–2002. The swarm episode is characterized by extremely high seismic activity compared to the preceding normal and the following gap episode (Table 5). The temporal pattern of events and their magnitude relationships within the elliptical preparation area (Fig. 5c) show seismicity fluctuations from low to high to low from 2003 forward in the three identified seismic episodes. Thus another well-defined earthquake swarm pattern exists in space and time in the western Nepal region.

Both the spatial and the focal depth distributions of swarm events exhibit clustering which may be used to approximate the probable site of an expected moderate sized earthquake in the preparatory area, depicted by an open star in Fig. 5a and b. Most events in the identified episodes are confined to a small area between longitude 81.2°–82.4°E within the 10–50-km-depth range (Fig. 5b). Except for the shallowest swarm event, all other events in this sequence have occurred in a narrow 15-km-depth range of 35–49 km. This suggests that the events in all the identified episodes may be caused by a single rupture zone and the expected main shock may occur in the 20–50-km-focal depth range. The occurrence of the earthquake swarm sequence followed by quiescence, which continues, is an indication of the existence of an anomalous seismicity gap in the region.

3.3. Eastern Nepal Himalaya and its Adjoining Region

3.3.1 27 March, 1964 earthquake of m_b 6.3

The precursory swarm pattern associated with this earthquake was discussed by Gupta and Singh (1986). They reported the occurrence of nine events in quick succession during 24 May, 1959 to 25 December, 1961 which constituted a precursory swarm for this main shock. The annual frequency of over three events per year was compared to extremely low annual frequency during the preceding background and following gap episodes. The preparatory area was estimated $\sim 1.4 \times 10^5$ km^2 in which the main shock was located at 27.2°N and 89.3°E and is oriented in a northwest–southeast direction. This earthquake was associated with a well-defined earthquake anomalous seismicity pattern and occurred some 5 years after the burst of swarm activity.

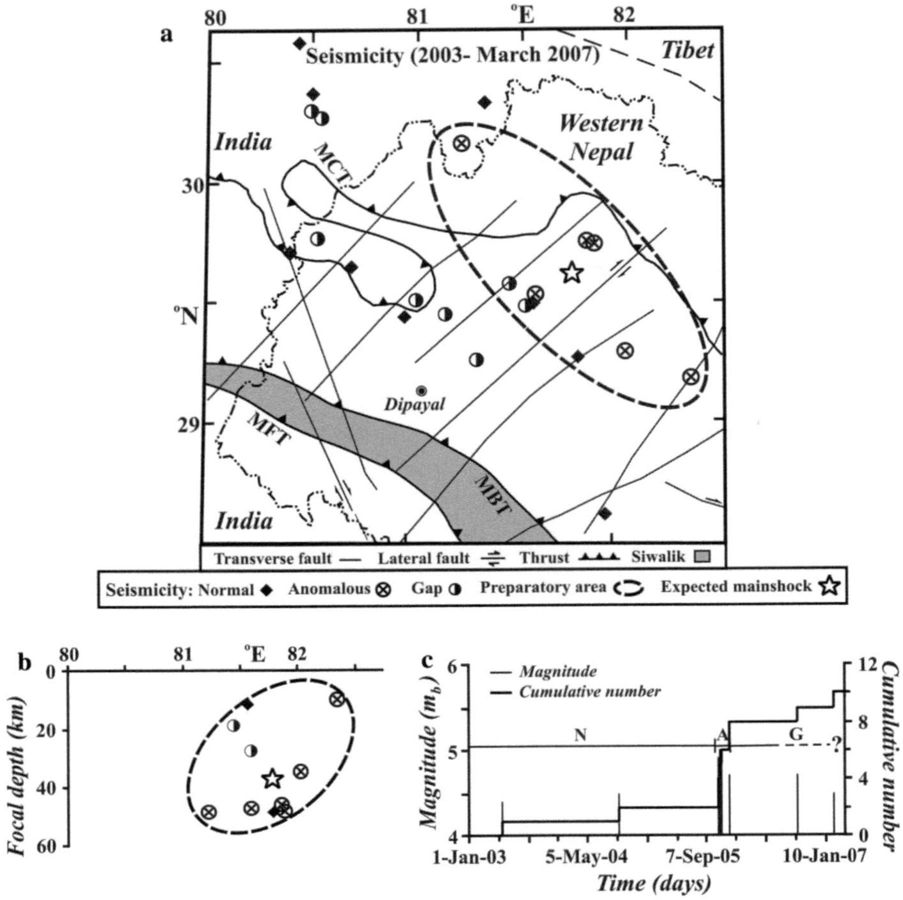

Figure 5
Spatial, focal depth and temporal distribution of events ($m_b \geq 4.3$) for the period 2003–2006 which constitute an anomalous seismicity/precursory swarm pattern for a possible future earthquake. The location of the expected mainshock is shown by an *open star*. The elliptical area delineated by a *dotted line* is the proposed preparatory zone (**a**, **b**) using spatial (**a**) and temporal (**c**) distribution of normal *N*, anomalous *A* and gap events *G*. **b** Represents the focal depth and longitude distribution of the events

Table 5

Characteristics of identified seismic phases in the preparatory area in western Nepal and its adjoining region during 2003–March 2007, possibly related to a future large earthquake

Seismic episodes	Duration	Days	Total events	Level of activity
Normal/background (N)	18 January 2003–24 October 2005	1,011	2	Extremely low
Anomalous/swarm (A)	25 October 2005–12 December 2005	47	6	Extremely high
Precursory gap (G)	13 December 2005–31 March 2007 onwards gap continues	476	2	Extremely low

3.3.2 19 November, 1980 Earthquake of m_b 6.1

Four anomalous episodes of seismic activity associated with this main shock were studied by Paudyal (2008) considering the seismicity data from 1971 to 1981 bounded by 26°–29°N and 86.2°–90°E and delineated the preparatory area ($\sim 3.5 \times 10^4$ km^2) orientated in the NE–SW direction. The background and gap seismic events show a similar nature of distribution whereas the swarm events occurred mostly along the transverse fault in the NE–SW direction. The anomalous seismicity started on 23 January, 1975 and continued for about 10 months during which eight earthquakes (m_b 4.7–5.2) occurred, of which six were within the delineated preparatory area. During the gap period three events with magnitudes ranging from 4.6 to 4.8 occurred in the identified preparatory area. The seismic activity thus fluctuated from low to high to low in the ratio 1:10:1 in the normal, anomalous and gap episodes, respectively (PAUDYAL, 2008). The events in the identified episodes within the preparatory area also show clusters in the focal depth range of 15–35 km. The main shock occurred at a focal depth of 18 km, after a precursory time period of 2,127 days, and was located to the east of Gangtok. The aftershock activity was not pronounced, even when considering the seismicity data for the next 1 year until 1981. The temporal pattern of events and their magnitude relationships within the preparation area demonstrate that this earthquake was preceded by a well-defined anomalous seismic activity which commenced about 6 years before the main shock.

3.3.3 Udaypur Earthquakes of 20 August 1988 (m_b 6.4) and 20 April 1988 (m_b 5.4)

The main shock of 20 August 1988 is the largest recent earthquake in southeastern Nepal. It caused widespread devastation in the adjoining areas of Nepal and India. The earthquake of 20 April 1988 occurred ~30 km to the north of the epicenter of the Udaypur earthquake. These two earthquakes occurred in the epicentral tract of the 1934 great earthquake. In order to investigate anomalous seismic patterns associated with these two earthquakes, the seismicity data for the period 1975–1988 within an area bounded by 26°–29°N and 84.6°–88.4°E is considered (PAUDYAL, 2008). The spatial distribution of events, including both the main shocks, is oriented in the northwest–southeast direction. A similarly oriented elliptical area is delineated as the preparatory area ($\sim 4.7 \times 10^4$ km^2) for the 1988 Udaypur earthquake. This area encloses most of the events during the period. The temporal pattern indicates a significant increase in the seismic activity during two periods, i.e., from 10 February, 1978 to 19 June, 1979, and 23 December, 1985 to 02 February, 1986. The first high activity phase is found to be related with the 20 August, 1988 main shock and the second one with the 20 April, 1988 main shock. The second high activity phase occurred well within the preparatory area of the 20 August, 1988 main shock during its gap episode. Four anomalous episodes associated with the 20 August, 1988 main shock show that the seismicity preceding the main shock fluctuated from low to high to low in the ratio 1:4:1 during the normal, anomalous and gap episodes, respectively. The anomalous seismic activity started about 10½ years prior to the main shock and continued for 495 days.

The seismic events in all the episodes are widely distributed, but mostly concentrated in the northwestern portion, leaving a seismic gap in the southeastern part of the preparatory area. The magnitudes of events in the swarm sequence range from 4.3 to 5.2. An event with M 5.4, which occurred on 20 April, 1988 during the gap period, is the second main shock which was itself found to be preceded by anomalous swarm activity. Four of five events in the swarm sequence are located between 84.5° and 86°E at about 33 km depth, whereas the remaining one is located remotely within the preparatory area. The main shock occurred on 20 August, 1988, after a precursory gap period of 3,349 days, in the southern part of the preparatory area at a depth of 57 km, close to the Main Frontal Thrust and to the south of the largest gap event. Aftershocks are confined to the depth range of 30–45 km. Based on this information, it is inferred that the main shock of 20 August, 1988 was preceded by anomalous seismic activity which commenced about 10½ years before the main shock.

The earthquake of 20 April, 1988 occurred close to the surface trace of the MCT at 27°N and 86.7°E. This earthquake was preceded by a swarm sequence which occurred within the continuing precursory gap period associated with 20 August, 1988 Udaypur main shock (PAUDYAL, 2008). This is a clear case of an earthquake swarm in which another swarm sequence occurs at a later stage, principally in the continuing gap episode of the first one. The delineated preparatory area ($\sim 2 \times 10^4$ km^2) for this earthquake falls totally within the preparatory area of the 20 August, 1988 main shock, and has a similar northwest–southeast orientation. The temporal pattern of events and their magnitude relationships within the preparation area show that the seismicity preceding the main shock in the normal, swarm and gap episodes fluctuated from low to high to low in the ratio of 1:41:2, respectively. The precursory swarm events are distributed in the western part of the preparatory area, and their magnitudes range from 4.5 to 4.6. The main shock occurred in the southeast portion of the preparatory area after a quiescence of 807 days. The swarm and the gap events are clustered in the depth range 30–45 km; whereas the main shock was deeper at 54 km. Based on these observations, it was reported that the main shock of 20 April, 1988 was preceded by anomalous seismic activity which commenced about 2 years and 4 months before the main shock.

4. *Predictive Regressions*

A generalized precursory swarm hypothesis was introduced by EVISON (1982), who found that $\bar{M}_p$, M_m and T_p are correlated in the following forms:

$$M_m = a\bar{M}_p + b, \quad (1)$$

and

$$\log_{10} T_p = c\bar{M}_p + d, \quad (2)$$

where M_m is the magnitude of the main shock; $\bar{M}_p$ the average magnitude of the three largest swarm earthquakes; T_p the time interval (in days) from the onset of the swarm sequence to the occurrence of the main shock; and a, b, c and d are constants, characteristic for a region and possibly different for different regions. EVISON (1982) reported the values of the above constants a, b, c and d for the Japan region as: 0.72, 2.73, 0.37 and 1.61; and for New Zealand as: 1.04, 1.52, 0.51 and 0.64, respectively.

M_m, $\bar{M}_p$ and T_p are parameters identified from each precursory sequence. In the present work, $\bar{M}_p$ is defined as the mean of the two largest events in the anomalous seismicity/swarm sequence. The parameters derived from the anomalous precursory seismicity activity for certain main shocks in the Himalayan Frontal Arc and its adjoining region are furnished in Table 6. This list includes six cases from the present work and the remaining nine from the published works of SINGH and SINGH (1984), GUPTA and SINGH (1986), and SINGH *et al.*, (2005a). The values of the parameters from the present work pertain mainly to the Central Himalaya; whereas the values included from previously published work pertain to other parts of the Himalayan Frontal Arc. The precursory parameters reveal interdependence between M_m, $\bar{M}_p$ and T_p. The predictive regressions are established here for the Himalayan compression zone by considering M_m, $\bar{M}_p$ and T_p for the 15 main shocks (Table 6). The plots of M_m versus $\bar{M}_p$, log T_p versus $\bar{M}_p$ and log T_p versus M_m are shown in Fig. 6. The predictive regressions thus obtained from the best fit are:

$$M_m = 1.05\bar{M}_p + 0.69 \quad \left(R^2 = 0.80 \quad \text{for } M_m \text{ estimation}\right) \quad (3)$$

$$\log T_p = 0.59\bar{M}_p + 0.08 \quad \left(R^2 = 0.59 \quad \text{for } T_p \text{ estimation}\right) \quad (4)$$

$$\log T_p = 0.52M_m - 0.05 \quad \left(R^2 = 0.63 \quad \text{for } T_p \text{ estimation}\right) \quad (5)$$

or,

$$M_m = 1.92 \log T_p + 0.10.$$

The expressions (3) and (5) are predictive regression equations, which give the estimation of the magnitude of the main shock (M_m), if $\bar{M}_p$ and T_p are known from the anomalous seismic (swarm) activity well before the main shock. These equations provide reliable results only if no earthquake with magnitude greater than or equal to the magnitude of the largest swarm event (M_p) occurs during the gap episode within the delineated preparation area. Based on the fault-plane solutions, SINGH and SINGH (1984) reported that the sequence of swarm events occur due to the converging trend of tectonic stresses towards the pending focal region. The occurrence of an event of magnitude greater than or equal to the largest swarm event, in the preparation area during the gap period, may divert the converging trend of stress accumulation. This situation may enhance the duration of the gap episode and also the magnitude of the main shock. In addition, it may be necessary to incorporate certain other considerations in these predictive

Table 6

Source parameters of the main shocks that were preceded by anomalous seismic activity in the Himalayan region along with related precursory parameters

Sl. Nos.	Date dd/mm/yyyy	Location °N	Location °E	Depth (km)	M_m	T_p (days)	$\bar{M}_p$	Preparatory area (km^2)	References
1	27/03/1964	27.2	89.30	32	6.3	1,769	5.0	1.2×10^5	Gupta and Singh (1986)
2	03/09/1972	35.94	73.33	45	6.2	1,448	5.2	9.4×10^3	Singh and Singh (1984)
3	11/08/1974	39.34	73.76	7	6.2	1,807	5.4	1.2×10^4	Singh and Singh (1984)
4	28/12/1974	35.06	72.91	24	5.9	845	5.0	7.9×10^3	Singh and Singh (1984)
5	19/01/1975	32.39	78.50	1	6.2	1,064	5.1	6.8×10^4	Singh and Singh (1984)
6	05/12/1975	33.10	76.13	24	5.3	385	4.8	3.7×10^3	Singh and Singh (1984)
7	12/08/1976	26.7	97.10	27	6.2	440	5.3	3.9×10^4	Gupta and Singh (1986)
8	29/07/1980	29.60	81.09	18	6.1	1,043	5.0	5.2×10^3	Paudyal (2008)
9	19/11/1980	27.34	88.75	17	6.1	2,127	5.2	3.6×10^4	Paudyal (2008)
10	18/05/1984	29.58	81.87	33	5.6	617	4.8	7.4×10^3	Paudyal (2008)
11	30/12/1984	24.6	92.90	22	5.6	1,084	5.0	8.4×10^4	Gupta and Singh (1986)
12	20/04/1988	27.04	86.69	54	5.4	859	4.6	2.0×10^4	Paudyal (2008)
13	06/08/1988	25.12	95.17	115	7.5	9,199	6.6	1.3×10^5	Singh *et al.* (2005a)
14	20/08/1988	26.75	86.62	57	6.4	3,844	5.2	4.7×10^4	Paudyal (2008)
15	28/03/1999	30.51	79.42	23	6.6	1,217	5.0	4.0×10^3	Paudyal (2008)

M_m and $\bar{M}_p$ are the magnitude of main shocks and the mean of two largest events in the anomalous (swarm) sequence respectively, and T_p is the preparatory time period measured from the onset of anomalous sequence to the occurrence of the main shocks. The values of parameters M_m, $\bar{M}_p$ and T_p are used to establish predictive regression among these parameters for the Himalayan Frontal Arc

equations prior to estimating the values of M_m and T_p for a pending earthquake. Further, it is evident from Eqs. (3) and (4) that both the parameters M_m and T_p are dependent on $\bar{M}_p$, and hence, the corrections if any, need to be applied for re-estimation of $\bar{M}_p$..

Rikitake (1978, 1981) defined "precursors of the first kind" on the basis of dependence of the logarithm of precursory time on the magnitude of the main shock. He added that precursors of this kind generally have large spatial extent and are scattered such that the precursor time increases with the increase in the magnitude of the main shock. Equation (5) and Fig. 6c support the above view, and show a linear correlation between the magnitude of the main shock and the logarithm of precursory time period. The anomalous seismicity pattern is thus a precursor of the first kind which may be used for long-range earthquake forecasting.

5. *Estimates of a Future Earthquake in Western Nepal Himalaya*

As described above, two anomalous seismic activity sequences in the western Nepal region occurred during (i) 15 April 2001–2004, June 2002 and (ii) 25 October–12, December 2005 with $\bar{M}_p$ values of 5.6 and 5.0, respectively (Table 7). For these sequences, no main shock occurred up to 2007 and the gap episodes continue. Here, it is estimated that the magnitudes of the expected earthquake and the precursor time are M 6.6 and T_p 6.7 years for case (i) and M 6.0 and T_p 3 years for case (ii) using the corresponding $\bar{M}_p$ values in the predictive equations (3, 4) established for the Himalayan Frontal Arc. According to the first estimates, an earthquake with M 6.6 should have occurred by the end of 2007, however no such event has been observed yet. According to the second estimates, an earthquake with M 6 should occur by September 2008 in the same region. The anomalous activities associated with these two cases were in the same locality. A large portion of their delineated preparatory areas are in common, although with different orientations. Evidently, this is a case of a repeated swarm sequence as pointed out by Evison (1982), in which the second swarm has occurred in the gap episode of the first. In view of the above, it may be considered that these two cases probably have a common origin. The occurrence of the repeated swarm sequence may

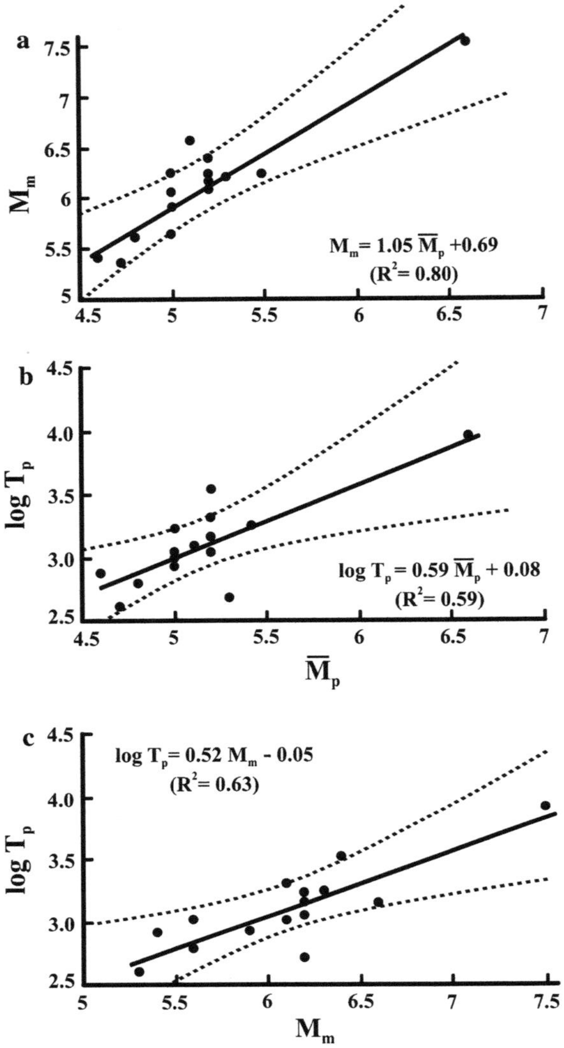

Figure 6
Predictive regressions among M_m (main shock magnitude), $\bar{M}_p$ (mean magnitude of the two largest events in the anomalous sequence/swarm) and T_p (precursory time period, in days, from the onset of the anomalous sequence to the main shock occurrence). The *solid straight lines* are the least-squares fits and the *dotted curves* are 99% confidence limits

indicate that a wider area is under threat which requires re-estimation of the magnitude and time of occurrence of the impending earthquake (Evison, 1982) after applying the required corrections. Such a situation may also change the rate of stress accumulation in the pending focal region (Singh and Singh, 1984) which evidently increases the duration of the preparatory period and the magnitude associated with the impending earthquake. Now, it is thought here that the delay in the occurrence of an expected earthquake is probably due to the interruption of the gap episode of the first sequence by the second one, enhancing both the preparatory period and the magnitude. In view of the above, and of the inherent error in the magnitude estimation, the magnitude of the impending earthquake may be M 6.5 ± 0.5. In this condition, a precursory time period of 10.7 years measured from 15 April 2001 for an earthquake with M 7 is estimated using the predictive equation (5). If so, the expected earthquake could occur at any time up to December 2011. Gupta and Singh (1986) invoked similar considerations to estimate the probable time of occurrence of the 6 August, 1988 earthquake in Arakan Yoma.

The same region of western Nepal was one of the seismogenic sources in which the probability is estimated to be 85% for the 10 years starting in 2005 for an earthquake with magnitude 6.4 ± 0.2 using the time and magnitude predictable model (Paudyal *et al.*, 2008b). It is noted that the estimates of time and magnitude associated with the impending earthquake using two different methods, namely the anomalous seismic activity and the time–magnitude predictable model, are in good agreement with each other.

The above estimates indicate that western Nepal is a potential zone for a forthcoming shallow-focus large earthquake. The impending earthquake may lie within the delineated preparatory area bounded by 29.3°–30.5°N and 81.2°–81.9°E and its focus would be 10–30 km deep. The anomalously low precursory quiescence period continues, and hence this observation may be used to search for other premonitory phenomena, if any.

6. Discussion

The spatial and temporal changes in seismic activity may be causally related to the time of occurrence and the magnitude of the main shocks. In view of this, an attempt has been made here to search for the pattern of the seismicity changes in space and time domains prior to main shocks with $m_b \geq 5.4$ during 1963–2006 in the Central Himalaya region.

Table 7

Characteristics of anomalous seismic activity/earthquake swarm observed for two cases during 2001–2005 in the western Nepal region along with approximately preparatory areas for the expected main shocks

Sl. nos.	Characteristics of anomalous seismic activity/earthquake swarm					Duration of data examined	Remarks
	Date of onset	Study grid	Preparatory area (km^2)	Depth range (km)	$\bar{M}_p$		
1	04.15.2001	28.0°–31.0°N 79.5°–82.2°E	3.0×10^4	5–35	5.6	1999–2006	Main shocks not occurred until 2008. Gap episodes continues
2	10.25.2005	28.5°–30.7°N 80.0°–82.3°E	1.1×10^4	10–50	5.0	2003–2006	

The magnitude of main shocks is reported to be 1–2 units higher than the magnitude of the largest swarm event in most of the anomalous sequences, as observed in different regions (Evison, 1977a, b; Singh *et al.*, 1982, 2005a; Shankar *et al.*, 1995). In the case of the Nepal Himalaya, such differences are 0.8–1.6 units. The difference between the magnitudes of the two largest swarm events is observed to range between 0.0 and 0.6.

The preparatory areas associated with six main shocks and of the expected one main shock indicate, in general, that they are oriented in the direction of the local tectonic features and can be of different size for similar-magnitude main shocks in the same area. In western Nepal and its environs, the preparatory areas were oriented in different directions, e.g., NW–SE (1999), NE–SW (1984) and E–W (1980), with the main shocks located in the northern part of the respective preparatory zones. In the case of the expected main shocks, a large part of their preparatory areas is found to be in common; however, they are oriented in the N–S and NW–SE directions. The preparatory areas of the main shocks in eastern Nepal and its adjoining region have two prominent orientations, NW–SE and NE–SW (Paudyal, 2008).

It is observed here that during the preparatory phases of six main shocks (Table 6), the seismic events exhibited clustering in the space and time domains with a high annual frequency during the anomalous episode compared to the preceding normal episode and the following gap episode, as shown in Fig. 7a. In some cases, e.g., the main shocks of 1984 (M 5.6) and 1988 (5.4), the existence of an extremely high annual frequency is due to a spurt of events within the short span of the anomalous episodes. In

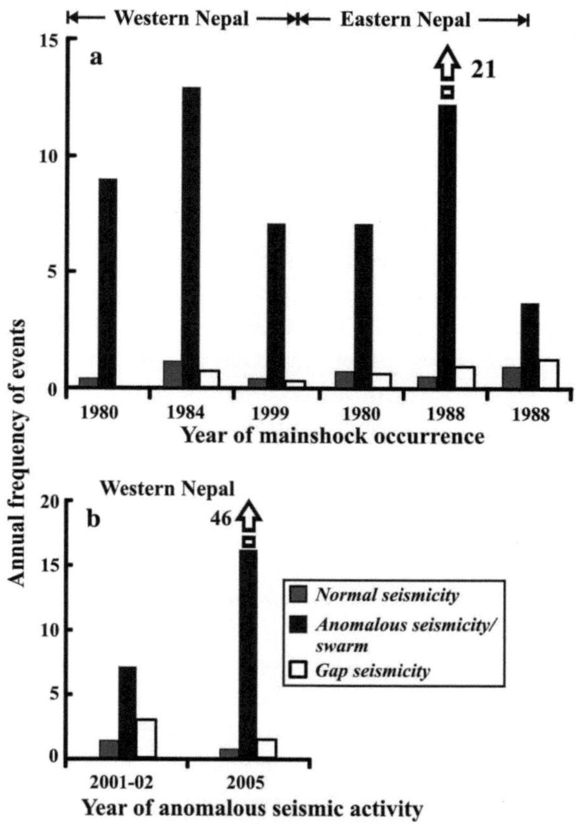

Figure 7
Annual frequency of events in the identified normal, anomalous (swarm) and gap episodes within the delineated preparatory areas for the main shocks that occurred (**a**) during 1963–2006 in western and eastern Nepal Himalaya and its adjoining Indian region; (**b**) during the period 2001–2006 for the expected main shocks in western Nepal region. It is evident that the anomalous episode is characterized by an extremely high annual frequency as compared to the preceding normal and the following gap episodes for all the main shocks. The *numbers* above the bars show the total annual frequency of events

the remaining cases, a longer extent of anomalous episode was observed, which provided a comparatively lower, but still elevated, annual frequency. It is evident that for a shorter duration anomalous episode, the associated main shock is smaller, and vice versa. It is also observed that a low level of seismic activity is quite common during the gap episodes with longer extent (Fig. 7a). This suggests that the anomalous seismic pattern may be a useful parameter in understanding the future trend in the seismic activity of a region. In light of these observations, an attempt has been made here to evaluate the future seismic hazards in the Nepal Himalaya and its adjoining region.

While searching for the anomalous seismicity pattern, it is observed that there are two well-defined cases of such features in the western Nepal (Figs. 4, 5) region, where the repeated anomalous seismic patterns with appreciably high annual frequency have occurred without any main shock through the end of 2008. It is evident that the anomalous episode is characterized by an extremely high annual frequency of events compared to that of the preceding normal and the following gap episodes, which has been depicted in Fig. 7b. This characteristic is similar to the anomalous pattern associated with the main shocks that have already occurred, as described above (Fig. 7a), and hence this observation may be useful to evaluate the magnitude and the time of occurrence of a pending earthquake.

7. Conclusion

Based on the data from 1963 to 2006, the anomalous seismicity (earthquake swarms) prior to the medium-sized earthquakes in the Nepal Himalaya and its adjoining region has been studied here. It is observed that the anomalous seismicity episodes follow episodes of relatively very low seismic activity, and it is an important finding to visualize that an area might be preparing for the occurrence of a forthcoming main shock. Such anomalous seismic patterns were observed prior to the six medium-sized main shocks that occurred from 1963 to 2006. From these observations it is inferred that the patterns of anomalous seismicity may be a useful tool for the forecasting of long-range earthquake hazards in the Nepal Himalaya region.

The following conclusions have been drawn:

1. Nepal Himalaya and its adjoining region are characterized by the occurrence of anomalous seismic activity prior to the medium-sized earthquakes. The events in the anomalous seismic episodes are causally related with the magnitude and the time of occurrence of the forthcoming main shocks, and the shorter the preparatory time period, the smaller the main shock, and vice versa.
2. The values of the precursory parameters were used to establish the predictive regressions for the Himalayan Frontal arc: $M_{\mathrm{m}} = 1.05\bar{M}_{\mathrm{p}} + 0.69$; $\log T_{\mathrm{p}} = 0.59\bar{M}_{\mathrm{p}} + 0.08$ and $M_{\mathrm{m}} = 1.92 \log T_{\mathrm{p}} + 0.10$ for the estimation of the magnitude (M_{m}) and the precursory time period (T_{p}) of the forthcoming main shocks provided the anomalous seismicity pattern is identified and $\bar{M}_{\mathrm{p}}$ is known.
3. The present study suggests that parts of the western Nepal region have potential for future medium-sized earthquakes. It has been estimated here that an earthquake with M 6.5 ± 0.5 may occur at any time from now up to 2011 in western Nepal, within an area bounded by 29.3°–30.5°N and 81.2°–81.9°E, in the focal depth range 10–30 km.

Acknowledgments

The authors are grateful to Dr. M. Banerjee, Department of Geophysics, Banaras Hindu University, Varanasi, India and Dr. T. Radhakrishna, Centre for Earth Science Studies, Thiruvanathapuram, India for critically examining the manuscript and offering useful comments and suggestions. Professor B. R. Arora, Director, WIHG, Dehradun, India and Professor Zhu Chuanzhen, Beijing, China, two anonymous reviewers for offering very valuable comments and suggestions and David Rhoades for his efforts throughout publication are acknowledged. The author HP is indebted to the University Grant Commission, Nepal for financial support in the form of a fellowship. Third author (DS) is grateful to the Head, Department of Earthquake Engineering Indian

Institute of Technology Roorkee for providing excellent computational facilities.

References

Bakun, W, H. and McEvilly, T. V. (1984), *Recurrence models and Parkfield, California earthquake*, J Geophys Res *89*, 3051–3058

Bullen, K. E. and Bolt, B. A., *An Introduction to the Theory of Seismology* (Cambridge University Press, Cambridge 1985), 499 pp

Evison, F. F. (1982), *Generalized precursory swarm hypothesis*, J Phy Earth *30*, 155–170

Evison, F. F. (1977a), *Fluctuation of seismicity before major earthquakes*, Nature *266*, 710–712

Evison, F. F. (1977b), *The precursory earthquake swarm*, Phys Earth Planet Inter *15*, 19–23

Evison, F. F. and Rhoades, D. A. (1999), *The precursory earthquake swarm in Japan: Hypothesis tests*, Earth Planets Space *51*, 1267–1277

Evison, F. F. and Rhoades, D. A. (1997), *The precursory earthquake swarm in New Zealand: Hypothesis tests II.* NZ J Geol. Geophys *40*, 537–547

Evison, F. F. and Rhoades, D. A. (1993), *The precursory earthquake swarm in New Zealand: Hypothesis tests*, NZ J Geol Geophys *36*, 51–60

Gupta, H. K. and Singh, H. N. (1989), *Earthquake swarms precursory to moderate to great earthquakes in the northeast India region*, Tectonophysics *167*, 285–298

Gupta, H. K. and Singh, H. N. (1986), *Seismicity of northeast India region: Part II: Earthquake swarms precursory to moderate magnitude to great earthquakes*, J Geol Soc India *28*, 367–406

Habermann, R. E. (1988), *Precursory seismic quiescence: Past, present, and future*, Pure and Appl Geophys. *126*, 279–318

Habermann, R. E. and Wyss, M. (1987), *Reply to "Comment on Habermann's method for detecting seismicity rate changes" by M.W. Matthews and P. Reasenberg*, J Geophys Res *92*, 9446–9450

Habermann, R. E. and Wyss, M. (1984), *Earthquake triggering during preparation for great earthquake*, Geophys. Res. Lett. *11*, 291–294

Ishida, M. and Kanamori, H. (1977), *The spatio-temporal variations of seismicity before the 1971 San Fernando earthquake, California*, Geophys Res Lett *4*, 345–346

Khattri, K. N. and Wyss, M. (1978), *Precursory variation of seismicity rate in the Assam area, India*, Geology *6*, 85–688

Lay, T. and Wallace, T. C., *Modern Global Seismology*, Vol. *58*, *International Geophysics Series* (Academic Press, San Diego 1995)

Marza, V. I. (1979), *A seismicity pattern of March 4, 1977 Vrancea, Romania earthquake: An. earthquake prediction insight*, Tectonophysics *53*, 217–222

Mogi, K., *Earthquake Prediction* (Academic Press, Tokyo 1985) 355 pp

Mogi, K. (1969), *Some features of recent seismic activity in and near Japan (2), Activity before and after great earthquake*, Bull Earthq Res Inst Univ Tokyo *47*, 395–417

Ohtake, M., Matumoto, T., and Latham, G. V. (1977a), *Seismicity gap near Oaxaca, Southern Mexico, as a probable precursor to a large earthquake*, Pure Appl Geophys *115*, 375–385

Ohtake, M., Matumoto, T., and Latham, G. V. (1977b), *Temporal changes in seismicity preceding some shallow earthquakes in Mexico and Central America*, Bull Int Inst Seismol Earthq Eng *15*, 105–123

Paudyal, H. (2008), *Seismicity and seismotectonics of Nepal and its adjoining region*, Ph.D. Thesis submitted to Banaras Hindu University, Varanasi, India (unpublished)

Paudyal, H., Singh, H. N., Shanker, D., and Singh, V. P. (2008a), *Stress pattern in two seismogenic sources in Nepal-Himalaya and its vicinity*, Acta Geophysica *56 (2)*, 313–323

Paudyal, H., Shanker, D., Singh, H. N., and Singh, V. P. (2008b), *Application of time—and magnitude—predictable model in the Central Himalaya and vicinity for estimation of seismic hazard,* Acta Geod Geophys Hung (accepted)

Rikitake, T. (1982), *Earthquake Forecasting and Warning,* Center for Academic Publications Japan *3*, 402 pp

Rikitake, T. (1981), Practical approach to earthquake prediction and warning. In *Current Research in Earthquake Prediction* (ed. T. Rikitake) Center for Academic Publications Japan *2*, 1–56

Rikitake, T. (1978), *Classification of earthquake precursors*, Tectonophysics *54*, 293–309

Rikitake, T., *Earthquake Prediction* (Elsevier, Amsterdam, 1976) 357 pp

Scholz, C. H., Sykes, L. R., and Aggarwal, Y. P. (1973), *Earthquake prediction: A physical basis*, Science *181*, 803–810

Sekiya, H. (1977), *Anomalous seismic activity and earthquake prediction*, J. Phys. Earth *25* (Suppl):S85–S93

Shankar, D., Singh, H. N., and Singh, V. P. (1995), *Anomalous seismic activity and long-range earthquake prediction in Himachal Pradesh, India*, Acta Geod. Geophys. Hung *30 (2–4)*, 379–395

Singh, H. N., Shanker, D., Neelakandan, V. N., Mathai, J., Singh, V. P., and Banerjee, M. (2007), *Spurt of geo-signatures signifying possible precursors to a major earthquake in southeastern Indian Peninsula*, ICFAI J Earth Sci *2* (2), 7–40

Singh, H. N., Shanker, D., and Singh, V. P. (2005a), *Occurrence of anomalous seismic activity preceding large to great earthquakes in northeast India region with special reference to 06 August, 1988*, Phys Earth Planet Inter *148*, 261–284

Singh, H. N., Mathai, J., Neelakandan, V. N., Shankar, D., and Singh, V. P. (2005b), *A database on occurrence patterns of unusual geological incidents in Southwest Peninsular India and its implication on future seismic activity*, Acta Geod Geophys Hung *40 (1)*, 69–88

Singh V. P. and Singh H. N. (1986), *Swarm hypothesis for occurrence of medium size earthquakes*, Earthq Predict Res *4*, 83–94

Singh, V. P. and Singh, H. N. (1985), *Stress distribution pattern and swarm occurrences*, Tectonophysics *113*, 295–306

Singh, V. P. and Singh, H. N. (1984), *Precursory swarm and medium size earthquake occurrences in Pamirs and its adjoining regions*, Earthq Predict Res. *2*, 245–258

Singh, V. P., Singh, H. N., and Singh, J. (1982), *On the possibilities of premonitory swarms for three sequences of Earthquakes of Burma-Szechwan region*, Tectonophysics *85*, T21–T29

Van Wormer, J. D. and RYALL, A. D. (1980), *Sierra Nevada–Great Basin Boundary Zone: Earthquake hazard related to structures, active tectonic processes, and anomalous patterns of earthquake occurrence*, Bull Seismol Soc Am *70*, 1557–1572

Wyss, M. (1997a), *Nomination of precursory seismic quiescence as a significant precursor*, Pure Appl Geophys *149*, 79–113
Wyss, M. and Burford, R. O. (1985), *Current episodes of seismic quiescence along the San Andreas Fault between San Juan Bautista and Stone Canyon, California: Possible precursors to local moderate main shock*, US Geol Survey Open-file Report *85-754*, 367–426

(Received August 28, 2008, revised January 30, 2009, accepted March 10, 2009, Published online March 10, 2010)

Pure Appl. Geophys. 167 (2010), 685–692

DOI 10.1007/s00024-010-0073-5

Pure and Applied Geophysics

Short- and Long-Term Earthquake Forecasts for California and Nevada

Y. Y. Kagan[1] and D. D. Jackson[1]

Abstract—We present estimates of future earthquake rate density (probability per unit area, time, and magnitude) on a 0.1-degree grid for a region including California and Nevada, based only on data from past earthquakes. Our long-term forecast is not explicitly time-dependent, but it can be updated at any time to incorporate information from recent earthquakes. The present version, founded on several decades worth of data, is suitable for testing without updating over a five-year period as part of the experiment conducted by the *Collaboratory for Study of Earthquake Predictability* (CSEP). The short-term forecast is meant to be updated daily and tested against similar models by CSEP. The short-term forecast includes a fraction of our long-term one plus time-dependent contributions from all previous earthquakes. Those contributions decrease with time according to the Omori law: proportional to the reciprocal of the elapsed time. Both forecasts estimate rate density using a radially symmetric spatial smoothing kernel decreasing approximately as the reciprocal of the square of epicentral distance, weighted according to the magnitude of each past earthquake. We made two versions of both the long- and short-term forecasts, based on the *Advanced National Seismic System* (ANSS) and *Preliminary Determinations of Epicenters (PDE)* catalogs, respectively. The two versions are quite consistent, but for testing purposes we prefer those based on the ANSS catalog since it covers a longer time interval, is complete to a lower magnitude threshold and has more precise locations. Both forecasts apply to shallow earthquakes only (depth 25 km or less) and assume a tapered Gutenberg-Richter magnitude distribution extending to a lower threshold of 4.0.

Key words: Earthquake regional forecasts, time-independent and time-dependent forecasts, California.

1. Introduction

Earthquake forecasts play an important role in hazard and risk estimation, informed decisions in risk management, and emergency response. California has a large vulnerable population, so there has been intense research on forecasting in that state. As a means of improving the forecasts, and perhaps more importantly vetting the ideas behind them, a collaborative effort known as RELM, for "Regional Earthquake Likelihood Models," was initiated in 2006. Published ideas used to gauge earthquake potential in California vary dramatically, including assumptions that earthquakes occur primarily on known faults, near recent earthquakes, or where geodetic strain rate is highest; that fault length limits or does not limit earthquake size, that large earthquakes behave very similarly or differently from smaller ones, and that large earthquakes deter or make more likely similar ones in the future. The importance of the assumptions has been highlighted in a series of earthquake probability reports (Working Group on Southern California Earthquake Probabilities, 1995; Working Group on California Earthquake Probabilities, 1990, 2003, 2009). These reports were intended to produce consensus, which could only be reached by including conflicting models with various weights. To help resolve the conflicts, the RELM project invited researchers to express their ideas as comprehensive forecasts in the form of earthquake rate density on a pre-assigned grid for prospective testing. About a dozen such models were described in a special issue of Seismological Research Letters (SRL), with an introduction by Field (2007). A set of likelihood tests for assessing the models after one and then five years was described by Schorlemmer *et al.* (2007) and Schorlemmer and Gerstenberger (2007). While the tests are still underway, the project has been subsumed under a larger international program: The Collaboratory for Study of Earthquake Predictability (CSEP). It adapts

[1] Department of Earth and Space Sciences, University of California, Los Angeles, CA 90095-1567, USA. E-mail: ykagan@ucla.edu; djackson@ucla.edu

the RELM philosophy and methods to other parts of the world and tests short-term models, updated daily. One short-term model (Gerstenberger *et al.*, 2005), nicknamed STEP for "Short Term Earthquake Probability," has been operating and made available on USGS web site (http://earthquake.usgs.gov/eqcenter/step//) for a few years.

Here we present long-term (5-year) and short-term (1-day) forecast models in the CSEP format with the intention that they will be tested against others in the CSEP project.

Our techniques have been extensively described in several of our publications (Kagan and Jackson, 1994, 2000, 2010; Jackson and Kagan, 1999). In those papers we used the Central Moment Tensor (CMT) earthquake catalog, described by Ekström *et al.* (2005), as our source of information on past earthquakes. That catalog has several advantages: Uniform definitions and stable methodology since 1977, provision of a seismic moment tensor for each event, and quite precise estimates of moment magnitude. The completeness threshold has improved from about magnitude 5.8 in the beginning to about 5.4 now. In Figure 1 we display the CMT earthquakes in the California-Nevada region.

Our forecast for two west Pacific regions (Kagan and Jackson, 1994, 2000, 2010; Jackson and Kagan, 1999) is subdivided into two programs: a long- and a short-term forecast. In the former program we use the first part of a catalog to smooth the seismicity level, and the second part of the catalog is used to validate and optimize the prediction. In the short-term forecast, Omori's law type of dependence is used to temporally extrapolate earthquake rate into the future.

We used the focal mechanisms to estimate the orientation of the fault systems controlling seismic behavior, indicating a preferred direction for future earthquakes. However, assumptions that we made in that process hold reliably only in subduction zones, so the focal mechanism information is less useful in locations like California and Nevada that are not on subduction zones. Thus, for this work we have modified our procedures to allow use of the PDE (Preliminary determinations of epicenters, 2008) and ANSS (Advanced National Seismic System ANSS Catalog Search, 2008) catalogs, with lower

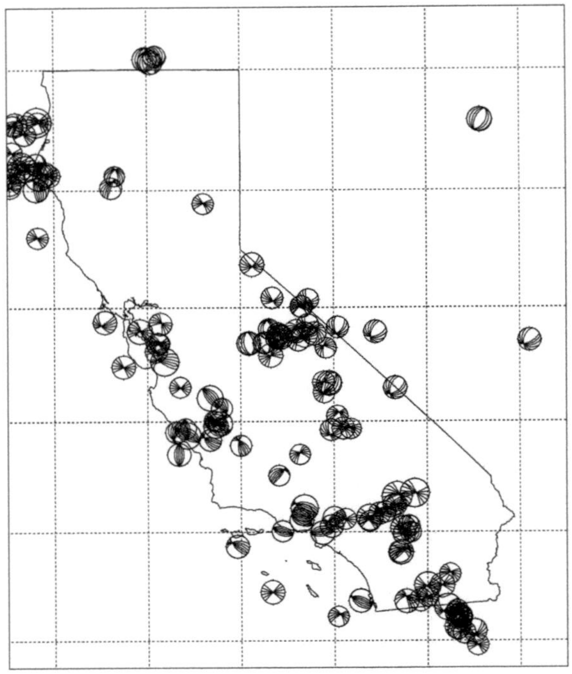

Figure 1
Earthquakes from the CMT catalog in California and Nevada during 1977/01/01–2008/08/01 in the window: Latitude limits 43.0°*N*–31.5°*N*, Longitude limits 113.0°*W*– 125.0°*W*. Magnitude threshold $m_t = 5.0$, the total number of earthquakes 122

completeness thresholds but without focal mechanism reports. We then compared the resulting forecasts.

2. *Catalogs*

The ANSS composite is a world-wide earthquake catalog which merges the master earthquake catalogs from contributing ANSS institutions but removes duplicate or non-unique solutions for the same event (ANSS Catalog Search, 2008). The ANSS catalog for the California and Nevada region extends back to about 1800, however reliable reporting of magnitude 4.0 earthquakes starts in southern California from 1932 (Hileman *et al.*, 1973) and in northern California and parts of Nevada from 1942 (Bolt and Miller, 1975, p. 24). Preliminary results are usually available within a few minutes of each earthquake, and revisions are made as needed, sometimes months later.

The PDE world-wide catalog is published by the USGS (Preliminary Determinations of Epicenters,

2008). Locations and magnitudes are determined by analysis of data from a single global network. Contrary to its name, the PDE catalog is published in final form with a latency of a few months. At the time of writing, the PDE catalog was in final form through the end of 2007. However, the temporary catalog is usually available with a delay of a few minutes.

The PDE catalog measures earthquake size using several magnitude scales, including the body-wave (m_b) and surface-wave (M_S) magnitudes for most moderate and large events since 1965 and 1968, respectively. The moment magnitude (m_w) estimate has been added recently. For our purposes the PDE catalog is useful from 1969 through 2008/08/01, with a completeness magnitude of about 4.7. Forecasting and testing would be improved if there were one standard magnitude, but no one magnitude is reported consistently for all relevant events in the PDE catalog. Kagan (1991) experimented with a weighted average of several magnitudes to estimate one standard magnitude scale, but results were inconclusive. In this work we decided to use the maximum magnitude among those shown for each earthquake. For moderate earthquakes it is usually the m_b or M_S magnitude scale, while for larger recent earthquakes the maximum magnitude is most likely m_w.

The ANSS catalog has some advantages over the PDE. It covers a longer time interval and has a lower magnitude threshold, thus it contains more events in any particular region. ANSS locations are generally more accurate because they are based on local seismic networks, while PDE locations are based on a sparser global network. However, ANSS has the disadvantage that the magnitudes are not defined consistently, because the catalog is a composite of data from several networks. For this reason, the magnitude estimates are less accurate. The inconsistencies in magnitude determination are not so severe in California and Nevada as in other places, because the three relevant networks supplying data to ANSS have been standardized to a certain extent. Advantages of the PDE catalog for our purposes are that the magnitudes are more consistent, as they are determined from a global network, and the catalog is not altered after publication in final form. Because both catalogs have advantages, we made separate forecasts from each before settling on a choice for use in the CSEP tests.

We computed rate density within a "test area" having latitude limits from 31.5°N to 43.0°N and longitude limits from 113.0°W to 125.0°W. This window encloses the California polygon proposed by Schorlemmer and Gerstenberger (2007, Table 2) for the RELM forecast testing. Our calculations show that earthquakes as much as 500 km outside the test window can influence the seismicity inside. Thus we compiled and analyzed earthquakes within a larger "data region" having latitude and longitude bounds 5 degrees outside those of the test region. We tested the obtained subcatalogs for the presence of duplicate solutions and absent earthquakes. Within the test region there are 569 $m \geq 4.7$ events in the PDE subcatalog and 4497 $m \geq 4.0$ in the ANSS subcatalog. The beginning time is 1932/01/01 for the ANSS catalog and 1969/01/01 for the PDE catalog, the end date of both subcatalogs is 2008/08/01.

3. *Forecast*

We have developed a long-term forecast procedure based on earthquake data alone (Kagan and Jackson, 1994, 2000, 2010). This procedure is based on the smoothing of past earthquake locations. The degree of spatial smoothing is controlled by the function

$$f(r) = \frac{1}{\pi} \times \frac{1}{r^2 + r_s^2}, \qquad (1)$$

where r is epicentroid distance, and r_s is the scale parameter.

In our previous investigations we optimized r_s to predict best the second part of a catalog from the first part (Kagan and Jackson, 2010). Here we apply the same value, r_s = 5 km to both catalogs using an isotropic kernel function. The rate density is calculated on a $0.1° \times 0.1°$ grid for shallow earthquakes (depth less or equal to 25 km); thus the forecast is estimated on a 125×117 grid.

Figures 2 and 3 show the time-independent forecasts for California and Nevada, based on the PDE catalog and on the ANSS catalog, respectively. The maps are similar in appearance, any difference being

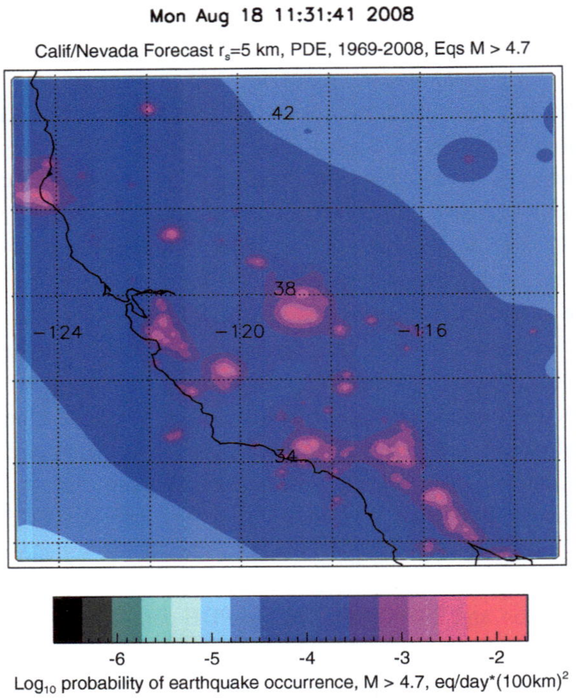

Figure 2
California and Nevada and its surrounding long-term seismicity forecast. Colors show the rates of shallow earthquake occurrence calculated using the PDE 1969-2008/08/01 Catalog. The forecast window is the same as in Figure 1

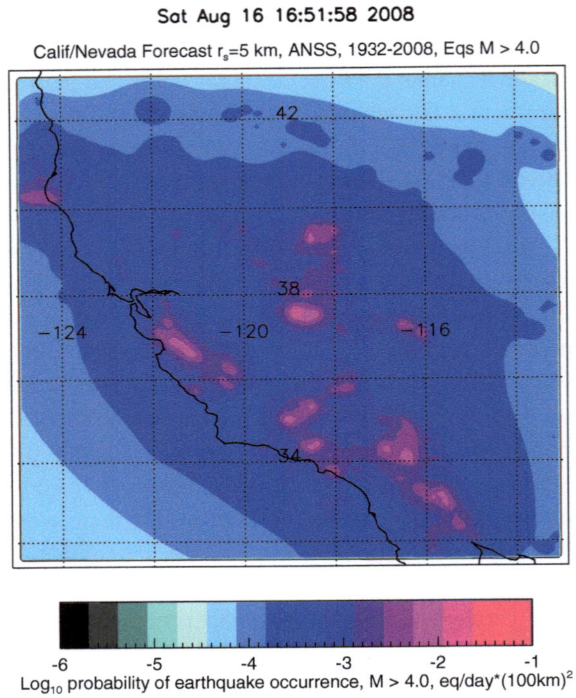

Figure 3
California and Nevada and its surrounding long-term seismicity forecast. Colors show the rates of shallow earthquake occurrence calculated using the ANSS 1932-2008/08/01 Catalog. The forecast window is the same as in Figure 1

due to the lower magnitude threshold and longer duration of the ANSS catalog.

The short-term forecast in this work was carried out according to the technique described by KAGAN and JACKSON (2000, Section 3.1.2, see also KAGAN and KNOPOFF, 1987). Using the branching model (*ibid.*), we obtained a likelihood function for the earthquake process. We compare the likelihood function for the branching process with a process based on the spatially inhomogeneous Poisson model of seismicity. We obtained the statistical estimates of the model parameters (KAGAN *et al.*, 2010) by maximizing the likelihood ratio of the two models using a Newton-Raphson technique.

For the PDE forecast we used the values of parameters obtained during the likelihood function search (KAGAN *et al.*, 2010, Table 5, $m \geq 4.7$, obtained for 'Active Continent' zones): The branching coefficient $\mu = 0.152$, the parent productivity exponent $a = 0.64$, the time decay exponent $\theta = 0.18$, and the horizontal location error $\varepsilon_\rho = 11.9$ km.

For the forecast based on the ANSS catalog we used the values of parameters obtained during the likelihood function search (KAGAN *et al.*, 2010, Table 7, $m \geq 4.0$): $\mu = 0.172$, $a = 0.64$, $\theta = 0.22$, $\varepsilon_\rho = 6.08$ km. The earthquake size distribution is assumed to follow the tapered Gutenberg-Richter law with the corner moment $M_c = 1.16 \times 10^{24}$ Newton m ($m_w \approx 8.05$) and the exponential falloff rate for the seismic moment distribution $\beta = 0.65$ (BIRD and KAGAN, 2004, Table 5).

In our forecasts we use combined final and temporary (preliminary) ANSS and PDE catalogs, updated every few days or weeks. Figures 4 and 5 display the short-term forecast for California and Nevada, based on the PDE and ANSS catalogs, respectively. The map for the PDE forecast is smoother because of larger location errors than those in the ANSS catalog. These location errors (ε_ρ) describe the spread of triggered events in the forecasting procedure.

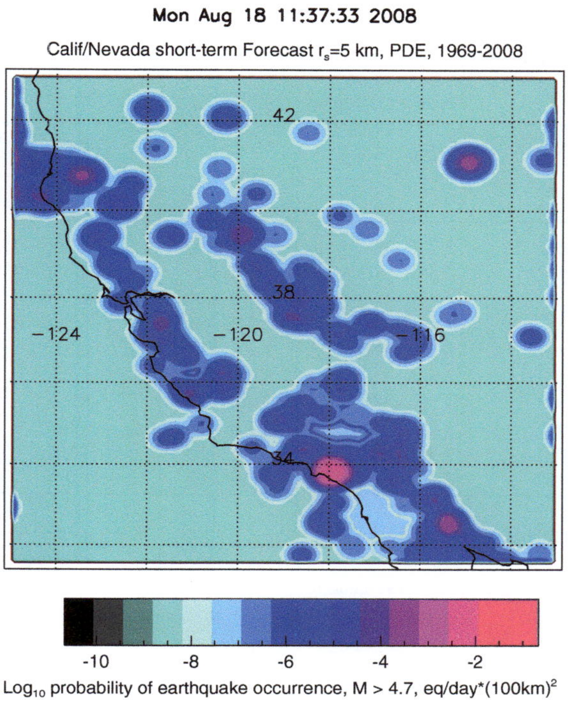

Figure 4
California and Nevada and its surrounding short-term seismicity forecast. Colors show the rates of shallow earthquake occurrence calculated using the PDE 1969-2008/08/01 Catalog. The forecast window is the same as in Figure 1

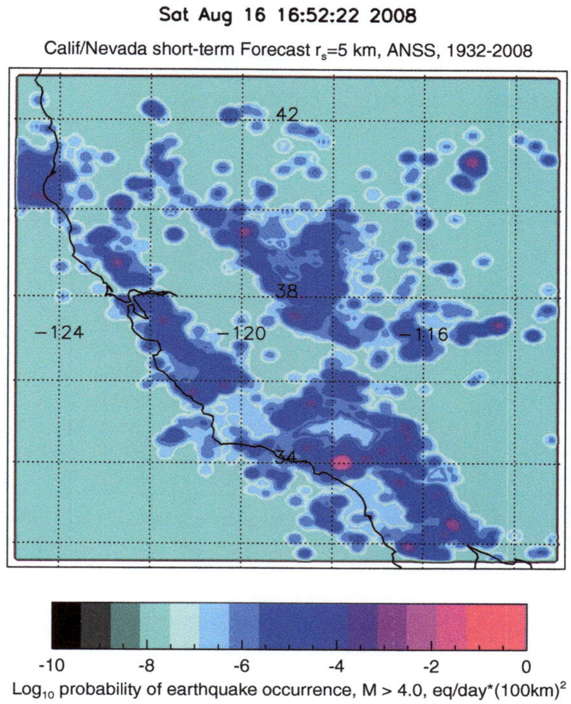

Figure 5
California and Nevada and its surrounding short-term seismicity forecast. Colors show the rates of shallow earthquake occurrence calculated using the ANSS 1932-2008/08/01 Catalog. The forecast window is the same as in Figure 1

In Table 1 we display the forecast for the area surrounding the recent Chino Hills, CA earthquake (2008/07/29 m 5.4, Hauksson *et al.*, 2008). According to the PDE temporary catalog, its coordinates were 33.734°N, 117.906°W, depth 13 km. The ANSS coordinates are 33.953°N, 117.7613°W, depth 14.7 km. We expect the ANSS location to be more accurate because it is determined from a denser, local network.

In the short-term forecast maps (Figs. 4 and 5) red spots in the east Los Angeles area correspond to the earthquake rate increase caused by the Chino Hills event. The PDE map patch is more extended due to the lower accuracy of earthquake locations in the catalog.

The long-term forecasts (Table 1 columns 3 and 6) are similar, taking into account different magnitude thresholds ($m_t = 4.7$ for the PDE and $m_t = 4.0$ for the ANSS catalog) and slightly different epicentral coordinates (see above). The long-term earthquake rate sum for the PDE catalog is 78.7×10^{-9} Eq/day $\times$ km^2, while for the ANSS catalog, it is equal to 339×10^{-9} Eq/day $\times$ km^2. The ratio of the two rates is 4.3, but to compare these rates we need to take into account different magnitude thresholds for both forecasts.

Although the b values for magnitudes in the ANSS and PDE catalogs may differ from the parameters obtained for the seismic moment tensor (Bird and Kagan, 2004), if we use $b = 1.5 \times \beta = 1.5 \times 0.65 = 0.975$, the ratio shown above approximately equals that expected from the Gutenberg-Richter relation: $10^{b \times 0.7} = 10^{0.68} = 4.81$. Similarly, the short-term rates (columns 4 and 7) differ by a factor of 4.51.

The experimental short- and long-term forecasts for California and Nevada based on the ANSS catalog are implemented at our Web site http://scec.ess.ucla.edu/~ykagan/cnpred_index.html. We update the forecasts every day around midnight Los Angeles time. To make our forecast compatible with other forecasts tested by CSEP we have slightly modified

Table 1

Example of long- and short-term forecast: August 1, 2008, Los Angeles Area

Long. °W	Lat. °N	PDE Catalog ($m \geq 4.7$)			ANSS catalog ($m \geq 4.0$)		
		Long-term rate[1]	Short-term rate[1]	Rate ratio[2]	Long-term rate[1]	Short-term rate[1]	Rate ratio[2]
1	2	3	4	5	6	7	8
118.4	33.9	7.45E−08	6.97E−09	9.36E−02	2.73E−07	9.28E−10	3.40E−03
118.3	33.9	7.55E−08	7.94E−08	1.05	3.58E−07	1.08E−09	3.02E−03
118.2	33.9	8.29E−08	5.42E−07	6.53	4.18E−07	2.00E−09	4.79E−03
118.1	33.9	9.10E−08	1.99E−06	21.8	4.16E−07	2.31E−09	5.56E−03
118.0	33.9	8.95E−08	3.65E−06	40.8	3.77E−07	1.93E−07	0.510
117.9	33.9	8.45E−08	4.05E−06	48.0	3.44E−07	8.07E−06	23.5
117.8	33.9	9.15E−08	3.56E−06	38.9	3.90E−07	3.40E−05	87.2
117.7	33.9	7.78E−08	1.52E−06	19.5	3.24E−07	2.59E−05	79.9
117.6	33.9	6.31E−08	3.47E−07	5.50	2.51E−07	3.09E−06	12.3
117.5	33.9	5.70E−08	5.22E−08	0.916	2.36E−07	3.76E−08	0.159

[1] Rate of earthquakes above the magnitude threshold ($\text{day}^{-1}\ \text{km}^{-2}$)

[2] Rate ratio = short-term rate/long-term rate

Two catalogs are used: 1969/01/01-2008/08/01 PDE catalog and 1932/01/01-2008/08/01 ANSS catalog

the forecast window and the magnitude threshold. The latitude limits are from 31.45°N to 43.05°N and the longitude limits are from 113.05°W to 125.45°W; and the magnitude cutoff is taken to be 3.95.

4. *Discussion*

Several long-term earthquake forecasts for California and southern California have been published recently (Bird and Liu, 2007; Shen *et al.*, 2007; Helmstetter *et al.*, 2007; Kagan *et al.*, 2007). Bird and Liu (2007) and Shen *et al.* (2007) forecasts are based on geodetic, geologic and tectonic measurements. They translated tectonic deformation into earthquake rates, using the results by Bird and Kagan (2004) of earthquake size distribution in the continental transform boundaries. Helmstetter *et al.* (2007) and Kagan *et al.* (2007) obtain the estimate of earthquake rates by using principally the same technique as in this paper—by smoothing past seismicity. Helmstetter *et al.* (2007) used small ($m \geq 2.0$) earthquake locations from the ANSS catalog in the time period 1981/01/01–2005/08/23. Contrary to that, Kagan *et al.* (2007) smoothed locations of large and moderate ($m \geq 5.0$) historical and instrumental events starting with year 1800, to obtain a forecast map for southern California. Moreover, they replaced all earthquakes of magnitude 6.5 and larger by an extended source representation: The whole rupture area is used in the forecast calculation.

How can we compare these and other forecasts collected in the special issue of SRL (Field, 2007) to the present forecast? A forecast based on tectonic deformation would most likely perform best for very long time intervals, whereas shorter term earthquake clustering effects, such as aftershocks, would not be effectively forecasted. Similarly, the forecast based on historical data (Kagan *et al.*, 2007) can be expected to perform less skillfully for shorter time intervals. We would expect that the forecast by Helmstetter *et al.* (2007) which has a high spatial resolution and an extensive data set of small earthquakes, would perform best in the short term. In this respect, the present forecast would fill the gap between these two programs. Kagan and Jackson (1994) conjectured that the forecast based on past earthquakes would be effective at the time scales comparable to the length of the catalog used in smoothing. However, the time intervals of the best predictive skills for all these forecasts are not known. The CSEP testing project (Field, 2007) should help us to quantitatively evaluate these bounds.

5. Conclusions

Forecasts based on the ANSS and PDE catalogs give very similar results for earthquake rate density. We prefer the ANSS forecast because it has a longer history, lower magnitude threshold and greater location accuracy than the PDE. Magnitude inconsistencies still present problems in both catalogs. For example, local and body wave magnitudes cannot be converted accurately to moment magnitudes. However, these problems do not seem to influence the forecasts severely. For example, differences between our smoothed seismicity forecasts and fault-based ones will dwarf any variations caused by magnitude inconsistencies.

The forecasts presented here are ready for testing by CSEP. The long-term forecast is based on earthquake data since 1932, and the rate density estimates vary little from year to year. Thus, the rate densities can be treated as constant in time for the duration of a five-year test. The short-term forecast is implemented by an open-source computer program with standard earthquake catalog input. It can be automatically updated on a daily basis in response to evolving aftershock sequences or other rapid changes in seismicity.

Acknowledgments

The authors appreciate support from the National Science Foundation through grants EAR 04-09890 and EAR-0711515, as well as from the Southern California Earthquake Center (SCEC). SCEC is funded by NSF Cooperative Agreement EAR-0529922 and the U.S. Geologic Survey (USGS) Cooperative Agreement 07HQAG0008. We thank Kathleen Jackson for significant improvements in the text. Comments by M. Murru, an anonymous reviewer and the Associate Editor D. Rhoades have been helpful in the revision of the manuscript. Publication 1246, SCEC.

References

Advanced National Seismic System (ANSS) Catalog Search (2008), Northern California Earthquake Data Center (NCEDC), http://www.ncedc.org/anss/catalog-search.html.

Bird, P. and Kagan, Y. Y. (2004), *Plate-tectonic analysis of shallow seismicity: Apparent boundary width, beta, corner magnitude, coupled lithosphere thickness, and coupling in seven tectonic settings*, Bull. Seismol. Soc. Am. *94*(6), 2380-2399 (plus electronic supplement).

Bird, P. and Liu, Z. (2007), *Seismic hazard inferred from tectonics: California*, Seism. Res. Lett. *78*(1), 37–48.

Bolt, B. A., and Miller, R. D. (1975), *Catalogue of Earthquakes in Northern California and Adjoining Areas; 1 January 1910 – 31 December 1972*, Seismographic Stations, Berkeley, Calif., 567 pp.

Ekström, G., Dziewonski, A. M., Maternovskaya, N. N., and Nettles, M. (2005), *Global seismicity of 2003: Centroid-moment-tensor solutions for 1087 earthquakes*, Phys. Earth Planet. Inter. *148*(2–4), 327–351.

Field, E. H. (2007), *Overview of the Working Group for the Development of Regional Earthquake Likelihood Models (RELM)*, Seism. Res. Lett. *78*(1), 7–16.

Gerstenberger, M. C., Wiemer, S., Jones, L. M., and Reasenberg, P. A. (2005), *Real-time forecasts of tomorrow's earthquakes in California*, Nature *435*, 328–331, doi:10.1038/nature03622.

Hauksson, E., K. Felzer, D. Given, M. Giveon, S. Hough, K. Hutton, H. Kanamori, V. Sevilgen, Wei, S. and Yong, A. (2008), Preliminary Report on the 29 July 2008 M_w 5.4 Chino Hills, Eastern Los Angeles Basin, California, Earthquake Sequence, *Seismol. Res. Lett.*, *79*(6), 855–866.

Helmstetter, A., Kagan, Y. Y., and Jackson, D. D. (2007), *High-resolution time-independent grid-based forecast for $M \geq 5$ earthquakes in California*, Seism. Res. Lett. *78*(1), 78–86.

Hileman, J. A., Allen, C. R., and Nordquist, J. M. (1973), *Seismicity of the Southern California Region, 1 January 1932 to 31 December 1972*, Cal. Inst. Technology, Pasadena.

Jackson, D. D. and Kagan, Y. Y. (1999), *Testable earthquake forecasts for 1999*, Seism. Res. Lett. *70*(4), 393–403.

Kagan, Y. Y. (1991), *Likelihood analysis of earthquake catalogues*, Geophys. J. Int. *106*(1), 135–148.

Kagan, Y. Y., Bird, P., and Jackson, D. D. (2010), *Earthquake patterns in diverse tectonic zones of the Globe*, *Pure Appl. Geophys.*, this issue.

Kagan, Y. Y. and Jackson, D. D. (1994), *Long-term probabilistic forecasting of earthquakes*, J. Geophys. Res. *99*(B7), 13,685–13,700.

Kagan, Y. Y. and Jackson, D. D. (2000), *Probabilistic forecasting of earthquakes*, Geophys. J. Int. *143*(2), 438–453.

Kagan, Y. Y. and Jackson, D. D. (2010), *Earthquake forecasting in diverse tectonic zones of the Globe, Pure Appl. Geophys.*, this issue.

Kagan, Y. Y., Jackson, D. D., and Rong, Y. F. (2007), *A testable five-year forecast of moderate and large earthquakes in southern California based on smoothed seismicity*, Seism. Res. Lett. *78*(1), 94–98.

Kagan, Y. Y. and Knopoff, L. (1987), *Statistical short-term earthquake prediction*, Science *236*(4808), 1563–1567.

Preliminary Determinations of Epicenters (PDE) (2008), U.S. Geological Survey, U.S. Dep. of Inter., Natl. Earthquake Inf.

Cent., http://neic.usgs.gov/neis/epic/epic.html and http://neic.usgs.gov/neis/epic/code_catalog.html.

SCHORLEMMER, D. and GERSTENBERGER, M. C. (2007), *RELM Testing Center*, Seism. Res. Lett. *78*(1), 30–35.

SCHORLEMMER, D., GERSTENBERGER, M. C., WIEMER, S., JACKSON, D. D., and RHOADES, D. A. (2007), *Earthquake likelihood model testing*, Seism. Res. Lett. *78*(1), 17–29.

SHEN, Z.-K., JACKSON, D. D., and KAGAN, Y. Y. (2007), *Implications of geodetic strain rate for future earthquakes, with a five-year forecast of M 5 earthquakes in southern California*, Seism. Res. Lett. *78*(1), 116–120.

WORKING GROUP ON CALIFORNIA EARTHQUAKE PROBABILITIES (WGCEP) (1990), *Probabilities of large earthquakes in the San Francisco Bay Region, California*, USGS Circular 1053, 51 pp.

WORKING GROUP ON SOUTHERN CALIFORNIA EARTHQUAKE PROBABILITIES, (WGCEP) (1995), JACKSON, D. D., AKI, K., CORNELL, C. A., DIETERICH, J. H., HENYEY, T. L., MAHDYIAR, M., SCHWARTZ, D., and WARD, S. N., *Seismic hazards in southern California: Probable earthquakes, 1994-2024*, Bull. Seism. Soc. Am. *85*, 379–439.

WORKING GROUP ON CALIFORNIA EARTHQUAKE PROBABILITIES (WGCEP) (2003), *Earthquakes probabilities in the San Francisco Bay region: 2002 to 2031*, USGS, Open-file Rept. 03-214.

WORKING GROUP ON CALIFORNIA EARTHQUAKE PROBABILITIES (WGCEP) (2009), FIELD, E. H., DAWSON, T.E., FELZER, K. R., FRANKEL, A. D., GUPTA, V., JORDAN, T. H., PARSONS, T., PETERSEN, M. D., STEIN, R. S., WELDON II, R. J., and WILLS, C. J., *Uniform California earthquake rupture forecast, Version 2 (UCERF 2)*, Bull. Seism. Soc. Am. *99*, 2053–2107.

(Received August 3, 2008, revised December 14, 2008, accepted March 6, 2009, Published online March 4, 2010)

Pure Appl. Geophys. 167 (2010), 693–707
© 2010 Birkhäuser / Springer Basel AG
DOI 10.1007/s00024-010-0068-2

Pure and Applied Geophysics

Retrospective Forecasting of $M \geq 4.0$ Earthquakes in New Zealand

R. Console,[1] M. Murru,[1] and G. Falcone[1]

Abstract—We have applied a variation of the Epidemic Type Aftershock Sequence (ETAS) model, which is a stochastic triggering epidemic model incorporating short-term clustering, to data collected by the New Zealand Seismological Observatory-Wellington (Geonet) for forecasting earthquakes of moderate and large magnitude in the New Zealand region. The model uses earthquake data only, with no explicit use of tectonic, geologic, or geodetic information. In this epidemic-type model every earthquake is regarded, at the same time, as being triggered by previous events and triggering following earthquakes. A maximum likelihood estimate of the model parameters has been performed on the learning period from 1960 to 2005 for earthquakes of magnitude 4.0 and larger. Forecast verification procedures have been carried out in a forward-retrospective way on the January 2006 to April 2008 data set, making use of statistical tools as the log-likelihood ratio, the Relative Operating Characteristics (ROC) diagrams, the Molchan error diagrams, the probability gain and the *R*-score. These procedures show that the clustering epidemic model achieves a log-likelihood ratio per event of the order of some units, and a probability gain up to several hundred times larger than a time-independent spatially uniform random forecasting hypothesis. The results show also that a significant component of the probability gain is linked to the time-independent spatial distribution of the seismicity used in the model.

Key words: Earthquake forecasting, real time forecasting, epidemic model, hypothesis testing, error diagrams.

1. Introduction

Earthquake occurrence is usually represented as a stochastic point process (Kagan and Jackson, 2000; Daley and Vere-Jones, 2003), and the event occurrence rate is one of the most important parameters in the theory of point processes. Progress is being made with the Epidemic Type Aftershock Sequence (ETAS) branching model of seismicity that offers a quantification of earthquake interactions. This model uses earthquake data only, with no explicit use of tectonic, geologic, or geodetic information. Every earthquake can be regarded at the same time as being triggered by previous events and also to trigger subsequent earthquakes. The details of the model can be found in Ogata (1998); Console and Murru (2001); Console *et al.* (2003, 2006a, b, 2007, 2010) and Murru *et al.* (2008).

Epidemic Type Aftershock Sequence models have been used in many studies to describe or forecast the spatio-temporal distribution of seismicity and reproduce many properties of real seismicity (Ma and Zhuang, 2001; Ogata, 2001, 2004a, b, 2005, 2006a, b, 2007; Ogata *et al.*, 2003; Ogata and Katsura, 2006; Ogata and Zhuang, 2006: Felzer *et al.*, 2002; Helmstetter and Sornette, 2002, 2003; Saichev and Sornette, 2006; Zhuang *et al.*, 2004, 2005).

Console and Murru (2001) and Console *et al.* (2003, 2006a) showed that a simple clustering ETAS model exhibits a substantially higher likelihood than the time-invariant Poisson hypothesis. They tested this clustering model on real seismicity of Italy, California, Greece and Japan by comparison with a plain time-independent Poisson model, through likelihood-based methods, proving the validity of this model in a retrospective way. The most important contribution to the information score achieved by short-term epidemic models derives commonly from their ability to forecast a large number of small aftershocks (Console and Murru, 2001; Console *et al.*, 2003). Here we examine the ability of our specific version of the ETAS model to forecast (in statistical sense) the impending occurrence of medium-to large ($M \geq 4.0$) earthquakes in a forward-retrospective way on a catalogue of instrumental

[1] Istituto Nazionale di Geofisica e Vulcanologia, Rome, Italy.
E-mail: console@ingv.it

seismicity. This study is related to the preparation of the algorithm for its possible application to the real-time data at the New Zealand Earthquake Forecast Testing Centre. Moreover, this paper continues our analysis of stochastic point process forecast verification (MURRU *et al.*, 2009; CONSOLE *et al.*, 2010).

2. *Brief Outline of the Earthquake Occurrence Probability Model Adopted*

Here we consider the short-term clustering properties of earthquakes and give a brief outline of a statistical method for modelling the interrelation of any earthquake with any other, referring the reader to the papers mentioned above for further details. Our method belongs to the general category of the ETAS models. In fact, as explained in the following, it is based on the same basic principles, though applying them with a specific version of the algorithms.

In the ETAS model all earthquakes have identical roles in the triggering process. The main departures of this model from conventional aftershock models are (1) that it can be used to model full catalogues and not just aftershocks, (2) that the arbitrary definition of fore/main/aftershock is not used, and (3) that therefore the meaning of aftershock is very different (they can be larger than their parent events). The clustering of earthquakes, in the ETAS model, is described by a process in which every earthquake is a main shock with its own aftershock sequence decaying according to the modified Omori law, and with magnitude distribution of all the earthquakes in a sample, following the Gutenberg–Richter law with a constant b value OGATA and KATSURA (1993). Although the decay in time, described by the Omori law, may have temporal variations up to very large timescales (see e.g., FAENZA *et al.*, 2007), this model is conventionally considered to be a short-range forecasting model because the temporal variation of an earthquake's contribution to the subsequent seismicity rate is greatest immediately after the earthquake occurs.

Such a stochastic process is characterized by a limited number of free parameters, which permit the computation of the occurrence rate density as a continuous function in space and time, according to the definition (OGATA, 1998):

$$\lambda(x,y,t,m) = \lim_{\Delta x,\Delta y,\Delta t,\Delta m \to 0} \frac{P_{\Delta x,\Delta y,\Delta t,\Delta m}(x,y,t,m)}{\Delta x \Delta y \Delta t \Delta m}, \tag{1}$$

where Δx, Δy, Δt and Δm are increments of longitude, latitude, time and magnitude and $P_{\Delta x,\Delta y,\Delta t,\Delta m}(x,y,t,m)$ is the probability of occurrence of an event in the volume $\{x,\ x+\Delta x;\ y,\ y+\Delta y;\ t,\ t+\Delta t;\ m,\ m+\Delta m\}$.

Consequently, this formulation allows the estimate of the log-likelihood function of a realization of seismic events described by a catalog $\{x_j,y_j,t_j,m_j,\ j=1,\ldots,N\}$ under the given hypothesis:

$$\ln L = \sum_{j=1}^{N} \ln\left[\lambda(x_j,y_j,t_j,m_j)V_0\right] - \int_X\int_Y\int_T\int_M \lambda(x,y,t,m)\,dxdydtdm, \tag{2}$$

where V_0 is an arbitrary coefficient whose dimensions are equal to those of the inverse of the rate density $\lambda(x_j,y_j,t_j,m_j)$, and X,Y,T and M denote the range of integration of the four space, time and magnitude variables. Equation 2 can be used for the assessment of the maximum likelihood parameters of the hypothesis (KAGAN, 1991; KAGAN and JACKSON 1994).

The expected occurrence rate density of earthquakes, at any instant of time and geographical point, $\lambda(x,y,t,m)$, is modelled as the sum of the independent, or time-invariant "spontaneous", Poisson mean activity and the contribution of every previous event using a kernel function that takes in proper account: (1) the magnitude of the j-th triggering earthquake, (2) the spatial distance r from the triggering event, and (3) the time interval $(t-t_j)$ between the triggering event and the instant considered for the computation:

$$\lambda(x,y,t,m) = f_r \cdot \lambda_0(x,y,m) + \sum_{j=1}^{N} H(t-t_j)\cdot k\cdot(t-t_j+c)^{-p} \cdot \left(\frac{d_j^2}{r^2+d_j^2}\right)^q \cdot 10^{-b(m-m_0)}, \tag{3}$$

where $\lambda_0(x,y,m)$ is a time-independent smoothed spatial density distribution of the seismicity, obtained from an undeclustered catalogue (as explained in the next section), f_r, k, c, p and q are free parameters, and m_0 is a threshold magnitude.

The average triggering distance of the aftershock zone d_j is related to the magnitude of the main shock KAGAN (2002):

$$d_j = d_0 10^{\alpha(m_j - m_0)/2}. \quad (4)$$

Equations 3 and 4 characterize a model that is substantially the same as the standard ETAS model adopted by ZHUANG *et al.* (2004, 2005) except that we are not using normalized time and space distributions.

The set of free parameters for the ETAS model actually estimated in this study are the following:

- k (productivity coefficient),
- d_0 (characteristic triggering distance),
- q (exponent of the spatial distribution of triggered events),
- α (coefficient of the exponential relation between the magnitude of triggering earthquakes and their average triggering distance),
- c (time constant of the generalized Omori law), and
- p (exponent of the generalized Omori law).

The fraction of spontaneous events over the total number of events of the process, f_r, is constrained by the requirement that the total number of events expected for these free parameters must be equal to the total number of events observed in reality. The b value is estimated from the entire catalogue independently from the other parameters.

3. Probabilistic Seismic Background Estimate

A basic ingredient of our algorithm based on an epidemic model for the estimate of the rate density $\lambda(x,y,m,t)$, is a smooth background time-independent rate density function $\lambda_0(x,y,m)$.

A smooth geographical distribution can be computed at each node k of a regular grid through the method introduced by FRANKEL (1995):

$$\tilde{N}_k = \frac{\sum_l N_l \exp\left(-\Delta_{kl}^2/\Delta_0^2\right)}{\sum_l \exp\left(-\Delta_{kl}^2/\Delta_0^2\right)}, \quad (5)$$

where N_l is the number of events in each cell centre on the lth node, Δ_{kl} is the distance between node k and any other node l, and Δ_0 is a scaling factor.

The space distribution $\mu_0(x,y,m_0)$ of earthquakes of magnitude equal to or larger than a reference magnitude m_0 at any point is obtained by linear interpolation of $\tilde{N}_k$ among the four nearest grid nodes. The scaling factor Δ_0 is determined by maximizing the likelihood of the seismicity contained in half catalogue, under the time-independent model obtained from the other half. Here the seismicity is modelled as a pure Poisson process ($K = 0$), so that no information about the parameters characterizing the induced seismicity is necessary at this stage. The rate density distribution $\lambda_0(x,y,m)$ is then obtained from $\mu_0(x,y,m_0)$ through the Gutenberg–Richter magnitude distribution. The details of the method have been first described by CONSOLE and MURRU (2001), and also can be found in CONSOLE *et al.* (2010). In the latter paper the authors have discussed the stability and robustness of the parameter estimates.

We aim at a spatial distribution that does not include the triggered component of the seismicity and still preserves the total seismic moment released by the seismicity, which is proportional to the total number of earthquakes. With this purpose, we apply an iterative process based on a clustering (epidemic) time-dependent model to a learning data set constituted by an earthquake catalogue:

1. We start finding the maximum likelihood set of free parameters using the initial distribution of the smoothed seismicity $\lambda_0(x,y,m)$;
2. We compute the probability of independence w_i as the ratio between the independent component $f_r \cdot \lambda_0(x_i,y_i,m_i)$ and the composite rate density $\lambda(x_i,y_i,m_i,t_i)$ for any event i;
3. Then we compute a new distribution of $\lambda_0(x,y,m)$ introducing the weights w_i to count a fractional number of events for each cell in the FRANKEL (1995) algorithm;
4. The new smoothed distribution is used in a new maximum likelihood best fit of the free parameters;
5. We proceed again from step 2, and continue until a reasonable convergence has been reached.

4. Retrospective Testing of the Model Capability

According to widely recognized standards, after a first phase, during which an earthquake forecast method is formulated and set up by best fit on real data, and before any practical use, it must be validated assessing its results in a rigorous and objective way (CONSOLE, 2001).

4.1. Validation of a Probability-based Forecast Method

KAGAN and KNOPOFF (1977, see VERE-JONES, 1998) suggested measuring the effectiveness of the earthquake prediction algorithm by first evaluating the likelihood ratio to test how well a model approximates an earthquake occurrence. In particular, they estimated the information gain (or information score) for measuring the effectiveness of earthquake prediction algorithms for events that actually occurred in the test volume. The information score is obtained by dividing the log-likelihood ratio by the number of target earthquakes in a catalog. Recently the information score has been considered in several papers to test the effectiveness of earthquake prediction algorithms (see, e.g., DALEY and VERE-JONES, 2003, 2004; IMOTO, 2004; HARTE and VERE-JONES, 2005).

In previous papers we have applied the log-likelihood ratio criterion, comparing the performance of the time-dependent ETAS model with that of a time-independent, spatially variable Poisson model (CONSOLE *et al.*, 2003, 2006a, b, 2007). In those papers the likelihood computation was possible because the forecasts were expressed by an occurrence rate density function defined at any point in space and time. In this study, related to the preparation of the algorithm for its possible application to the real-time data at the New Zealand Earthquake Forecast Testing Centre, we estimate both the log-likelihood ratio and the information score through the estimate of the probability of occurrence of earthquakes exceeding a threshold magnitude in a set of time–space cells covering the region and the time span considered for the test. The likelihood of realisation of a process described by an earthquake catalogue, under a given model, is expressed by (KAGAN and JACKSON, 1995):

$$\log L = \sum_{i=1}^{P} \left[c_i \log\left(\frac{p_i}{1-p_i}\right) + \log(1-p_i) \right], \quad (6)$$

where P is the total number of cells, $c_i = 0$ denotes the cells in which no events have been observed, $c_i = 1$ those in which at least one event has occurred, and p_i is the probability of occurrence of at least one event in every cell, under the specific model. The probability p_i is estimated from the expected number of events λ_i through the following relation:

$$p_i = 1 - e^{-\lambda_i}. \quad (7)$$

The log-likelihood ratio is given by the difference of the log-likelihood for the model to be tested and the log-likelihood of another reference model (null hypothesis).

4.2. Alarm-based Scoring Methods

For the verification of this statistical forecasting ETAS model, we also use a binary forecast approach for the generation of the Relative Operating Characteristic (ROC) diagram. According to HOLLIDAY *et al.* (2005) and CHEN *et al.* (2006) the maximum likelihood test is overly sensitive to least probable events, and consequently the above method has significant advantages with respect to the maximum likelihood test in evaluating the performance of the forecast model relative to random chance.

In a binary forecast approach the events (earthquakes) of given magnitude are considered as being forecast either to occur or not to occur in a given time–space cell in which the whole space–time volume is divided. The result of every specific test for a given magnitude is summarized on a 2 × 2 contingency table.

Contingency table

Forecast	Observed	
	Yes	No
Yes	*a*	*b*
No	*d*	*c*

The entries for classifying the results in a 2 × 2 table, reading clockwise from the top left corner, are the following:

$a =$ number of cells containing successful forecasts of occurrence, given by forecast YES and observed YES. (e.g., a is the number of forecast events that really occurred),

$b =$ number of false alarms, given by forecast YES and observed NO,

$c =$ number of successful forecasts of non-occurrence, given by forecast NO and observed NO,

$d =$ number of failures to predict, given by forecast NO and observed YES (that is the cell containing events that occurred but not forecast, missed alarms).

These entries comply with the following constraints:

$a + b =$ total number of cells containing alarms

$a + d =$ total number of cells containing events that really occurred,

$b + c =$ total number of cells without any occurred events,

$c + d =$ total number of cells without any alarms,

$e =$ $a + b + c + d =$ total number of geographic cells multiplied by the number of time bins.

Following the terminology introduced by HOLLIDAY *et al.* (2005), we make use of the parameters Hit rate (H) and False alarm rate (F), defined as:

$H =$ $a/(a + d)$ (the fraction of successful forecasts issued where at least one event has occurred),

$F =$ $b/(b + c)$ (the fraction of false alarms issued where an event has not occurred).

The meaning of H corresponds to that of the Reliability (MATTHEWS and REASENBERG, 1988; RHOADES and EVISON, 1989) that is the probability that an event is preceded by a warning.

We also consider the probability gain G as a function of false alarm rate F to measure the effectiveness of a prediction technique. The probability gain is commonly defined as the ratio of the probability of one model over that of another model. In this study we consider a spatially-uniform time-independent Poisson model as null hypothesis, and adopt the definition of probability gain given by AKI (1981) as the ratio between the conditional and the unconditional (Poisson) rate, namely:

$$G = a/(a+d) \cdot e/(a+b) = H \cdot e/(a+b)$$
$$= \text{Success rate/average rate of occurrence,}$$

(or in other words the ratio between the conditional probability, success rate, and the unconditional probability, average rate or frequency of occurrence). G varies between 0 and ∞.

In case the prediction algorithm is expressed in terms of probabilities (or expected rates) as for our ETAS model, it is necessary to transform the probability forecasts into binary predictions defined by some probability threshold. The result of the test produces a single point on the ROC diagram. For different thresholds the corresponding Hit rates and False alarm rates can be computed. Applied to earthquake forecasting, a ROC diagram is a plot of the Hit rate H (the fraction of "hotspot" cells that have an earthquake forecast over the total number of cells with actual earthquakes) versus the False alarm rate F (the fraction of the forecast cells that do not have earthquakes over the total number of cells with no actual earthquakes in them). In the case of purely random forecasts, $H = 1 - F$, and the diagram consists of the diagonal joining the points (0,0) and (1,0).

MOLCHAN (1990) modified the Relative Operating Characteristic technique as an error diagram to predict random point processes. He was the first, in 1990, that used this method for evaluating the performance of a prediction program. MOLCHAN and KAGAN (1992) and MOLCHAN (1997, 2003) also reviewed the error diagram method and its applications. For any prediction algorithm, and for every possible threshold rate, the error diagrams plot the fraction of failures to predict, v, versus the fraction of alarm time, τ. Recently, MCGUIRE *et al.* (2005); KOSSOBOKOV (2006); BAIESI (2006); ZECHAR and JORDAN (2008) and others applied this method to evaluate their earthquake prediction algorithms.

It is important to note that v and τ can be connected to the Aki probability gain by this relation:

$$G = (1 - v)/\tau.$$

Another test statistic, commonly called the R score, can be derived from a 2×2 contingency table (see e.g., HARTE and VERE-JONES, 2005). The R score has been defined in two different forms. The first, defined by HANSSEN and KUIPERS (1965), is:

$R = a/(a + b) - d/(c + d)$ (number of cells in which earthquakes are successfully predicted/total number of cells containing alarms)—(number of failures to predict/total number of cells without any alarms.

The second definition (applied in this study) has been adopted by SHI *et al.* (2001) and nominated hereinafter as:

$R' = a/(a + d) - b/(b + c) = H - F$ (number of cells in which earthquakes are successfully predicted/total number of cells in which earthquakes occur)—(number of cells with false alarms/total number of cells without any earthquakes).

The two quantities R and R' have similar properties, however they may have significantly different values.

The previous form can be interpreted as the proportion of successful forecasts less the proportion of non-forecasts resulting in failures to predict.

R and R' vary between -1 and 1 with the following meanings:

$R, R' = -1$, all predictions are wrong.
$R, R' \approx 0$, random prediction scores.
$R, R' = 1$, all positive and negative predictions are correct, no false alarm.

A meaningful prediction must have $R, R' > 0$. A significant prediction needs R, R' larger than a marginal level.

Of course a relationship exists between G and R and/or 'R'': when G tends to ∞, R goes to 1, when $G = 1$ R is equal to 0 and when $G = 0$ $R = -1$.

5. Application to the New Zealand Seismicity

We use the shallow seismic ($h \leq 40$ km) catalogue collected by the New Zealand Seismological Observatory, Wellington (Geonet), from January 1960 to April 2008 in the area limited by the coordinates 48°N–34°N and 166°E–180°E. The data were used in a retrospective way to test the forecasting capability of the ETAS model. This data set, in which magnitudes are given with two decimal digits, can be considered complete for magnitudes equal to 3.95 and larger (RHOADES, personal communication), which yields a total of 9,881 events (Fig. 1).

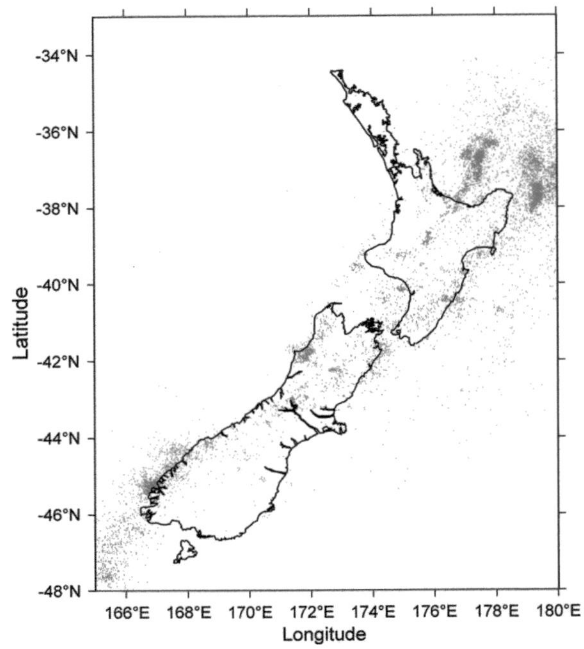

Figure 1
Epicenters of earthquakes in the New Zealand catalogue, 1960–2007, with magnitudes $M \geq 3.95$ and hypocentral depths $h \leq 40$ km

5.1. Learning Phase

The data set until December 2005 (9,594 events) has been used to estimate the model free parameter values in the learning phase. In this period, the maximum likelihood value for the b parameter of the Gutenberg–Richter frequency–magnitude relation (FMD) is 0.959 ± 0.008, with the error computed using the formula suggested by SHI and BOLT (1982) (Fig. 2).

In this study the best fit of the Δ_0 value has been carried out by maximizing the likelihood of the New Zealand earthquake catalogue ($M_l \geq 3.95$) from 1990 to 2005 (4,778 events) under the time-independent Poisson model obtained from the same catalogue from 1960 to 1989 (4,516 events). This search has lead to the value of $\Delta_0 = 23$ km (Fig. 3).

The smooth geographical distribution is shown in Fig. 4a. This is the distribution of all the observed earthquakes, which include both the spontaneous and the triggered seismicity of the epidemic model. As shown in Fig. 5 (upper line), the cumulative number

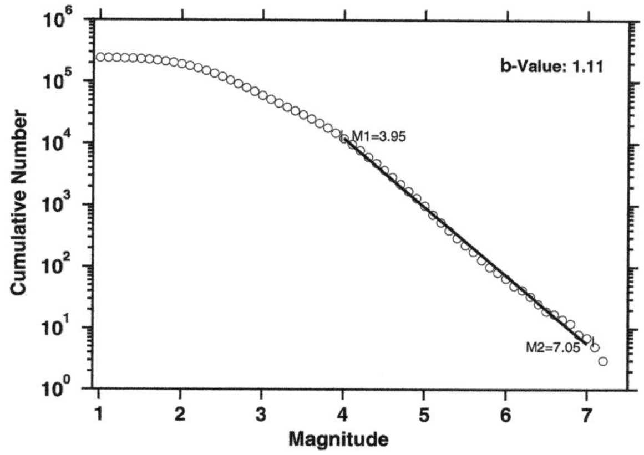

Figure 2
Gutenberg–Richter plot for the New Zealand earthquakes analysed in this study

of all the earthquakes is strongly irregular, because of the presence of aftershock sequences in the catalogue.

The best fit of the model parameters obtained by the maximum likelihood criterion is shown in Table 1.

As can be clearly noted in Fig. 5 (lower line), the cumulative distribution of w_i over the time spanned by the catalogue is considerably closer to a linear trend than the starting cumulative distribution. This favourably indicates the capacity of our algorithm in removing the contribution of the triggered seismicity, without really deleting any independent events from the catalogue.

The final geographical distribution of the background seismicity, after normalization to the total number of events, can be compared with the starting distribution in Fig. 4b. It is evident how the new distribution obtained by the weighting process of the seismicity exhibits a smaller spatial clustering than the original one.

5.2. *Testing Phase*

The capability of the model has been tested in a forward retrospective way for the earthquakes (274 events, $M_l \geq 4.0$) that occurred in the New Zealand region during the January 2006 to April 2008 period.

The test region covers the New Zealand land area plus a region extending about 50 km offshore (RHOADES, personal communication) (Fig. 6). The location grid consists of cells of area 0.1° squared on 1/10th degree coordinates of latitude and longitude which have their centers within the test region.

We first report the application of the log-likelihood ratio criterion, taking as reference model the spatial time-independent occurrence density distribution obtained from the learning phase (1960–2005). Following the procedure described in the previous section, we estimate the result of Eq. (6) by summation over the total number of space–time cells (in this study the number of cells is 5,396,191). Letting the target magnitude change from 4.0 to 8.0 in steps of

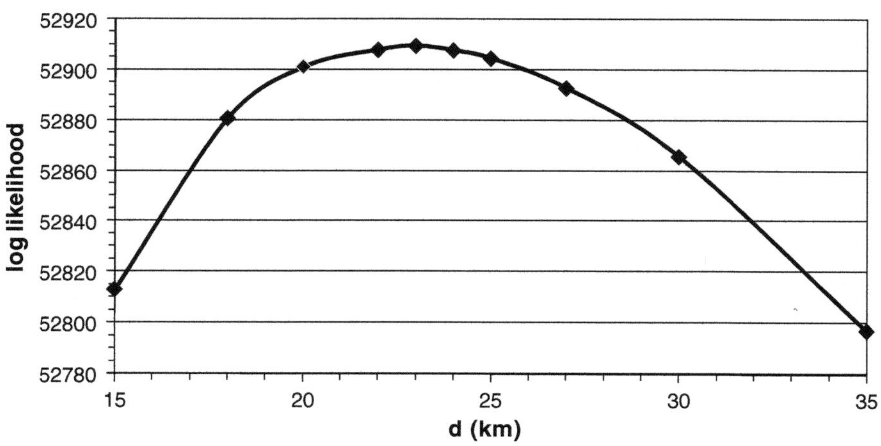

Figure 3
Plot of the log-likelihood of the earthquake catalogue recorded in the New Zealand region from 1990 to 2005 under the time-independent Poisson model obtained from the same catalogue from 1960 to 1989. This plot shows the dependence of the log-likelihood versus the correlation distance Δ_0 used in the smoothing algorithm of Eq. 5

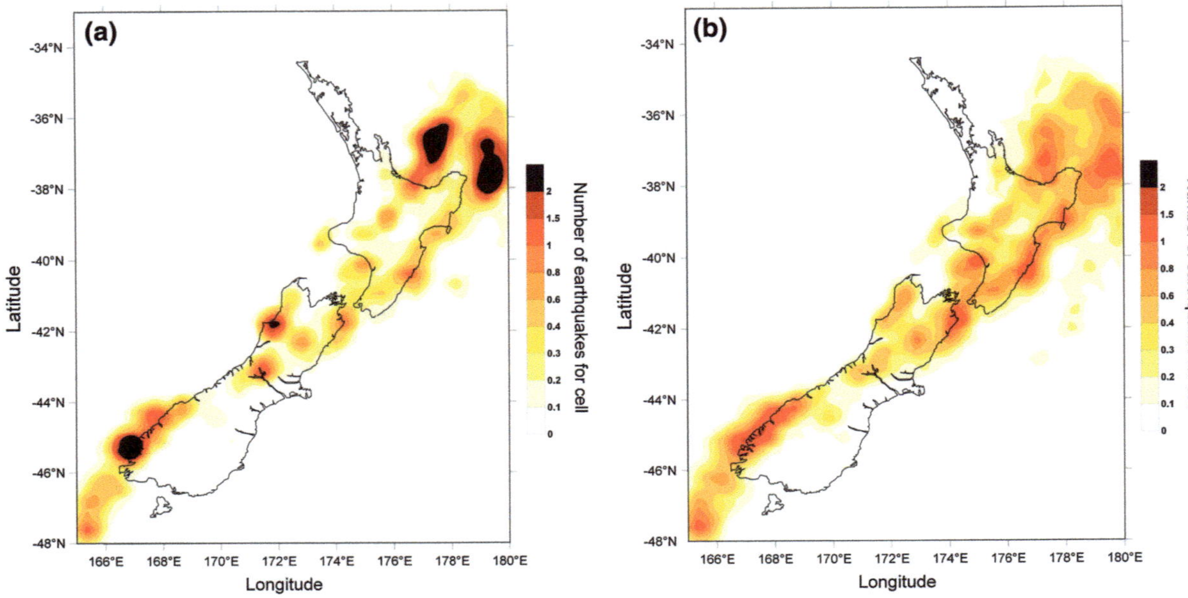

Figure 4
Comparison between the initial and the final geographical distribution of the space density. **a** New Zealand 1960–2005 smoothed seismicity obtained by the raw catalogue. **b** As in **a** but obtained using a weighted catalogue

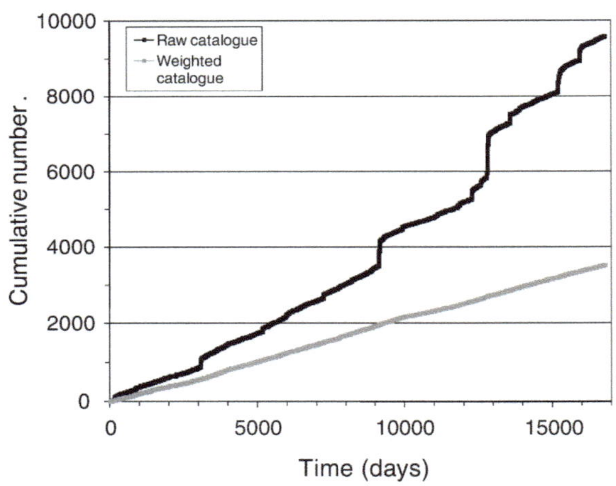

Figure 5
Cumulative seismicity for the period 1960–2005. The *gray* and *black lines* represent the weighted and raw series, respectively

0.1 units, we obtain different values of the log-likelihood ratio, shown in Fig. 7. The total likelihood ratio decreases with magnitude, due to the smaller number of target events for larger magnitudes.

In order to compensate for the decrease of the number of earthquakes with magnitude, we considered also the information score defined as the average log-likelihood ratio per event. Figure 8 shows that the information score has a general trend to increase with magnitude (larger earthquakes are more likely forecast) until magnitude 5.6, after which the information score seems to decrease dramatically. It must be noted, however, that the limited number of events with large magnitude makes the statistical significance of this test very low. Our results show that for all examined magnitudes the information score is significantly larger than zero, which would represent the information score pertaining to a random forecast based on a time-independent spatially heterogeneous model. This plot exhibits the largest values of the information gain in the magnitude range 5.2–5.6, with values exceeding 4.0, i.e., likelihood ratios exceeding 50 for each earthquake forecast.

For the following methods, we perform the test only on the occurrence or no occurrence of earthquakes of given magnitude. The first step has been the generation of the ROC diagrams for the verification of the statistical forecasting ETAS model with the construction of the 2 × 2 contingency tables with variable alarm thresholds.

Here, in the preparation of the contingency tables, we define forecasts when the expected occurrence rate of earthquakes with magnitudes greater than the

Table 1

Maximum log-likelihood parameters of the ETAS model (Learning phase)

Catalogue	New Zealand
Time span	1960–2005
Lower magnitude threshold of triggering events	3.95
Lower magnitude threshold of targets events	4.0
Productivity coefficient k (days^{p-1})	7.10E-04
Exponent of the spatial distribution q	1.77
Characteristic distance in the spatial distribution d_0 (km)	3.30
Coefficient of the exponential magnitude productivity law (α)	4.98E-01
Time constant in Omori law c (days)	8.66E-03
Exponent in Omori law p	1.77
Failure rate f_r	0.364
Log (likelihood ratio)	35,524

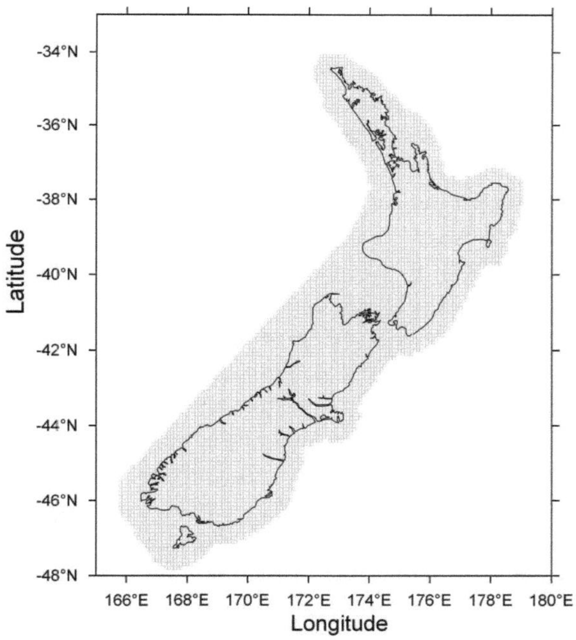

Figure 6
Map showing test region (*shaded region*), including both the New Zealand land area and a region extending approximately 50 km offshore zone (*lightly shaded*)

lower cutoff magnitude (equal to 4.0) exceeds a given threshold value r. Table 2 shows the results for three contingency tables, obtained for three occurrence rate thresholds of 1.00E-05, 7.00E-03 and 5.00E-05 (events/day/100 km^2), respectively.

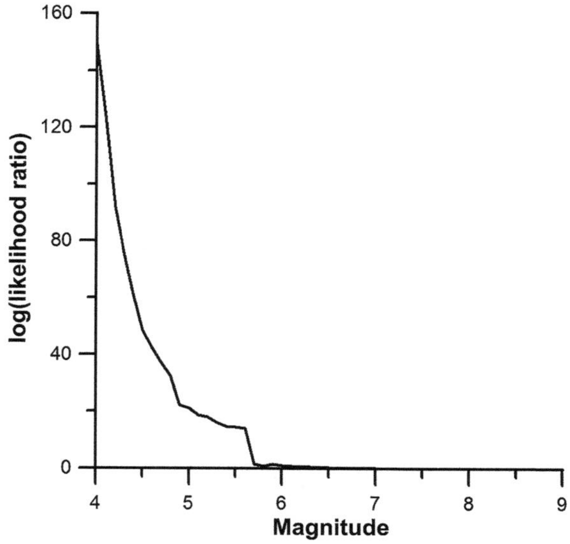

Figure 7
Log-likelihood ratio of the ETAS model with respect to a spatially variable time-independent Poisson model, versus the magnitude of target events. The zero-value pertains to a random forecast, based on the spatially heterogeneous time-independent Poisson model only

Varying the occurrence rate threshold for the verification of the ETAS forecast model, we have obtained the values of H (fraction of successful forecasts issued where at least an event has occurred) and F (fraction of false alarms issued where an event has not occurred) to be used for the preparation of the ROC diagram for the given (4.0) threshold magnitude of the target events. The same procedure has been repeated varying the threshold magnitude from 4.0 up to 5.5 in steps of 0.5 magnitude units. This produces changes in the entries of the contingency tables, and consequently in the corresponding ROC diagram (the higher the threshold magnitude, the smaller the number of successes, although the smaller the number of false alarms, will be). The results are summarized in Fig. 9a. This figure shows that the ETAS model yields a performance significantly better than the simple time-independent spatially uniform model for all the magnitude thresholds considered in this study. For comparable purposes we have also computed the ROC diagram for a time-independent spatially heterogeneous forecast model. Figure 9b shows that this spatially heterogeneous Poisson model still achieves a better performance with respect

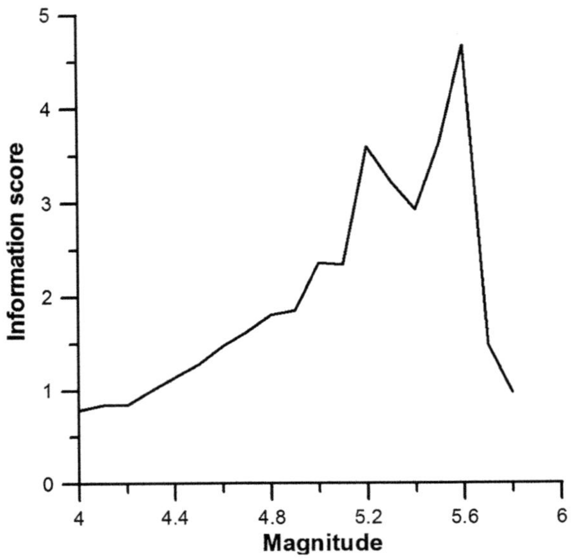

Figure 8
Information score of the ETAS model with respect to a spatially variable time-independent Poisson model, versus the magnitude of target events. Natural logarithms are used. The value 0 pertains to a random forecast, based on the spatially heterogeneous time-independent Poisson model

Table 2

Contingency tables for the ETAS model from January 1, 2006 to April 30, 2008 as a function of three occurrence rate thresholds (r), expressed as events/day/100 km²

(a) $r = 1.00\text{E-}01$

Forecast	Observed		Total
	Yes	No	
Yes	(*a*) 2	(*b*) 2	4
No	(*d*) 163	(*c*) 5,396,024	5,396,187
Total	165	5,396,026	(*e*) 5,396,191

(b) $r = 7.00\text{E-}03$

Forecast	Observed		Total
	Yes	No	
Yes	(*a*) 23	(*b*) 7,160	7,183
No	(*d*) 142	(*c*) 5,388,866	5,389,008
Total	165	5,396,026	(*e*) 5,396,191

(c) $r = 5.00\text{E-}05$

Forecast	Observed		Total
	Yes	No	
Yes	(*a*) 122	(*b*) 829,485	829,607
No	(*d*) 43	(*c*) 4,566,541	4,566,584
Total	165	5,396,026	(*e*) 5,396,191

to a spatially uniform model, although lower than that obtained from the ETAS model.

The same contingency tables have also allowed the computation of the respective values for the Molchan's error diagram. The results for Molchan's diagram are plotted in Fig. 10a and b for the ETAS and a spatially heterogeneous Poisson model, respectively. The Molchan's diagrams (Fig. 10) show nearly the same behaviour as the ROC diagrams (Fig. 10a), taking into account that in these diagrams the best performance is below the diagonal corresponding to the spatially homogeneous random forecasts. The comparison between Fig. 10a and b shows that an important component of the performance derives from the time-independent spatial distribution of the seismicity, even if the time-dependent ETAS model performs better than the time-independent Poisson model.

Figure 11a and b show the probability gain G for the ETAS and spatially heterogeneous Poisson model, respectively. We can note that the largest probability gain is obtained for low values of the false alarm rate F, i.e., when the alarm threshold r is high (only a very slight fraction of the time–space volume is occupied by alarms). In particular, in the enlargement of the F scale shown in Fig. 12 it can be observed that the probability gain can exceed values of 100 for false alarm rates smaller than 0.001. This result is stable with the threshold magnitude adopted in the computations. Also for the probability gain, the comparison between Fig. 11a and b puts clearly in evidence the important role of the time-independent spatial distribution of the seismicity in the performance of the ETAS model.

Finally, Fig. 13 shows the plots for the R score versus false alarm rate (F) according to the definition adopted by Shi *et al.* (2001) (named R' in Chapter 4). Again, these plots denote a better performance of the ETAS model with respect to the spatially uniform time-independent Poisson hypothesis (R') (Fig. 13a). However, unlike the parameters considered previously, they exhibit a maximum in the false alarm range between 0.1 and 0.3. We may interpret this behaviour as the capability of the R' score to make a sort of compromise between the need of maximizing the Hit rate (H) and minimizing the False alarm rate (F). Moreover,

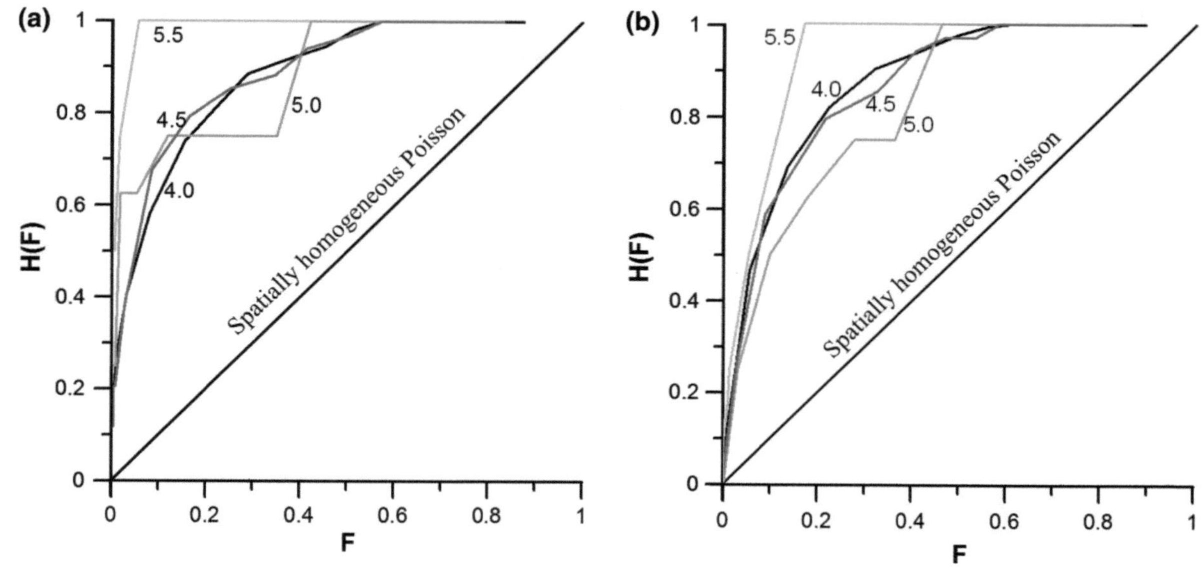

Figure 9
Relative operating characteristic diagrams (ROC) for **a** the ETAS model and **b** a spatially heterogeneous Poisson model. These plots show the hit rate H (fraction of cells where a forecast has been a successful), versus the false alarm rate F (ratio between the false alarms and the total number of alarms). Different magnitude thresholds are indicated above the respective lines. The *diagonal line* shows the results expected for a spatially homogeneous time-independent Poisson model

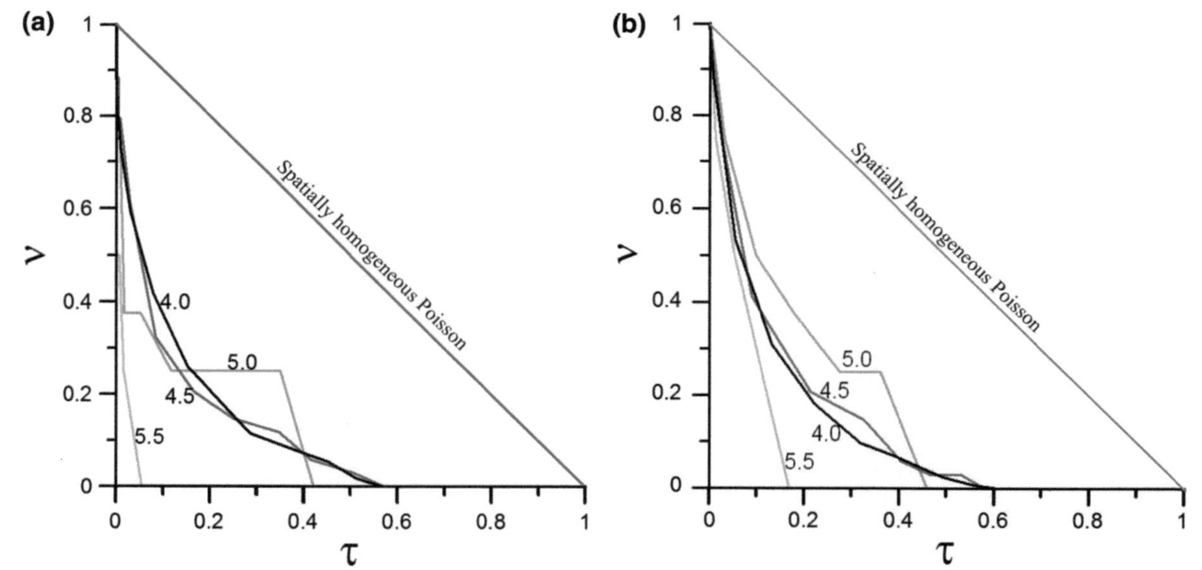

Figure 10
Molchan's error diagram for **a** the ETAS model and **b** a spatially heterogeneous Poisson model. These plots show the fraction of failures to predict, ν, versus the fraction of alarm targets, τ, for different magnitude thresholds. The *diagonal line* shows the results expected for a spatially homogeneous time-independent Poisson model

the comparison between the ETAS model and the spatially heterogeneous time-independent Poisson hypothesis was also made (Fig. 13b), showing the better performance of the ETAS model, although the contribution of the time-independent spatial distribution of the seismicity is quite relevant.

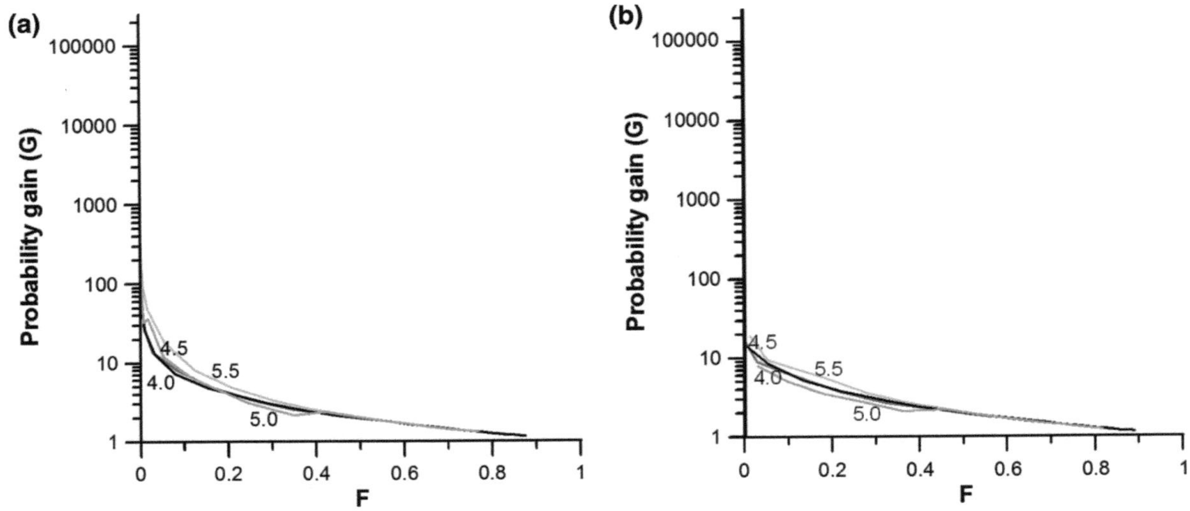

Figure 11
Probability gain versus false alarm rate for **a** the ETAS model and **b** a spatially heterogeneous Poisson model for different magnitude thresholds. The value 1.0 pertains to a complete random forecast, based on the spatially homogeneous time-independent Poisson model

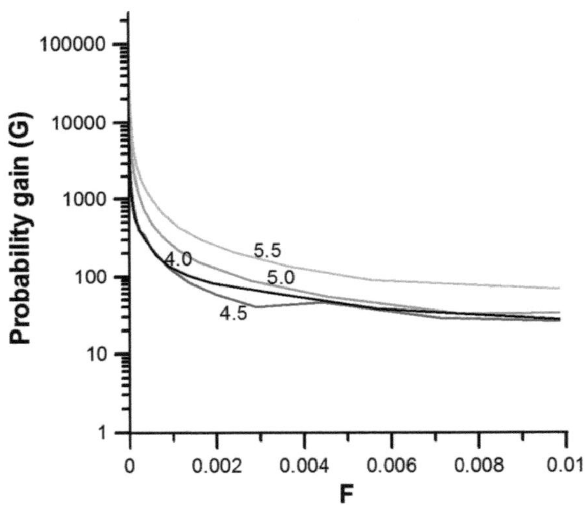

Figure 12
Probability gain versus false alarm rate (F) for the ETAS model. The scale of F parameter covers only the range between 0 and $1-10^{-2}$. The magnitude thresholds are reported above each line. The value 1.0 pertains to a complete random forecast, based on the spatially homogeneous time-independent Poisson model

6. Discussion and Conclusions

In this paper we adopt a simple model of clustered seismicity (ETAS model) characterized by isotropic spatial distribution, and positive triggering, so that the effect of previous earthquakes can only increase the occurrence rate of the following activity. The simple model adopted in the present study is to be considered in the context of a real time forecast system, even with a possible practical application for decision-making and public information. However, before its possible practical use, as is necessary for every forecast method, the method needs a suitable testing phase.

In this exercise we have applied different statistical validation methods. In consideration of the seismicity level characterizing the New Zealand region in the test period (January 2006 to April 2008) the magnitude of the target events for which forecasts are considered was chosen equal to or larger than 4.0 (274 events actually occurred in different space time boxes).

The likelihood ratio method has shown that forecasts based on the ETAS time-dependent model are more reliable than forecasts simply based on a random choice of the occurrence time and a heterogeneous spatial distribution based on the past seismicity. The average information score per event obtainable with the likelihood ratio method ranges from 1 to 4.5 for target events of magnitudes ranging from 4.0 to 5.5. The corresponding average likelihood ratio per event ranges from 3 to 50. For comparison, Helmstetter *et al.* (2006) found for California a likelihood ratio per event equal to 4 and 8 for magnitudes 5.0 and 4.0, respectively.

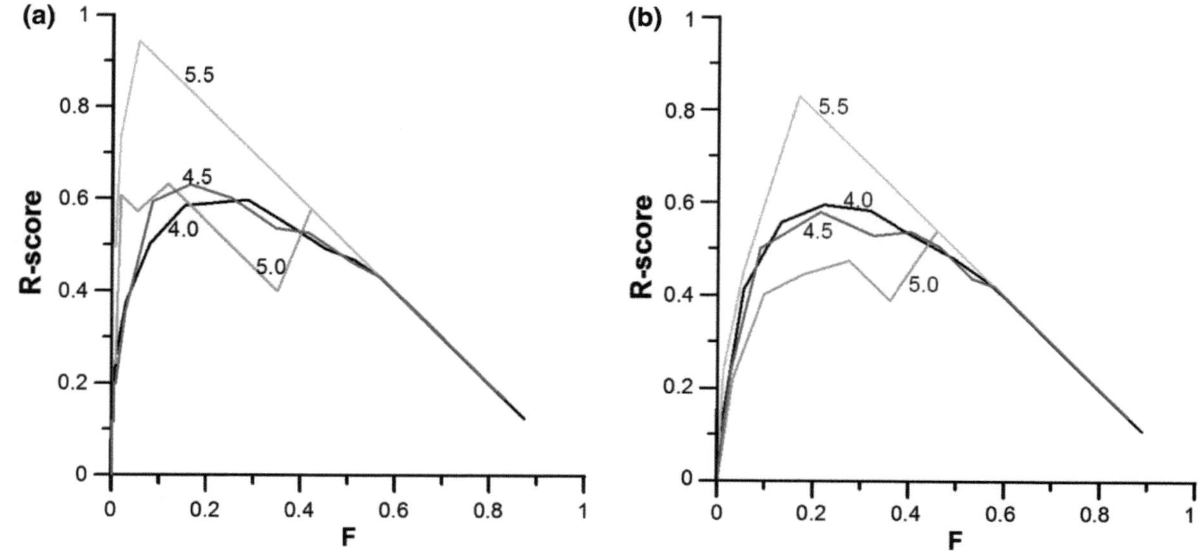

Figure 13
R score versus false alarm rate (F) for **a** the ETAS and **b** the spatially heterogeneous Poisson models. The scale of F parameter covers the whole range between 0 and 1. The zero-value pertains to a complete random forecast, based on the spatially homogeneous time-independent Poisson mode only

In agreement with these results, a more recent study by Kagan et al. (2010), found for California a likelihood ratio per event equal to 3 and 10 for magnitudes 4.7 and 4.0, respectively.

The likelihood ratio method is suitable for the comparison of the performance of two or more forecast algorithms. In this context, the ETAS algorithm proposed in this study could be taken as a reference model for testing the forecasting capability of other, more sophisticated models.

With respect to the likelihood criterion, the binary criteria (Forecast = yes or no, Occurrence = yes or no), such as the ROC and Molchan's diagrams, are more strictly related to the possible real use of the forecasting algorithm.

Taking the ROC diagram as an example, as shown in Table 2 and Fig. 9a, the Hit rate H provided by our algorithm is considerably larger than that of a random spatially uniform forecast for the whole range spanned by the False alarm rate F. For instance, for target events of magnitude 4.0 or larger, and for $r = 5.00\text{E-}05$ (one event expected every 20,000 alarms) we gained a real success rate equal to 0.000147 (122 successes over 829,607 alarms) against a random occurrence rate equal to 0.0000306 (165 cells containing events over 5,396,191 space–time cells). This yields a probability gain G equal to 4.8 ($F = 0.154$, $H = 0.74$). Raising the value of r to 0.007 (one event expected every 143 alarms), we attained a real success rate equal to 0.0032 (23 success over 7,183 alarms), the random occurrence rate remaining the same (165 cells containing events over 5,396,191 space–time cells). This yields a probability gain G equal to 105 ($F = 0.139$, $H = 0.00133$). Finally, the performance of our forecasting method becomes even better (however the results are less significant) raising the value of r to 0.1 (one event expected over 10 alarms). In this case we realized a real success rate equal to 0.5 (2 successes over 4 alarms), and a probability gain G equal to 16,340 ($F = 0.0121$, $H = 3.7\text{E-}07$).

Figure 11 shows that for smaller magnitudes of the target events (in the magnitude range between 4.0 and 5.5), and for false alarm rates smaller than 0.002, the probability gain assumes values ranging in the order of magnitude 100–1,000.

The results of the tests carried out with the ROC and the Molchan's error diagrams as well as with the probability gain, consistently show that the better performance is achieved with higher rate thresholds. It is, in fact, in the range of higher rate thresholds (very low false alarm rates) that the time-dependent

ETAS model demonstrates its superiority with respect to the time-independent spatially variable Poisson model (as shown by the comparison of the a and b panels in Figs. 9, 10 and 11. In contrast, the *R*-score method, with the definition proposed by SHI *et al.* (2001), shows a maximum for the performance at intermediate threshold values ($F \approx 0.1$–0.3) (Fig. 12). This is because the *R* score seeks a compromise between the maximization of the Hit rate and the minimization of the False alarm rate.

The test of the method at higher threshold rate values (e.g., $r \approx 0.5$) and for larger cut-off magnitudes (e.g. $M = 6.0$) would require observations over wider regions and/or longer periods of time.

Acknowledgments

The authors are grateful to two anonymous reviewers for their suggestions that greatly improved the quality of the manuscript, and to Matt Gerstenberger, who has carried out a careful editorial revision of the manuscript.

REFERENCES

AKI, K. (1981), *A probabilistic synthesis of precursory phenomena.* In *Earthquake Prediction*, Am. Geophys. Union. Washington, pp. 556–574.

BAIESI, M. (2006), *Scaling and precursor motifs in earthquake networks*, Physica A, *360*(2), 534–542.

CHEN, C.-C., RUNDLE, J.B., LI, H.-C., HOLLIDAY, J.R., NANJO, K.Z., TURCOTTE, D.L. and TIAMPO, K.F. (2006), *From tornados to earthquakes: forecast verification for binary events applied to the 1999 Chi-Chi, Taiwan, Earthquake*, Terr. Atmos. Ovean. Sci. *17*, 3, 503–516.

CONSOLE, R. (2001), *Testing earthquake forecast hypotheses*, Tectonophysics, *338*, 261–268.

CONSOLE, R. and MURRU, M. (2001), *A simple and testable model for earthquake clustering*, J. Geophys. Res. *106*, 8699–8711.

CONSOLE, R., MURRU, M. and LOMBARDI, A.M. (2003), *Refining earthquake clustering models*, J. Geophys. Res. *108*, 2468, doi: 10.1029/2002JB002130.

CONSOLE, R., MURRU, M. and CATALLI, F. (2006a), *Physical and stochastic models of earthquake clustering*, Tectonophysics *417*, 141–153.

CONSOLE, R., RHOADES, D.A., MURRU, M., EVISON, F.F., PAPADIMITRIOU, E.E. and KARAKOSTAS, V.G. (2006b), *Comparative performance of time-invariant, long-range and short-range forecasting models on the earthquake catalogue of Greece*, J. Geophys. Res. *111*, B09304, doi:10.1029/2005JB004113.

CONSOLE, R., MURRU, M., CATALLI, F. and FALCONE, G. (2007), *Real time forecasts through an earthquake clustering model constrained by the rate-and-state constitutive law: Comparison with a purely stochastic ETAS model*, Seismol. Res. Lett. *78*, 49–56.

CONSOLE, R., MURRU, M. and FALCONE, G. (2010), *Probability gains of an epidemic-type aftershock sequence model in retrospective forecasting of $M \geq 5$ earthquakes in Italy*, J. Seismol. *14*, 9–26.

CONSOLE, R., JACKSON, D.D. and KAGAN, Y.Y. (2010), *Using the ETAS model for catalog declustering and seismic background assessment*, Pure Appl. Geophys. Seismogenesis and Earthquake Forecasting: The Frank Evison Volume, this issue.

DALEY, D.J. and VERE-JONES, D. (2003), *An introduction to the Theory of Point Processes*, 2-nd ed., vol. 1 (Springer-Verlag, New York), p. 469.

DALEY, D.J. and VERE-JONES, D. (2004), *Scoring probability forecasts for point processes: The entropy score and information gain*, J. Appl. Prob. *41A*, 297–312.

FAENZA, L., HAINZL, S., SCHERBAUM, F. and BEAUVAL, C. (2007), *Statistical analysis of time dependent earthquake occurrence and its impact on hazard in the low seismicity region Lower Rhine Embayment*, Geophys. J. Int. *171*(2), 797–806.

FELZER, K.R., BECKER, T.W., ABERCROMBIE, R.E., EKSTRÓM, G. and RICE, J.R. (2002), *Triggering of the 1999 M_w 7.1 Hector Mine earthquake by aftershocks of the 1992 M_w 7.3 Landers earthquake*, J. Geophys. Res. *107*(B9), 2190, doi:10.1029/2001JB 000911.2002.

FRANKEL, A. (1995), *Mapping seismic hazard in the central and eastern United States*, Seismol. Res. Lett. *66*, 8–21.

HANSSEN, A.W. and KUIPERS, W.J.A. (1965), *On the relationship between frequency of rain and various meteorological parameters*, Mededelingen en Verhandelingen, *81*, 2–15.

HARTE, D. and VERE-JONES, D. (2005), *The entropy score and its uses in earthquake forecasting*, In (Ben-Zion, Y, R. Zuniga and D. Vere-Jones, Eds), *Statistical Seismology*, Pure Appl. Geophys. *162*, 1229–1253.

HELMSTETTER, A. and SORNETTE, D. (2002), *Subcritical and supercritical regimes in epidemic models of earthquake aftershocks*, J. Geophys. Res. *107*(B10), 2237, doi:10.1029/2001JB001580.

HELMSTETTER, A. and SORNETTE, D. (2003), *Predictability in the Epidemic-Type Aftershock Sequence model of interacting triggered seismicity*, J. Geophys. Res. *108*(B10), 2482, doi:10.1029/2003JB002485, 2003.

HELMSTETTER, A., Kagan, Y.Y. and Jackson, D.D. (2006), *Comparison of short-term and time-independent earthquake forecast models for southern California*, Bull. Seismol. Soc. Am. *96*(1), 90–106.

HOLLIDAY, J.R., NANJO, K.Z., TIAMPO, K.F., RUNDLE, J.B. and TURCOTTE, D.L. (2005), *Earthquake forecasting and its verification*, Nonlinear Proc. Geophys. *12*, 965–977.

IMOTO, M. (2004), *Probability gain expected for renewal models*, Earth Planets Space *56*, 561–571.

KAGAN, Y.Y. (1991), *Likelihood analysis of earthquake catalogues*, Geophys. J. Int. *106*, 135–148.

KAGAN, Y.Y. (2002), *Aftershock zone scaling*, Bull. Seismol. Soc. Am. *92*(2), 641–655.

KAGAN, Y.Y. and JACKSON, D.D. (1994), *Long-term probabilistic forecasting of earthquakes*, J. Geophys. Res. *99*, 13,685–13,700.

KAGAN, Y.Y. and JACKSON, D.D. (1995), *New seismic gap hypothesis: five years later*, J. Geophys. Res. *100*, 3,943–3,959.

KAGAN, Y.Y. and JACKSON, D.D. (2000), *Probabilistic forecasting of earthquakes*, Geophys. J. Int. *143*, 438–453.

KAGAN, Y.Y. and KNOPOFF, L. (1977), *Earthquake risk prediction as a stochastic process*, Phys. Earth Planet. Inter. *14* (2), 97–108.

KAGAN, Y.Y., BIRD, P. and JACKSON, D.D. (2010), *Earthquake patterns in diverse tectonic zones of the globe*, Pure Appl. Geophys. Seismogenesis and Earthquake Forecasting: The Frank Evison Volume, this issue.

KOSSOBOKOV, V.G. (2006), *Testing earthquake prediction methods: "The West Pacific short-term forecast of earthquakes with magnitude $M_wHRV \geq 5.8$"*, Tectonophysics *413*(1–2), 25–31.

MA, L. and ZHUANG, J. (2001), *Relative quiescence within the Jiashi Swarm in Xinjiang, China: An application of the ETAS point process model*, J. Appl. Prob. *38, Probability, Statistics and Seismology*, 213–221.

MATTHEWS, M.M. and REASENBERG, P.P. (1988), *Statistical methods for investigating quiescence and other temporal seismicity patterns*, Pure Appl. Geophys. *126*, 2–4, 357–372.

MCGUIRE, J.J., BOETTCHER, M.S. and JORDAN, T.H. (2005), *Foreshock sequences and short-term earthquake predictability on East Pacific Rise transform faults*, Nature *434*(7032), 457–461; Correction-Nature *435*(7041), 528.

MOLCHAN, G.M. (1990), *Strategies in strong earthquake prediction*, Phys. Earth Planet. Inter. *61* (1–2), 84–98.

MOLCHAN, G.M. (1997), *Earthquake prediction as a decision-making problem*, Pure Appl. Geophys. *149* (1), 233–247.

MOLCHAN, G.M. (2003), *Earthquake prediction strategies: A theoretical analysis*. In: (Keilis-Borok, V.I. and Soloviev, A.A., eds) *Nonlinear Dynamics of the Lithosphere and Earthquake Prediction* (Springer, Heidelberg, 2003), pp. 208–237.

MOLCHAN, G.M. and KAGAN, Y.Y. (1992), *Earthquake prediction and its optimization*, J. Geophys. Res. *97*, 4823–4838.

MURRU, M., CONSOLE, R. and FALCONE, G. (2009), *Real-time earthquake forecasting in Italy*, Tectonophysics. *3–4*, 214–223.

OGATA, Y. (1998), *Space–time point-process models for earthquake occurrences*, Ann. Inst. Statist. Math. *50*, 2, 379–402.

OGATA, Y. (2001), *Increased probability of large earthquakes near aftershock regions with relative quiescence*, J. Geophys. Res. *106*, B5, 8729–8744.

OGATA, Y. (2004a), *Space-time model for regional seismicity and detection of crustal stress changes*, J. Geophys. Res. *109*, B3, B03308, doi:10.1029/2003JB002621.

OGATA, Y. (2004b), *Seismicity quiescence and activation in western Japan associated with the 1944 and 1946 great earthquakes near the Nankai trough*, J. Geophys. Res. *109*, B4, B04305, doi: 10.1029/2003JB002634.

OGATA, Y. (2005), *Synchronous seismicity changes in and around the northern Japan preceding the 2003 Tokachi-oki earthquake of M 8.0*, J. Geophys. Res. *110*, B5, B08305, doi:10.1029/2004JB003323.

OGATA, Y. (2006a), *Monitoring of anomaly in the aftershock sequence of the 2005 earthquake of M 7.0 off coast of the western Fukuoka, Japan, by the ETAS model*, Geophys. Res. Lett. *33*, 1, L01303, doi:10.1029/2005GL024405.

OGATA, Y. (2006b), *Seismicity anomaly scenario prior to the major recurrent earthquakes off the east coast of Miyagi Prefecture, northern Japan*, Tectonophysics *424*, 291–306, doi:10.1016/j.tecto.2006.03.038.

OGATA, Y. (2007), *Seismicity and geodetic anomalies in a wide area preceding the Niigata-Ken-Chuetsu earthquake of 23 October 2004, central Japan*, J. Geophys. Res. *112*, B10301, doi: 10.1029/2006JB004697.

OGATA, Y. and KATSURA, K. (1993), *Analysis of temporal and spatial heterogeneity of magnitude frequency distribution inferred from earthquake catalogues*, Geophys. J. Int. *113*, 727–738.

OGATA, Y. and KATSURA, K. (2006), *Immediate and updated forecasting of aftershock hazard*, Geophys. Res. Lett. *33*, 10, L10305, doi:10.1029/2006GL025888.

OGATA, Y. and ZHUANG, J. (2006), *Space-time ETAS models and an improved extension*, Tectonophysics *413*, 13–23.

OGATA, Y., JONES, L.M. and TODA, S. (2003), *When and where the aftershock activity was depressed: Contrasting decay patterns of the proximate large earthquakes in southern California*, J. Geophys. Res. *108*, B6, 2318, doi:10.1029/2002JB002009 (1-12).

RHOADES, D.A. and EVISON, F.F. (1989), *On the reliability of precursors*, Phys. Earth Planet. Int. *58*, 137–140.

SAICHEV, A. and SORNETTE, D. (2006), *Renormalization of branching models of triggered seismicity from total to observable seismicity*, Eur. Phys. J. B *51*, 443–459, doi:10.1140/epjb/e2006-00242-6.

SHI, Y. and BOLT, B. (1982), *The standard error of the magnitude frequency b value*, Bull. Seism. Soc. Am. *72*, 1677–1687.

SHI, Y., LIU, J. and ZHANG, G. (2001), *An evaluation of Chinese annual earthquake predictions, 1990–1998*, J. Appl. Probab. *38A*, 222–231.

VERE-JONES, D. (1998), *Probabilities and information gain for earthquake forecasting*, Comput. Seismol., *30*, Geos, Moscow, 248–263.

ZECHAR, J.D. and JORDAN, T.H. (2008), *Testing alarm-based earthquake predictions*, Geophys. J. Int. *172*, 715–724.

ZHUANG, J., OGATA, Y. and VERE-JONES, D. (2004), *Analyzing earthquake clustering features by using stochastic reconstruction*, J. Geophys. Res. *109*, B5, B05301, doi:10.1029/2003JB002879.

ZHUANG, J., CHANG, C., OGATA, Y. and CHEN, Y. (2005), *A study on the background and clustering seismicity in the Taiwan region by using point process models*, J. Geophys. Res. *110*, B5, B05S18, doi:10.1029/2004JB003157.

(Received July 31, 2008, revised April 2, 2009, accepted October 7, 2009, Published online March 12, 2010)

Pure Appl. Geophys. 167 (2010), 709–719

DOI 10.1007/s00024-010-0074-4

Pure and Applied Geophysics

Earthquake Forecasting in Diverse Tectonic Zones of the Globe

Y. Y. Kagan[1] and D. D. Jackson[1]

Abstract—We present a simple method for long- and short-term earthquake forecasting (estimating earthquake rate per unit area, time, and magnitude). For illustration we apply the method to the Pacific plate boundary region and the Mediterranean area surrounding Italy and Greece. Our ultimate goal is to develop forecasting and testing methods to validate or falsify common assumptions regarding earthquake potential. Our immediate purpose is to extend the forecasts we made starting in 1999 for the northwest and southwest Pacific to include somewhat smaller earthquakes and then adapt the methods to apply in other areas. The previous forecasts used the CMT earthquake catalog to forecast magnitude 5.8 and larger earthquakes. Like our previous forecasts, the new ones here are based on smoothed maps of past seismicity and assume spatial clustering. Our short-term forecasts also assume temporal clustering. An important adaptation in the new forecasts is to abandon the use of tensor focal mechanisms. This permits use of earthquake catalogs that reliably report many smaller quakes with no such mechanism estimates. The result is that we can forecast earthquakes at higher spatial resolution and down to a magnitude threshold of 4.7. The new forecasts can be tested far more quickly because smaller events are considerably more frequent. Also, our previous method used the focal mechanisms of past earthquakes to estimate the preferred directions of earthquake clustering, however the method made assumptions that generally hold in subduction zones only. The new approach escapes those assumptions. In the northwest Pacific the new method gives estimated earthquake rate density very similar to that of the previous forecast.

Key words: Earthquake global forecasts, statistical analysis of earthquake occurrence, branching models of seismicity, forecasts testing.

1. Introduction

The importance of earthquake forecasting for seismic hazard and risk estimation and the difficulty of resolving basic differences in forecast models have motivated an international effort. How can we establish standards for reporting and testing earthquake forecasts? One significant effort began in California, wherein the Regional Earthquake Likelihood Models (RELM) project published a dozen models for earthquake rate density and a likelihood-based method for prospective testing (Field, 2007; Schorlemmer and Gerstenberger, 2007). The various models included ones with strongly varying emphasis based on past earthquakes, fault geometry and slip rates, and geodetic strain rates. Some assume that large earthquakes reduce the probability of similar ones soon thereafter, while others assume the opposite. While testing awaits further time and more events, the Collaboratory for Study of Earthquake Predictability (CSEP) is extending the tests to several natural laboratories around the globe. CSEP will also test short-term forecasts updated daily.

Our technique is to establish a statistical model which fits the catalog of earthquake times, locations, and seismic moments, and subsequently to base forecasts on this model. While most components of the model have been tested (Kagan and Knopoff, 1987; Kagan, 1991; Jackson and Kagan, 1999; Kagan and Jackson, 2000), some require further exploration and can be modified as our research progresses. Our purpose in this work is to extend a clustering model that we used to make testable forecasts over large regions of the western Pacific (*ibid.*) to short-term forecasts and additional test laboratories.

Our previous forecast model was based on constructing a map of smoothed rates of past earthquakes. We used the CMT catalog (Ekström *et al.*, 2005) because it employs relatively consistent methods and reports tensor focal mechanisms. The

[1] Department of Earth and Space Sciences, University of California, Los Angeles, CA 90095-1567, USA. E-mail: ykagan@ucla.edu; djackson@ucla.edu

focal mechanisms allow us to estimate the fault plane orientation for past earthquakes, through which we can identify a preferred direction for future events. Using the forecasted tensor focal mechanism, it is possible in principle to calculate an ensemble of seismograms for each point of interest on the earth's surface.

However, the CMT catalog imposes some restrictions: It only began in 1977, and from that date is complete only for earthquake magnitudes of about 5.8 and larger. Here we experiment with a forecast based on the US Geological Survey PDE catalog (US Geological Survey, 2008) instead. While that catalog does not report focal mechanisms routinely, it uses relatively stable methods and reliably reports earthquakes about one magnitude unit smaller than does the CMT. Preliminary tests indicate that the PDE catalog gives similar results to the CMT and can thus enable credible forecasts for tests in the CSEP effort.

Both the CMT and PDE are global catalogs employing data from worldwide seismic networks. We advocate use of global catalogs even in CSEP test laboratories where regional catalogs with lower magnitude thresholds are available. Why? Global catalogs are more homogeneous than local catalogs, and lack spatial boundary effects which greatly complicate analysis of local catalogs. Moreover, local seismicity may be dominated by a few aftershock sequences of strong events, such as the *m* 7.5 1952 Kern County and the *m* 7.3 1992 Landers earthquakes in southern California. Explosions and earthquakes caused by volcanic and geothermal activity are more likely to contaminate earthquake records in local and regional catalogs.

The PDE catalog has significant advantages over that of the CMT. The PDE has a longer observation period (the surface wave magnitude M_S was determined starting from the middle of 1968), and a lower magnitude threshold (m_t). Depending on time period and region, the threshold is of the order 4.5–4.7 (Kagan, 2003), i.e., much lower than the CMT catalog threshold (around 5.4–5.8). This means that the forecast estimates can be practically obtained for all global seismic areas. The PDE reports earthquake hypocenters, which can be estimated far more precisely than the moment centroid locations reported by the CMT catalog.

On the other hand, the PDE catalog has a few drawbacks when compared to the CMT data set. For one, the PDE catalog generally lacks the focal mechanism solutions. Also, the PDE reports a somewhat inconsistent mix of different magnitudes (local, body wave, surface wave, moment, etc.) with less accuracy than the moment magnitude inferred from the CMT catalog. Moreover, the PDE magnitudes are influenced by strong systematic effects and biases (Kagan, 2003). Another drawback is that the hypocenter, which the PDE catalog uses for representing location, could be at the edge of the rupture zone for a large earthquake. The moment centroid, reported by the CMT, more meaningfully describes the location even though the centroid is generally more uncertain than the hypocenter.

Our forecasting method involves several steps which we have performed here but which still deserve more research. These steps include the following.

- Select an earthquake catalog, test it for homogeneity in time and space, and investigate its accuracy and magnitude threshold (Kagan, 2003).
- Determine the distribution of hypocentral distances between all pairs of earthquakes used to normalize the spatial earthquake distribution in the branching model of earthquake occurrence (Kagan, 1991, Eq. (27); 2007).
- Using likelihood analysis, estimate the parameters of the clustering model and their uncertainties for use in the short-term forecast model (Kagan *et al.*, 2010).
- Optimize the smoothing kernel parameters of the long-term forecast using the first part of the catalog as a learning set and the second part as a control set (Kagan and Jackson, 1994).
- Use determined parameters of the branching model for a short-term forecast of earthquake occurrence.
- Write a computer program to combine the results of the steps above into an earthquake forecast algorithm, such that the program can be run by an independent party with a clearly identified source of future earthquake data (Jackson and Kagan, 1999; Kagan and Jackson, 2000).

- Finally, test the forecast using independent data to ensure that the program works as intended (KAGAN and JACKSON, 2000; KAGAN, 2009).

The program we propose in this paper should be considered as a scientific demonstration project; implementing the technique for the real-life earthquake forecast would require additional work.

2. *Methods and Data*

For our long-term model we factor the rate density into spatial and magnitude distributions, assuming they are mutually independent. We estimate the spatial distribution using all earthquakes above the assumed completeness threshold in the catalog. For the magnitude distribution, we assume a tapered Gutenberg–Richter distribution (BIRD and KAGAN, 2004; KAGAN *et al.*, 2010). At any place, the spatial rate serves as a multiplicative constant, proportional to the "a-value" of the magnitude distribution. Thus, any forecast based on a catalog with a given lower magnitude threshold applies as well to larger earthquakes. We test the models over all magnitudes above the threshold, and because large earthquakes are expected to be less frequent, they count more than smaller ones for the likelihood test. In principle this method would allow us to forecast earthquakes smaller than the threshold magnitude. But we choose not to do so because the smoothing kernels that determine our spatial distribution are influenced indirectly by earthquake magnitudes. Thus, catalogs with smaller earthquakes provide higher spatial resolution.

In our previous papers (KAGAN and JACKSON, 1994, 2000; JACKSON and KAGAN, 1999) we studied earthquake distributions and clustering for the global CMT catalog of moment tensor inversions compiled by EKSTRÖM *et al.* (2005). The CMT catalog now contains more than 28,000 earthquake entries for the period 1977/1/1 to 2007/12/31. This catalog characterizes earthquake size by the scalar seismic moment M. Here we use the scalar seismic moment directly, but for easy comparison we convert it into an approximate moment magnitude using the relationship

$$m_w = \frac{2}{3}(\log_{10} M - 9.0), \qquad (1)$$

(Hanks and Kanamori, 1979; HANKS, 1992), where M is measured in Newton m (Nm). Note that various authors use slightly different values for the final constant (here 9.0) so the magnitude values we use in tables and diagrams may not be fully consistent with values used by others.

The PDE (preliminary determinations of epicenters) worldwide catalog is published by the USGS (US GEOLOGICAL SURVEY, 2008). In spite of its name, the catalog is distributed in final form with a few months latency. When we wrote this paper, the catalog was in final form through the end of 2007. However, truly preliminary information on earthquakes is usually available with a delay of just a few minutes.

The PDE catalog reports earthquake size using several magnitude scales, providing the body wave (m_b) and surface wave (M_S) magnitudes for most moderate and large events since 1965 and 1968, respectively. The moment magnitude (m_w) estimate has been added recently. To construct our smoothed seismicity maps, we need a single standard magnitude estimate for each earthquake. Ideally, we would like to convert all other magnitude estimates into moment magnitude and use that as a standard. Unfortunately, we have not found a reliable way to do that (see KAGAN, 2003). KAGAN (1991) used a weighted average of several different magnitude scales but found the results not fully satisfying, especially now since several additional magnitudes are reported.

In this work we decided to use as standard the maximum magnitude among those shown for each earthquake. For smaller moderate earthquakes it is usually m_b, since other magnitude estimates are unavailable. For larger events M_S magnitude would be selected; for even larger recent earthquakes the maximum magnitude most likely is m_w.

3. *Forecast Models*

3.1. *Long-term Rate Density Estimates*

We have developed a long-term forecast procedure based on earthquake data alone (KAGAN and JACKSON, 1994, 2000). Our procedure is based on the smoothing of past earthquake locations. The degree of spatial smoothing is controlled by the kernel function

$$f(r) = \frac{1}{\pi} \times \frac{1}{r^2 + r_s^2}, \tag{2}$$

where r is epicentroid distance, r_s is the scale parameter. The scaling distance r_s may depend on many factors including tectonic environment, Earth structure, earthquake location accuracy, and so on. Thus, one should determine it in principle for each forecast region.

Recently similar techniques based on smoothing of earthquake locations have been applied to long-term forecasts in California and other regions (Rhoades and Evison, 2005, 2006; Rhoades, 2007; Helmstetter *et al.*, 2007; Schorlemmer and Gerstenberger, 2007; Console *et al.*, 2010; Kagan and Jackson, 2010).

For the CMT catalog, we use the same values for a long-term smoothing kernel as Kagan and Jackson (2000, see their Eq. 3): the spatial scale parameter, r_s is 15 km for the Northwest Pacific (NW), and the azimuthal concentration factor (δ in Eq. 6, *ibid.*) is 100.

In Figure 1 we display the NW Pacific long-term rate density for magnitudes of 5.8 and above based on the CMT catalog. A similar forecast based on the PDE catalog with the same time period and magnitude threshold as the CMT catalog is shown in Figure 2. For the PDE catalog we apply an isotropic kernel function with r_s = 15 km (2). Both forecasts are similar in appearance. Given that the magnitude threshold is lower for the PDE catalog (m_t = 5.0), the forecast in Figure 3 shows more details since many more earthquakes were used in computation. More detailed maps are, in principle, preferable; however, to evaluate the forecast performance quantitatively we need to perform a statistical analysis of the forecasts and earthquakes that occurred after the forecasts have been issued.

Our procedure allows us to optimize the parameters by choosing those r_s values which best predict the second part of a catalog from the first part. Kagan and Jackson (1994) subdivided the catalog into two equal parts, whereas Kagan and Jackson (2000) used the CMT catalog for 1977–1996 as the 'training' part and as the test catalog the data for 1997–1998. In this

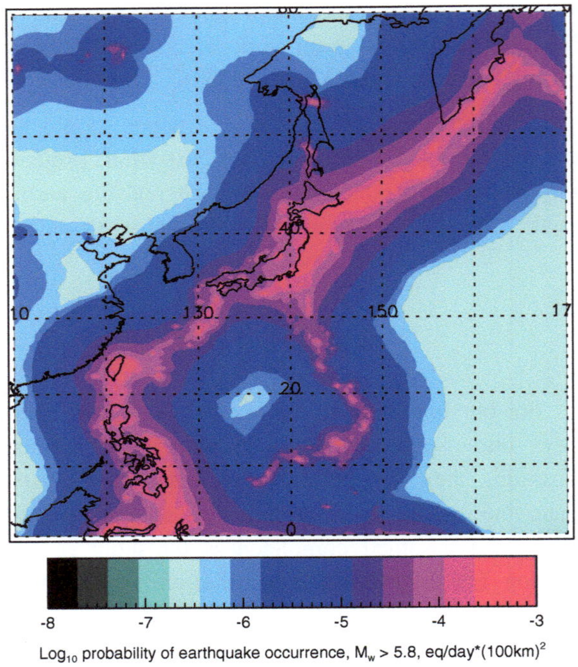

Figure 1
Northwest Pacific long-term seismicity forecast: Latitude limits from 0.25°S to 60.25°N, longitude limits from 109.75°E to 170.25°E. *Colors* show the rate-density of earthquake occurrence calculated using the CMT 1977–2008/08/06 catalog

paper earthquakes of 2004–2006 are used as the control set.

Based on the optimized parameters, we simulate thousands of earthquake catalogs to compare with the observed test catalog, and evaluate the likelihood function for both types of catalogs (Kagan and Jackson, 1994). In the simulations, earthquakes are assumed to occur at the centers of grid cells with the rates defined by the forecast. We normalize the cell rates and simulate a random number uniformly distributed in the interval [0, 1]. The random number corresponding to a particular segment of the cumulative normalized rate curve defines the cell where an event occurs. We obtain one synthetic catalog by repeating this procedure n times. The simulated catalogs, by construction, are consistent with the forecast model. If the observed catalog is also consistent, its likelihood score should be in the middle range of the simulated values.

Several statistics can be used to characterize a forecast and its fit to a future earthquake occurrence

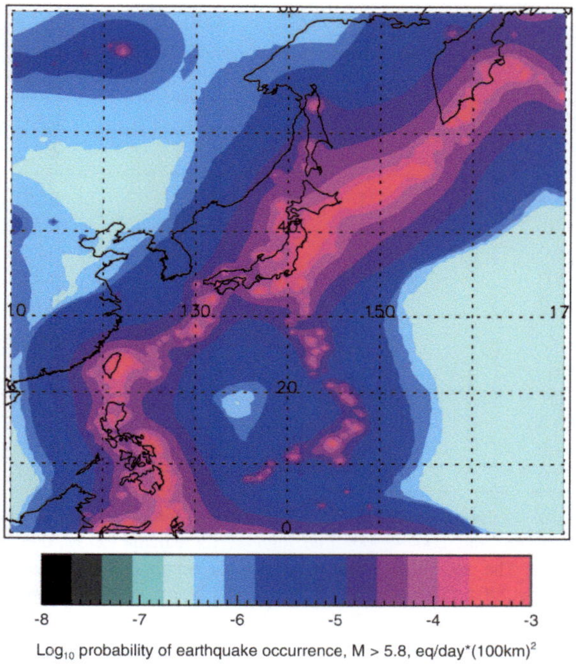

Figure 2
Northwest Pacific long-term seismicity forecast: Latitude limits from 0.25°S to 60.25°N, longitude limits from 109.75°E to 170.25°E. *Colors* show the rate-density of earthquake occurrence calculated using the PDE 1977–2008/08/06 $m \geq 5.8$ catalog

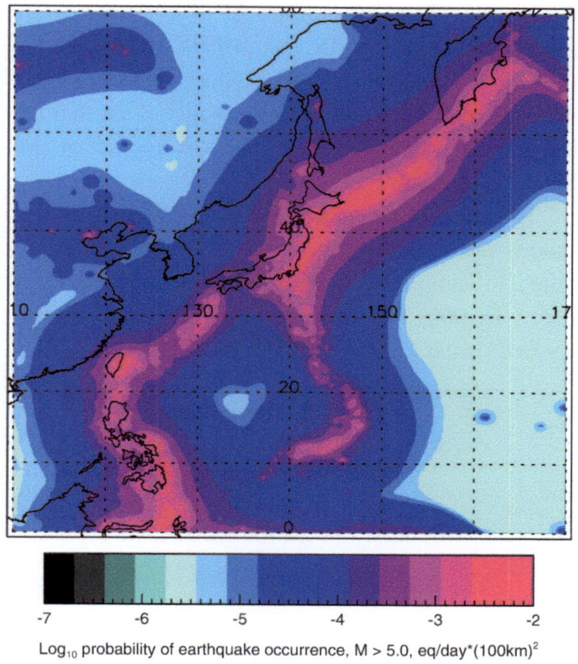

Figure 3
Northwest Pacific long-term seismicity forecast: Latitude limits from 0.25°S to 60.25°N, longitude limits from 109.75°E to 170.25°E. *Colors* show the rate-density of earthquake occurrence calculated using the PDE 1969–2008/08/06 $m \geq 5.0$ catalog

(Kagan, 2009). For an inhomogeneous Poisson process in which n points are observed $(x_1,\ldots,x_n)$ in a region A, the log-likelihood can be written as (Daley and Vere-Jones, 2003, Eq. 7.1.2)

$$\log L(x_1,\ldots,x_n) = \sum_{i=1}^{n} \log \lambda(x_i) - \int_A \lambda(x)\mathrm{d}x, \quad (3)$$

where $\lambda(x_i)$ is the process rate (density) at a point x_i.

The log-likelihood of an inhomogeneous Poisson process is normally compared to a similar log-likelihood, L_0, calculated for a Poisson process with constant intensity (ξ) to obtain the log-likelihood ratio (Daley and Vere-Jones, 2003, Ch. 7; Schorlemmer *et al.*, 2007)

$$\log(L/L_0) = \sum_{i=1}^{n} \log \lambda(x_i/\xi) - \int_A [\lambda(x) - \xi]\mathrm{d}x. \quad (4)$$

In our calculations we normalize both rates (λ, ξ) by the observed event number n, hence the integral term in (4) is zero.

Kagan and Knopoff (1977) suggested measuring the performance of the earthquake prediction algorithm by first evaluating the likelihood ratio to test how well a model approximates an earthquake occurrence. In particular, they estimated the information score, $\hat{I}$, per one event by

$$\hat{I} = \frac{\ell - \ell_0}{n} = \frac{1}{n}\sum_{i=1}^{n} \log_2 \frac{\lambda_i}{\xi}, \quad (5)$$

where $\ell - \ell_0$ is the log-likelihood ratio, n is the number of earthquakes in a catalog, $\log_2$ is used to obtain the score measured in the Shannon bits of information, and λ_i is the rate of earthquake occurrence according to a stochastic model.

In this work we compute the information score

$$I_2 = \frac{1}{n}\sum_{k=1}^{n} \log_2 \frac{\lambda_k}{\xi}, \quad (6)$$

where λ_k is the forecasted rate for the actual epicenter (epicentroid) locations. In the above formulas a spatially uniform Poisson process is assumed as the null hypothesis. This hypothesis, of course, cannot be considered 'sensible' (STARK, 1997), the actual distribution of seismicity is highly inhomogeneous. Thus, we use the actual forecast model as another null hypothesis and compare it with the record of predicted earthquakes.

To evaluate the distribution of events in a forecast model we simulate them by a Monte-Carlo procedure. In simulated catalogs we generate multiple ($\mathcal{N} = 10,000$) sets of n events and calculate the rate for cell centers as the earthquake location

$$I_3 = \frac{1}{n}\sum_{l=1}^{n} \log_2 \frac{\lambda_l}{\xi}, \tag{7}$$

and

$$\langle I_3 \rangle = \frac{1}{\mathcal{N}}\sum_{\ell=1}^{\mathcal{N}} (I_3)_\ell. \tag{8}$$

The I_3 values characterize the information gain which is obtained from smoothed seismicity maps, as compared to randomly selecting a point from a region on a spherical surface.

The standard deviation of the log-likelihood for the set of independent n events is (KAGAN, 2009)

$$\sigma_n = \sqrt{v_2/n}, \tag{9}$$

where v_2 is the second moment of the likelihood score and n is the earthquake number.

In Table 1 we display an example of r_s determination for the NW Pacific region. To compare the CMT and PDE catalogs, we selected the time period 1977–2003 for calculating a forecast table and next the earthquakes of 2004–2006 to test and optimize our spatial kernel (2).

The best r_s value for forecast purposes corresponds to $I_3 - I_2 = 0$, i.e., when the score for earthquakes which occurred in 2004–2006 equals the average score obtained by simulating events with the smoothed forecast spatial distribution. (KAGAN and JACKSON, 1994, Fig. 4 discuss in more detail interpretation of $I_3 - I_2$ differences.) Therefore, the optimal estimates of r_s for both catalogs are between 10 and 15 km.

We could test both forecasts to infer which is preferable on statistical grounds (SCHORLEMMER *et al.*, 2007). In such a test, simulated earthquake sets from one forecast may be tested for agreement with another model and *vice versa*. These more extensive tests are the subject of our further work and are not discussed here.

Table 1

Dependence of likelihood score on smoothing distance r_s for NW-Pacific region

r_s km	I_2	$I_3 - I_2$	σ_n
1	2	3	4
2.5	2.1223	0.9773	0.2714
5.0	2.2590	0.5845	0.2498
10.0	2.3667	0.1687	0.2249
15.0	2.4012	−0.0445	0.2151
25.0	2.4060	−0.2628	0.1963
2.5	2.2060	1.1533	0.2972
5.0	2.2551	0.4907	0.2454
7.5	2.2616	0.2425	0.2299
10.0	2.2560	0.1084	0.2197
15.0	2.2346	−0.0556	0.2058
25.0	2.1872	−0.2199	0.1861

Both catalogs cover 1977–2003 time interval for calculation of the I_3 score, and 2004–2006 time interval for calculation of the I_2 score. The magnitude threshold is 5.8, the number of earthquakes during 2004–2006 is $n = 108$ and $n = 114$ for the CMT (upper Table) and PDE (lower Table) catalogs, respectively

To illustrate how the long-term method can be applied to arbitrary areas of the Earth, in Figure 4 we show the forecast for Italy and surrounding areas and in Figure 6 that for Greece and surrounding areas. In Figure 5 we display the distribution of likelihood scores for observed and simulated catalogs for the forecast model as shown in Figure 4. We simulate earthquake locations with $r_s = 5.0$ km, each time calculating the likelihood function and comparing the function value with that obtained for the real catalog in 2004–2006. We normalize the distributions by dividing the values by the standard deviation [see Eq. (9) and Table 1] so the plotted cumulative curve has a unit standard deviation. From the plot it is clear that $r_s = 5.0$ km is close to optimal, since the curve for observed earthquakes intersects the solid vertical line near the middle.

3.2. *Short-term Rate Density Estimates*

The assumptions we make in constructing our initial branching model of earthquake occurrence have been summarized in KAGAN and KNOPOFF

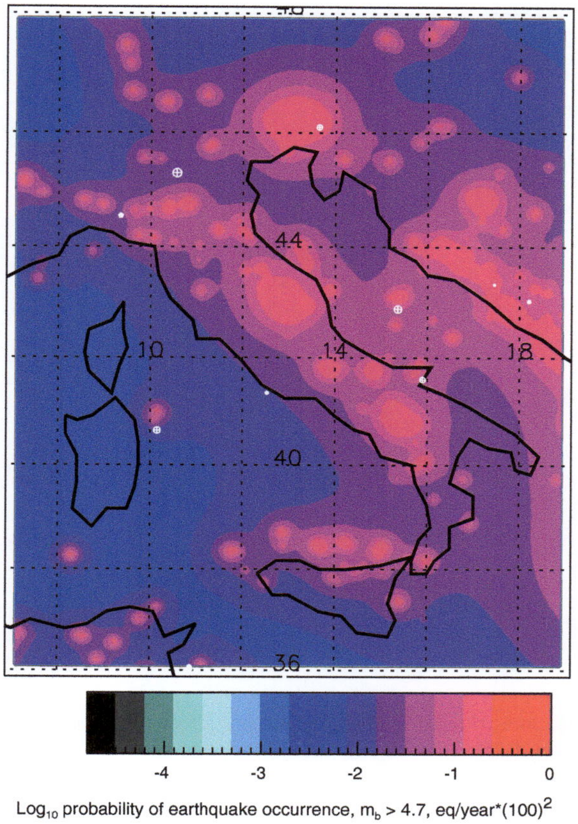

Figure 4
Italy and its surrounding seismicity forecast: Latitude limits from 36.0°N to 48.0°N, longitude limits from 7.0°E to 19.0°E. The forecast is calculated at 121 × 121 grid. *Colors* show the rate density of shallow (depth less or equal to 70 km) earthquake occurrence calculated using the PDE 1969–2003 catalog; nine $m \geq 4.7$ earthquakes for 2004–2006 are shown in *white*

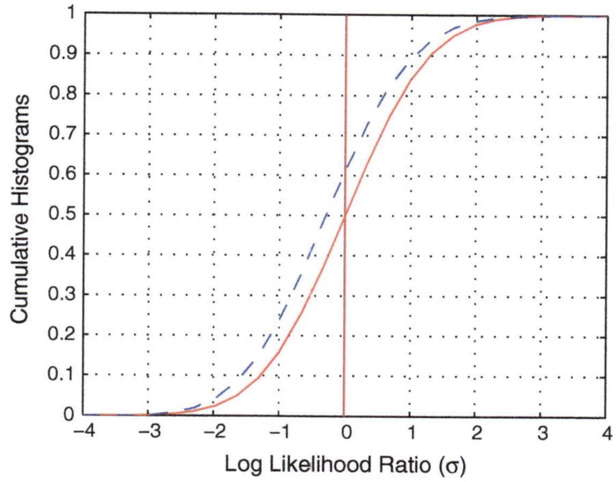

Figure 5
Dashed line is a cumulative curve of the log-likelihood function differences for 2004–2006 simulated earthquakes (see Fig. 4). The function is normalized to have a unit standard deviation. The *red solid line* is the Gaussian curve with a zero mean and unit standard deviation. *Curves on the right* from the Gaussian curve correspond to simulations that are worse than the observed earthquake distribution; *curves on the left* correspond to simulations that are better than the observed earthquake distribution

(1987), KAGAN (1991), and KAGAN and JACKSON (2000). A similar model called the ETAS (Epidemic Type Aftershock Sequence) was proposed by OGATA (1988, 1998), as well as by OGATA and ZHUANG (2006).

In both our model and ETAS, seismicity is approximated by a Poisson cluster process, in which clusters or sequences of earthquakes are statistically independent, although individual earthquakes in the cluster are triggered. The clusters are assumed to form a spatially inhomogeneous Poisson point process with a constant temporal rate. The major assumption regarding the interrelationships between events within a cluster is that the propagation of an earthquake rupture is closely approximated by a

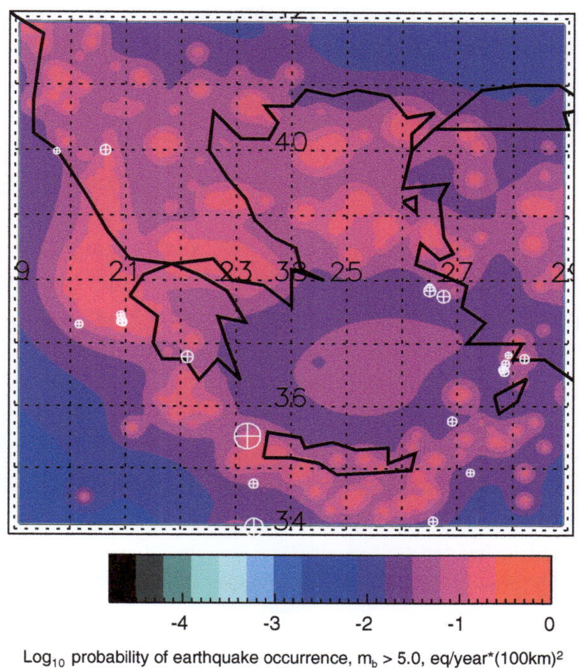

Figure 6
Greece and its surrounding seismicity forecast: Latitude limits from 34.0°N to 42.0°N, longitude limits from 19.0°E to 29.0°E. The forecast is calculated at 121 × 121 grid. *Colors* show the rate density of shallow (depth less or equal to 70 km) earthquake occurrence calculated using the PDE 1969–2003 catalog; 15 $m \geq 5$ earthquakes for 2004–2006 are shown in *white*

stochastic space–time critical branching process. Under this assumption there is a sole trigger for any given dependent event. As shown below, the space–time distribution of interrelated earthquake sources within a sequence is controlled by simple relations justified by analyzing available statistical data on seismicity.

Recently Console and Murru (2001), Console *et al.* (2006a, b, 2007) and many other researchers have applied the ETAS branching models to Japan, Greece, California, and Italy. Lombardi and Marzocchi (2007) and Marzocchi and Lombardi (2008) have applied the ETAS model to two global catalogs of large earthquakes. Kagan *et al.* (2010, Section 6.2) discuss certain drawbacks of the ETAS model in its parametrization of seismicity parameters. These defects may influence the implementation and quality of the forecast programs based on the ETAS model.

The short-term forecast in this work was carried out according to the technique described by Kagan and Jackson (2000, Section 3.1.2). Examples of the short-term forecast are shown in several of our papers (*ibid.*, Kagan and Knopoff, 1987; Molchan and Kagan, 1992; Jackson and Kagan, 1999; Helmstetter *et al.*, 2006; Kagan *et al.*, 2003, 2007).

We used the values of parameters obtained during the likelihood function search (Kagan *et al.*, 2010, Table 4) obtained for the full PDE catalog, $m \geq 5.0$: the branching coefficient $\mu = 0.141$, the parent productivity exponent $a = 0.63$, the time decay exponent $\theta = 0.28$, and the horizontal location error $\varepsilon_\rho = 9.5$ km. From other tables by Kagan *et al.* (2010) we see that the values of the earthquake clustering parameters are similar for subduction zones (trenches), active continental zones and plate interiors. These regions are of interest for evaluating seismic hazard.

Figure 7 shows a short-term forecast for the Northwest Pacific region. The map is more detailed than similar ones obtained with the help of the CMT catalog such as in Fig. 5 by Jackson and Kagan (1999) or Fig. 5.4 by Kagan *et al.* (2003). More general comparison of both forecasts based on the PDE and the CMT catalogs is complicated by the issue of close-in-time aftershocks (Kagan *et al.*, 2010, Section 6.1). As with our long-term forecast,

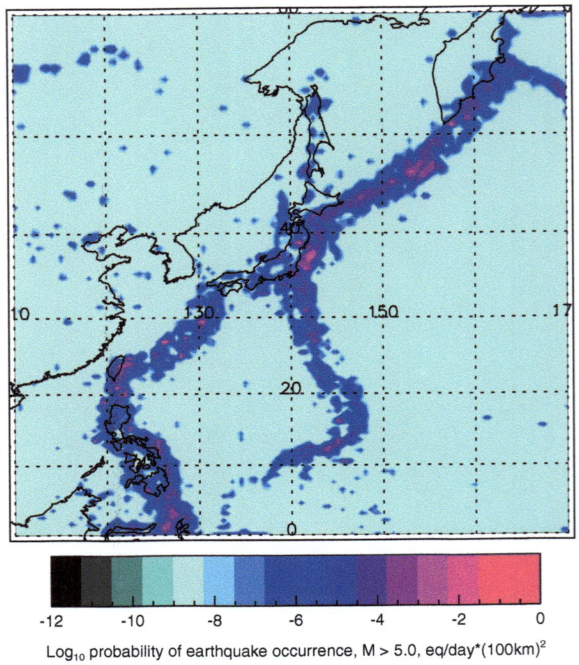

Figure 7
Northwest Pacific short-term seismicity forecast: Latitude limits from 0.25°S to 60.25°N, longitude limits from 109.75°E to 170.25°E. *Colors* show the rate-density of earthquake occurrence calculated using the PDE 1969–2008/08/06 $m \geq 5.0$ catalog

these tests will be subjects of our future work and are not discussed here.

3.3. Forecasts Comparison

In Table 2 we display numerically the forecasts based on the CMT and PDE catalogs for an East-West profile north of the Phillipines. Kagan and Jackson (2000) showed the same area in their Table 1. Columns 3 and 6 show the rate of events per unit area at or above the magnitude thresholds of two catalogs, $m_t = 5.8$ for the CMT and $m_t = 5.0$ for the PDE catalog. The earthquake rate sum for the PDE catalog is 670×10^{-9} Eq/day $\times$ km^2, while for the CMT catalog it is equal to 94×10^{-9} Eq/day $\times$ km^2. The ratio of the two rates is 7.1, close to the value of 6.3 expected for a magnitude threshold difference of 0.8 from the Gutenberg–Richter magnitude distribution with a b value of 1.0. The same sum for Table 1 of Kagan and Jackson (2000) is 117×10^{-9} Eq/day $\times$ km^2, close to the

Table 2

Example of long- and short-term forecast: August 6, 2008, north of Phillipines.

Long.	Lat.	PDE catalog ($m \geq 5.0$)			CMT catalog ($m \geq 5.8$)		
°E	°N	Long-term Rate[a]	Short-term Rate[a]	Rate Ratio[b]	Long-term Rate[a]	Short-term Rate[a]	Rate Ratio[b]
1	2	3	4	5	6	7	8
119.5	19.5	3.61E-08	3.24E-09	8.96E-02	4.20E-09	1.51E-14	3.61E-06
120.0	19.5	5.72E-08	2.77E-10	4.84E-03	6.11E-09	6.46E-12	1.06E-03
120.5	19.5	9.53E-08	2.54E-10	2.67E-03	2.68E-08	3.86E-10	1.44E-02
121.0	19.5	1.27E-07	1.32E-09	1.04E-02	2.09E-08	3.16E-09	0.152
121.5	19.5	1.30E-07	2.05E-09	1.57E-02	1.67E-08	1.62E-09	9.68E-02
122.0	19.5	1.00E-07	4.85E-10	4.84E-03	7.87E-09	1.63E-10	2.07E-02
122.5	19.5	5.44E-08	1.10E-13	2.02E-06	4.63E-09	2.69E-12	5.80E-04
123.0	19.5	3.61E-08	8.34E-22	2.31E-14	3.55E-09	1.16E-15	3.28E-07
123.5	19.5	3.03E-08	4.52E-11	1.49E-03	2.95E-09	1.05E-20	3.56E-12

[a] Rate of earthquakes above the magnitude threshold ($\text{day}^{-1}\text{km}^{-2}$)

[b] Rate ratio = short-term rate/long-term rate

Two catalogs are used: 1969–2008/08/06 PDE catalog and 1977–2008/08/06 CMT catalog

present value of 94×10^{-9} Eq/day $\times$ km^2. This slight difference may result from the fact that here we are using data up to the end of 2007 for estimating rate density.

The short-term forecast rates are more variable. The PDE and CMT catalogs produce similar rate densities (columns 4 and 7 of Table 2), but the 2008 forecast using the CMT differs from the corresponding value from 1999. Also, the ratio of long- and short-term rates is dissimilar for both catalogs, a difference mostly due to lower occurrence rates for the CMT catalog because of its higher magnitude threshold. Why is the ratio of long- and short-term rates higher in Table 1 by Kagan and Jackson (2000)? At that time (*ibid.*) we selected the region of high current activity to demonstrate high short-term rates due to nearby earthquakes also close-in-time.

4. Conclusions

We demonstrated a technique for calculating and evaluating long- and short-term earthquake rate forecasts in practically any seismically active region of the Earth. This advance was possible after our statistical analysis of the global earthquake occurrence (Kagan *et al.*, 2010) that yielded parameter values necessary for a forecast program. The forecast tables can be tested both retrospectively and prospectively, using the methods developed by us and other researchers. In using the PDE rather than the CMT catalog, we give up moment tensor information but gain a lower magnitude threshold, a longer observation period, and consequently more earthquakes on which to base the forecast. Our comparison of forecasts based on the CMT and PDE in the NW Pacific revealed that the PDE forecasts are quantitatively similar to and produce higher resolution than the CMT. Thus, our present forecast program represents a significant progress when compared to our Pacific forecast (Kagan and Jackson, 2000).

Acknowledgments

The authors appreciate support from the National Science Foundation through grants EAR 04-09890 and EAR-0711515, as well as from the Southern California Earthquake Center (SCEC). SCEC is funded by NSF Cooperative Agreement EAR-0529922 and the US Geologic Survey (USGS) Cooperative Agreement 07HQAG0008. We thank Kathleen Jackson for significant improvements in the text. Comments by two anonymous reviewers and by the Associate Editor D. Rhoades have been helpful in the revision of the manuscript. Publication 1245, SCEC.

References

Bird, P. and Kagan, Y.Y. (2004), *Plate-tectonic analysis of shallow seismicity: Apparent boundary width, beta, corner magnitude, coupled lithosphere thickness, and coupling in seven tectonic settings*, Bull. Seismol. Soc. Am. *94*(6), 2380–2399 (plus electronic supplement).

Console, R. and Murru, M. (2001), *A simple and testable model for earthquake clustering*, J. Geophys. Res. *106*, 8699–8711.

Console, R., Murru, M. and Catalli, F. (2006a), *Physical and stochastic models of earthquake clustering*, Tectonophysics, *417*, 141–153.

Console, R., Rhoades, D.A., Murru, M., Evison, F.F., Papadimitriou, E.E. and Karakostas, V.G. (2006b), *Comparative performance of time-invariant, long-range and short-range forecasting models on the earthquake catalog of Greece*, J. Geophys. Res. *111*, B09304, doi:10.1029/2005JB004113.

Console, R., Murru, M., Catalli, F. and Falcone, G. (2007), *Real-Time forecasts through an earthquake clustering model constrained by the rate-and- state constitutive law: Comparison with a purely stochastic ETAS model*, Seism. Res. Lett. *78*(1), 49–56.

Console, R., Jackson, D.D. and Kagan, Y.Y. (2010), *Using the ETAS model for catalog declustering and seismic background assessment*, Pure Appl. Geophys., this issue.

Daley, D.J. and Vere-Jones, D. An Introduction to the Theory of Point Processes, Springer-Verlag, New York (2003), 2nd ed., vol. 1, 469 pp.

Ekström, G., Dziewonski, A.M., Maternovskaya, N.N. and Nettles, M. (2005), *Global seismicity of 2003, Centroid-moment-tensor solutions for 1087 earthquakes*, Phys. Earth Planet. Inter., *148*(2–4), 327–351.

Field, E.H. (2007), *Overview of the Working Group for the Development of Regional Earthquake Likelihood Models* (RELM), Seism. Res. Lett. *78*(1), 7–16.

Hanks, T.C. (1992), *Small earthquakes, tectonic forces*, Science *256*, 1430–1432.

Hanks, T.C. and Kanamori, H. (1979), *A moment magnitude scale* J. Geophys. Res. *84*, 2348–2350.

Helmstetter, A., Kagan, Y.Y. and Jackson, D.D. (2006), *Comparison of short-term and time-independent earthquake forecast models for southern California*, Bull. Seismol. Soc. Amer. *96*(1), 90-106.

Helmstetter, A., Kagan, Y. Y. and Jackson, D. D. (2007), *High-resolution time-independent grid-based forecast for $M \geq 5$ earthquakes in California*, Seism. Res. Lett. *78*(1), 78–86.

Jackson, D.D. and Kagan, Y.Y. (1999), *Testable earthquake forecasts for 1999*, Seism. Res. Lett. *70*(4), 393–403.

Kagan, Y.Y. (1991), *Likelihood analysis of earthquake catalogs*, Geophys. J. Int. *106*(1), 135–148.

Kagan, Y.Y. (2003), *Accuracy of modern global earthquake catalogs*, Phys. Earth Planet. Inter. (PEPI) *135*(2–3), 173–209.

Kagan, Y.Y. (2007), *Earthquake spatial distribution: The correlation dimension*, Geophys. J. Int. *168*(3), 1175–1194.

Kagan, Y.Y. (2009), *Testing long-term earthquake forecasts: Likelihood methods and error diagrams*, Geophys. J. Int., *177*(2), 532–542, doi:10.1111/j.1365-246X.2008.04064.x.

Kagan, Y.Y., Bird, P. and Jackson, D.D. (2010), *Earthquake patterns in diverse tectonic zones of the Globe*, Pure Appl. Geophys., this issue.

Kagan, Y.Y. and Jackson, D.D. (1994), *Long-term probabilistic forecasting of earthquakes*, J. Geophys. Res. *99*(B7), 13,685–13,700.

Kagan, Y.Y. and Jackson, D.D. (2000), *Probabilistic forecasting of earthquakes*, Geophys. J. Int. *143*(2), 438–453.

Kagan, Y.Y. and Jackson, D.D. (2010), *Short- and long-term earthquake forecasts for California and Nevada* Pure Appl. Geophys., this issue.

Kagan, Y.Y., Jackson, D.D. and Rong, Y.F. (2007), *A testable five-year forecast of moderate and large earthquakes in southern California based on smoothed seismicity*, Seism. Res. Lett. *78*(1), 94–98.

Kagan, Y.Y., Rong, Y.F. and Jackson, D.D. *Probabilistic forecasting of seismicity*, Chapter 5.2 in Earthquake Science and Seismic Risk Reduction, NATO Science Series IV: Earth and Environmental Sciences, vol. 32, eds. F. Mulargia and R.J. Geller, pp. 185–200 (Kluwer, Dordrecht, The Netherlands. 2003).

Kagan, Y.Y. and Knopoff, L. (1977), *Earthquake risk prediction as a stochastic process*, Phys. Earth Planet. Inter., *14*(2), 97–108.

Kagan, Y.Y. and Knopoff, L. (1987), *Statistical short-term earthquake prediction*, Science *236*(4808), 1563–1567.

Lombardi, A.M. and Marzocchi, W. (2007), *Evidence of clustering and nonstationarity in the time distribution of large worldwide earthquakes*, J. Geophys. Res. *112*, B02303, doi:10.1029/2006JB004568.

Marzocchi, W. and Lombardi, A.M. (2008), *A double branching model for earthquake occurrence*, J. Geophys. Res. *113*, B08317, doi:10.1029/2007JB005472.

Molchan, G.M. and Kagan, Y.Y. (1992), *Earthquake prediction and its optimization*, J. Geophys. Res. *97*(B4), 4823–4838.

Ogata, Y. (1988), *Statistical models for earthquake occurrence and residual analysis for point processes*, J. Am. Statist. Assoc. *83*, 9–27.

Ogata, Y. (1998), *Space–time point-process models for earthquake occurrences*, Ann. Inst. Statist. Math. *50*(2), 379–402.

Ogata, Y. and Zhuang, J.C. (2006), *Space–time ETAS models and an improved extension*, Tectonophysics *413*(1–2), 13–23.

Rhoades, D.A. (2007), *Application of the EEPAS Model to Forecasting Earthquakes of Moderate Magnitude in Southern California*, Seism. Res. Lett. *78*(1), 110–115.

Rhoades, D.A. and Evison, F.F. (2005), *Test of the EEPAS forecasting model on the Japan earthquake catalog*, Pure Appl. Geophys. *162*(6–7), 1271–1290.

Rhoades, D.A. and Evison, F.F. (2006), *The EEPAS forecasting model and the probability of moderate-to-large earthquakes in central Japan*, Tectonophysics, *417*(1–2), 119–130.

Schorlemmer, D. and Gerstenberger, M.C. (2007), *RELM testing Center*, Seism. Res. Lett. *78*(1), 30–35.

SCHORLEMMER, D., GERSTENBERGER, M.C., WIEMER, S., JACKSON, D.D. and RHOADES, D.A. (2007), *Earthquake likelihood model testing*, Seism. Res. Lett. *78*(1), 17–29.

STARK, P.B., (1997), *Earthquake prediction: The null hypothesis*, Geophys. J. Int. *131*(3), 495–499.

US GEOLOGICAL STURVEY, Preliminary determinations of epicenters (PDE), (2008), US Dep. of Inter., Natl. Earthquake Inf. Cent., http://neic.usgs.gov/neis/epic/epic.html or http://neic.usgs.gov/neis/epic/code_catalog.html.

(Received August 30, 2008, revised December 14, 2008, accepted March 6, 2009, Published online March 10, 2010)

Pure Appl. Geophys. 167 (2010), 721–741

DOI 10.1007/s00024-010-0075-3

Pure and Applied Geophysics

Earthquake Patterns in Diverse Tectonic Zones of the Globe

Y. Y. KAGAN,[1] P. BIRD,[1] and D. D. JACKSON[1]

Abstract—We extend existing branching models for earthquake occurrences by incorporating potentially important estimates of tectonic deformation and by allowing the parameters in the models to vary across different tectonic regimes. We partition the Earth's surface into five regimes: trenches (including subduction zones and oceanic convergent boundaries and earthquakes in outer rise or overriding plate); fast spreading ridges and oceanic transforms; slow spreading ridges and transforms; active continental zones, and plate interiors (everything not included in the previous categories). Our purpose is to specialize the models to give them the greatest possible predictive power for use in earthquake forecasts. We expected the parameters of the branching models to be significantly different in the various tectonic regimes, because earlier studies (BIRD and KAGAN in Bull Seismol Soc Am 94(6):2380–2399, 2004) found that the magnitude limits and other parameters differed between similar categories. We compiled subsets of the CMT and PDE earthquake catalogs corresponding to each tectonic regime, and optimized the parameters for each, and for the whole Earth, using a maximum likelihood procedure. We also analyzed branching models for California and Nevada using regional catalogs. Our estimates of parameters that can be compared to those of other models were consistent with published results. Examples include the proportion of triggered earthquakes and the exponent describing the temporal decay of triggered earthquakes. We also estimated epicentral location uncertainty and rupture zone size and our results are consistent with independent estimates. Contrary to our expectation, we found no dramatic differences in the branching parameters for the various tectonic regimes. We did find some modest differences between regimes that were robust under changes in earthquake catalog and lower magnitude threshold. Subduction zones have the highest earthquake rates, the largest upper magnitude limit, and the highest proportion of triggered events. Fast spreading ridges have the smallest upper magnitude limit and the lowest proportion of triggered events. The statistical significance of these variations cannot be assessed until methods are developed for estimating confidence limits reliably. Some results apparently depend on arbitrary decisions adopted in the analysis. For example, the proportion of triggered events decreases as the lower magnitude limit is increased, possibly because our procedure for assigning independence probability favors larger earthquakes. In some tests we censored earthquakes occurring near and just after a previous event, to account for the fact that most such earthquakes will be missing from the catalog. Fortunately the branching model parameters were hardly affected, suggesting that the inability to measure immediate aftershocks does not cause a serious estimation bias. We compare our branching model with the ETAS model and discuss the differences in the models parametrization and the results of earthquake catalogs analysis.

Key words: Classification of the Earth's tectonic zones, global and regional earthquake catalogs, magnitude-frequency relation, statistical analysis of earthquake occurrence, branching models of seismicity.

1. *Introduction*

In our previous work, we have already proven some important variations in seismicity determined by tectonic setting. Bird *et al.* (2002) confirmed that mid-ocean spreading ridges have very low corner magnitude ($m_c \approx 5.8$) and low seismic coupling, which declines further with increasing relative plate velocity. Oceanic transform faults also showed an order-of-magnitude decline in coupling with velocity, and a corner magnitude that declines from ~ 7.1 to ~ 6.3 with increasing velocity. BIRD and KAGAN (2004) extended the study to all shallow earthquakes, and showed additional variations in coupling and corner magnitude between subduction, continental collision, continental transforms, and continental rifts. Given this proven variability, it would not be surprising if different tectonic zones had different earthquake-clustering and/or -triggering behaviors. BOETTCHER and JORDAN (2004) and MCGUIRE *et al.* (2005) have already found that earthquakes on oceanic transform faults have more foreshocks and fewer aftershocks. Therefore, our first hypothesis is that interesting

[1] Department of Earth and Space Sciences, University of California, Los Angeles, CA 90095-1567, USA. E-mail: kagan@moho.ess.ucla.edu; pbird@ess.ucla.edu; djackson@ucla.edu

variations will be discovered if the empirical clustering parameters in the earthquake branching model (Kagan, 1991) are redetermined in tectonically-defined zones. (Previous global studies, which did not distinguish between tectonic settings, presumably gave results dominated by the behavior of subduction zones.) However, we do not propose to divide as finely as Bird and Kagan (2004), because their code for classification of earthquakes into subcatalogs by "plate boundary class" is complex and not easy to incorporate into forecasts testable by independent agencies. Instead, we will define geographically-contiguous areas of related tectonic style, and define the union of all areas with the same style as a "tectonic zone" (see Fig. 1, below).

These zones are similar in concept to the Flinn–Engdahl zones (Flinn *et al.*, 1974; Young *et al.*, 1996); but they are much larger, and they will be based on the detailed classification of plate-boundary steps (mean length 43 km) in the PB2002 model of Bird (2003). The spatial contiguity of each area within each zone is an important advantage:

- Epicenter location errors become relatively unimportant.
- No complex algorithm is needed to decide which actual earthquakes belong to a zone.
- Testing can include earthquakes below the Harvard CMT threshold (~5.6) in cases where focal mechanisms are not required.

We start by investigating how different branching models approximate the behavior of global earthquake catalogs, then consider the proposition that different tectonic regimes may have different clustering behaviors. Global catalogs are free of spatial boundary effects and considerably more homogeneous than regional catalogs. Moreover, regional seismicity may be dominated by aftershock sequences of a few strong events, like the *m*7.5 1952 Kern County and the *m*7.3 1992 Landers earthquakes in southern California. Explosions and earthquakes caused by volcanic and geothermal activity are more likely to contaminate earthquake records in local and regional catalogs.

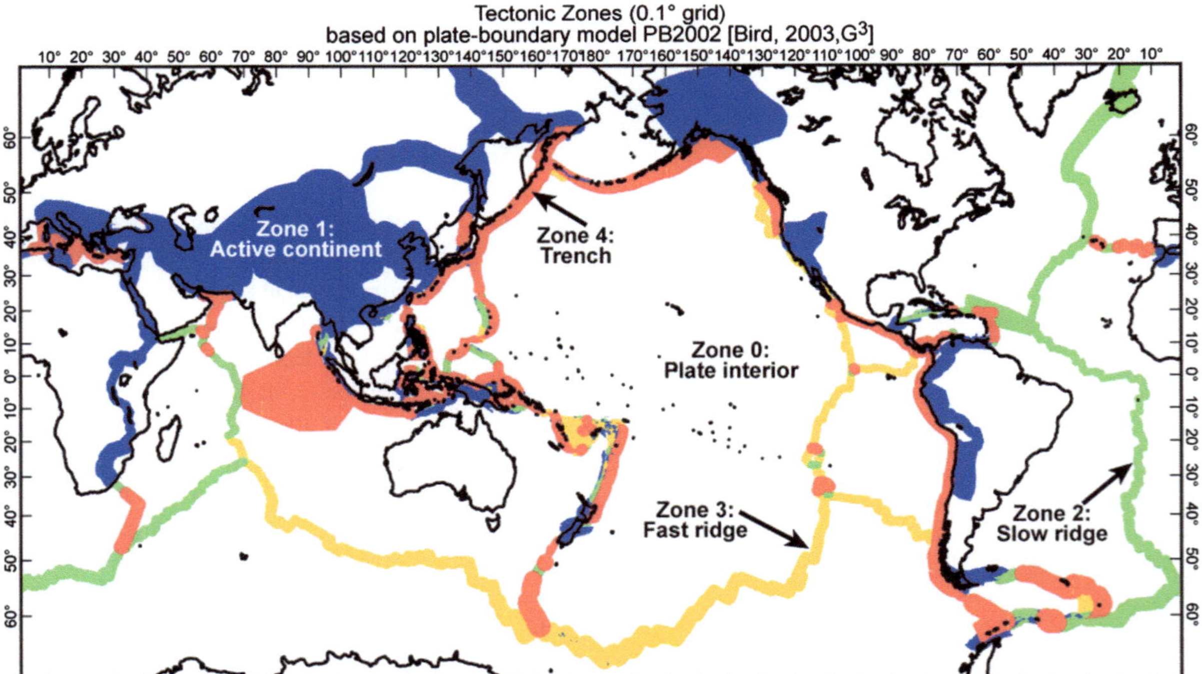

Figure 1
Global tectonic zones

On the other hand, it is important to analyze regional catalogs as well. If we see that model parameter values are similar for worldwide and regional catalogs (KAGAN, 1991), then we may conclude that the models are relatively robust and therefore suitable for earthquake forecasting. It is also important to investigate various parameterizations of the branching models, especially the spatial and temporal fitting of seismicity patterns (KAGAN, 1991) in order to find the best algorithms. KAGAN and JACKSON (2010) used the results of the present study to develop a technique for calculating and evaluating long- and short-term earthquake rate forecasts in practically any seismically active region of the Earth.

Recently many researchers have applied ETAS (Epidemic Type Aftershock Sequence) branching models (OGATA, 1988, 1998; OGATA and ZHUANG, 2006; ZHUANG *et al.*, 2005; OGATA *et al.*, 2003; CONSOLE and MURRU, 2001; CONSOLE *et al.*, 2003) to Japan, California, and Italy. The present ETAS programs do not calculate the distance between earthquake locations based on spherical geometry. It should be noted that even the regional studies were conducted with no attempts to compare the results for different tectonic areas or catalogs. LOMBARDI and MARZOCCHI (2007) and MARZOCCHI and LOMBARDI (2008) have applied the ETAS model to two global catalogs of large earthquakes.

This work is dedicated to the late Frank Evison, a superb exemplar of careful attention to detail and a passionate advocate of rigorous testing in earthquake prediction research.

2. *Defining Tectonic Zones*

2.1. *Objective*

Our goal is to divide the Earth surface into a modest number of zones of different tectonic style, defined by objective criteria, which are known from previous research (e.g., BIRD and KAGAN, 2004) to contain interesting variations in seismicity parameters such as seismic coupling and/or corner magnitude. It is reasonable to anticipate that these zones might have different branching behaviors, and our present definition of zones is designed with consideration for practical matters that will permit such testing to be conducted relatively easily. Zones are defined here as surface areas into which epicenters (and/or epicentroids) of shallow earthquakes may fall. Precise depth of the earthquake will not be a criterion because it is usually poorly known (within the 0–70-km-depth range) unless there is local station control. Focal mechanism will not be a criterion because this is not always available for the smaller earthquakes ($m < 5.6$) which make up large portions of the aftershock swarms we wish to include in this study. The number of zones we have defined is small (5) so that sufficient earthquakes fall into each zone within a few decades, allowing for reliable optimizations of branching models. Therefore, no zone has a local geographic name like "Aleutian;" instead, each zone has a generic name like "Trench." We allow one tectonic zone to be the union of many non-contiguous patches. The preceding choices make it impossible to separate strike-slip from normal-faulting earthquakes on mid-ocean spreading ridges, or in continental rift zones. Also, along many trenches it will be impossible to separate subduction-related earthquakes from back-arc-spreading earthquakes. Therefore, the tectonic zones are not equivalent to the seven plate-boundary classes of BIRD (2003), and they have been given new, distinct names that will not be confused with plate-boundary classes.

We propose the following short list of tectonic zones (with identifying integers for compact representation in computer files):

- (4) Trench (including incipient subduction, and earthquakes in outer rise or upper plate);
- (3) Fast-spreading ridges (oceanic crust, spreading rate $\geq$40 mm/a; includes transforms);
- (2) Slow-spreading ridges (oceanic crust, spreading rate <40 mm/a, includes transforms);
- (1) Active continent (including continental parts of all orogens of PB2002, plus continental plate boundaries of PB2002); and
- (0) Plate-interior (the rest of the Earth's surface).

Fast-spreading versus slow-spreading ridges are divided at 40 mm/a because (a) BIRD and KAGAN (2004) found a change in OTF (Oceanic Transform Fault) earthquakes at $\sim$39 mm/a, and (b) there is a change in ridge morphology around this rate, from central-graben to peaked (MACDONALD, 1986). "Active continent" is not subdivided by style of faulting, as this would require focal mechanisms (not always available) and obscure the basic definition that a zone is the union of a set of predefined geographic regions.

2.2. General Method

Our tectonic zones are defined by objective rules, implemented in a computer program, because this is more reproducible, easier to explain, easier to revise, and less subject to procedural errors. Our program assigns tectonic zone integers to grid points rather than drawing curves along boundaries of tectonic zones. It loops through latitudes (in 0.1° steps) and also loops through longitudes (in 0.1° steps), and creates a grid of integers to identify the tectonic zone at each point.

The output is a relatively compact representation (1801 latitudes $\times$ 3601 longitudes $\times$ 1 to 2 bytes $=$ 6 to 12 Mb). The necessary data sets are available in digital form: elevation from ETOPO5 (ANONYMOUS, 1988), age of seafloor from MUELLER *et al.* (1997); plate boundaries and Euler poles from BIRD (2003). In the Appendix we describe details of zone definition implementation. The result of the global tectonic zones classification is shown in Fig. 1 and is available at http://bemlar.ism.ac.jp/wiki/index.php/Bird%27s_Zones.

3. Earthquake Catalogs

We applied the likelihood technique to several available global and regional earthquake catalogs.

We studied the earthquake distributions and clustering for the global CMT catalog of moment tensor inversions compiled by EKSTRÖM *et al.* (2005). The present catalog contains more than 28,000 earthquake entries for the period 1977/1/1 to 2007/12/31, earthquake size is characterized by a scalar seismic moment M.

The PDE worldwide catalog is published by the USGS (U.S. GEOLOGICAL SURVEY, 2008); the catalog available at the time this article was written ended on January 1, 2008. The catalog measures earthquake size, using several magnitude scales, and provides the body-wave (m_b) and surface-wave (M_S) magnitudes for most moderate and large events since 1965 and 1968, respectively. Recently, the moment magnitude (m_w) estimate has been added.

Determining one measure of earthquake size for the PDE catalog entails a certain difficulty: for example, KAGAN (1991) calculates a weighted average of several magnitudes to use in the likelihood search. KAGAN (2003) also analyzes systematic and random errors for various magnitudes in the PDE catalog. At various times many magnitudes have been listed in the PDE catalog, and their relationships are difficult to establish. Therefore, in this work we chose a palliative solution: to use the maximum magnitude among those magnitudes shown for each earthquake. This solution is easier to carry out and the results are easily reproducible. For moderate earthquakes usually m_b or M_S magnitude is selected, for larger recent earthquakes the maximum magnitude is most likely m_w.

The ANSS (ADVANCED NATIONAL SEISMIC SYSTEM, 2008) composite catalog is a world-wide earthquake catalog that is created by merging the master earthquake catalogs from contributing ANSS institutions and then removing duplicate solutions for the same event.

The CalTech (CIT) data set (HILEMAN *et al.*, 1973; HUTTON and JONES, 1993) was the first instrumental local catalog to include small earthquakes ($m \geq 3$), beginning in 1932. In recent years even smaller earthquakes have been included in the catalog.

4. Corner Magnitudes in the Tectonic Zones

Corner magnitude (m_c) is, by definition, associated with corner moment (M_c). We use the moment-magnitude conversion formula

$$m_w = \frac{2}{3}(\log_{10} M - C), \qquad (1)$$

(HANKS and KANAMORI, 1979; HANKS, 1992), where M is measured in Newton m (Nm), however the calculations in our programs have been made with the seismic moment tensor which is the appropriate measure of earthquake size. In order to keep the available computer programs unchanged, in this section we take C = 9.05 (BIRD and KAGAN, 2004, Eq. 1), but in Sects. 5 and 6 C is taken to be 9.0.

Table 1

Parameters of the tapered Gutenberg–Richter frequency/magnitude relations of the tectonic zones, with 95%-confidence ranges, $M \geq 10^{17.45} = 2.818 \times 10^{17}$ Nm *(C is taken 9.05 in Eq. 1).*

	Tectonic zone	N	N/N_{All}	β	m_c
4.	Trenches	4,234	0.648	$0.639^{+0.024}_{-0.024}$	$8.75^{+?}_{-0.35}$
1.	Active continent	862	0.132	$0.647^{+0.060}_{-0.055}$	$7.59^{+0.72}_{-0.25}$
0.	Plate-interior	251	0.038	$0.639^{+0.105}_{-0.101}$	$8.15^{+?}_{-0.53}$
2.	Slow-spreading ridges	490	0.075	$0.812^{+0.098}_{-0.094}$	$7.38^{+?}_{-0.33}$
3.	Fast-spreading ridges	701	0.107	$0.767^{+0.089}_{-0.095}$	$6.79^{+0.30}_{-0.18}$
	Global total	6,538	1.0	$0.671^{+0.021}_{-0.021}$	$8.73^{+?}_{-0.34}$

Table 2

Parameters for various subdivisions of CMT catalog, 1982/01/01–2007/03/31, $m_w \geq 5.55$ *or* $M \geq 10^{17.325} = 2.113 \times 10^{17}$ Nm *(C is taken 9.0 in Eq. 1).*

Zones#		All	Trench (4)	Act-Con. (1)	Inter. (0)	Slow (2)	Fast (3)	All/d	Trench/d (4)
1	2	3	4	5	6	7	8	9	10
1.	N	7,720	5,022	1,004	279	584	831	7,471	4,845
2.	N/N_{All}	1.0	0.651	0.130	0.036	0.076	0.108	1.0	0.649
3.	m_{max}	9.07	9.07	8.28	8.15	7.67	7.15	9.07	9.07
4.	I/N	1.03	1.18	1.11	0.59	0.30	0.44	0.857	0.997
5.	$\lambda T/N$	0.745	0.690	0.819	0.866	0.935	0.941	0.758	0.707
6.	μ	0.131	0.169	0.093	0.099	0.042	0.035	0.131	0.168
7.	b	1.02	0.98	0.96	1.05	1.28	1.23	1.01	0.98
8.	a	0.64	0.59	0.56	0.45	0.52	0.59	0.62	0.57
9.	θ	0.12	0.11	0.27	*0.1	0.27	*1.0	0.1*	0.1*
10.	s_r	0.30	0.29	0.15*	0.47	0.17	0.27	0.28	0.29
11.	ϵ_p	21.8	22.1	21.6	18.0	19.5	17.7	21.9	22.1
12.	ϵ_h	3.5	4.4	3.0*	5.1	3.0*	3.0*	3.4	4.2

This Table is similar in format to Table 3 in KAGAN (1991). Briefly, N is the earthquake number, m_{max} maximum magnitude in a subcatalog, I/N information score per event in bits/eq (1 bit means that the statistical uncertainty can be reduced on average by a factor of 2), $\lambda T/N$ ratio of spontaneous events to total, μ branching ratio (more branching—more triggered events), b b-value, a parent productivity exponent, $\delta = a/1.5$ (see Eq. 9), θ temporal exponent, see Eq. 6 ($1 + \theta$ is similar to "p" in Omori's law), s_r focal size for $m4$ earthquake in km (12), ϵ_p horizontal error in km (12), ϵ_h vertical error in km. The values preceded by an asterisk (*) mean that the parameters could not be evaluated and are assigned the upper limit, the values followed by (*) are assigned the lower limit. */d means that immediate aftershocks have been removed from a catalog (as in the last two columns).

These changes in C and the moment threshold values explain the differences in the earthquake numbers for the CMT catalog shown in Tables 1 and 2 below.

Corner moment and asymptotic spectral slope (β) are the two parameters of the tapered Gutenberg–Richter model for the cumulative frequency/moment relation:

$$G(M, M_t, \beta, M_c) = (M_t/M)^{\beta} \exp[(M_t - M)/M_c] \quad \text{for} \quad M_t \leq M \leq \infty, \qquad (2)$$

where G is the fraction of earthquakes (by event count) in the catalog with moment exceeding M, and M_t is the threshold moment for completeness of the catalog (JACKSON and KAGAN, 1999; KAGAN and JACKSON, 2000; BIRD and KAGAN, 2004). The corner moment (M_c) can be thought of as the earthquake size which is rarely exceeded (e.g., about once per century for subduction earthquakes), and in maximum-likelihood fitting of (2) to actual large (sub)catalogs it is typically not very different from the second-largest magnitude. Although the form of (2) is chosen for simplicity and acceptable fit, a manner of upper magnitude limit is required to keep seismic moment rates finite. BIRD and KAGAN (2004) reported corner magnitudes for seven types of plate boundaries, and these ranged from a low of $5.86^{+0.19}_{-0.16}$ for normal-faulting earthquakes on Ocean Spreading Ridges (OSR) to a high of $9.58^{+0.48}_{-0.46}$ for shallow events in Subduction zones (SUB). (Ranges are 95%-confidence.) The spread of corner magnitudes among tectonic zones is less, because each tectonic zone (except zone 0) merges the seismicity of different plate-boundary types as defined by BIRD and KAGAN (2004). At the low end, the corner magnitude for zones 2 and 3 (slow- and fast-spreading ridges) is higher than that of OSR/normal because of the inclusion of OTF seismicity. At the high end, the corner magnitude of zone 4 (Trenches) is less than that of SUB because of dilution by Oceanic Convergent Boundary (OCB) and oceanic convergent orogen seismicity, and that of some overlying backarc OSRs.

To determine the corner magnitudes of the tectonic zones, we divided the shallow ($\leq$70-km-depth) events with $m_w \geq 5.6$ (scalar seismic moment $M = 10^{17.45}$ Nm or $M = 2.818 \times 10^{17}$ Nm, see Eq. 1) from the Global CMT catalog (1982/01/01-2008/03/31) into 5 zone subcatalogs using the gridded zone assignments of Fig. 1. In Fig. 2 we compare the tapered Gutenberg–Richter model (2) to actual cumulative distribution for two tectonic zones: 'Active continent' (1) and 'Slow-spreading ridges' (2).

Then each subcatalog was analyzed with program BetaCorner.f90 (published by BIRD and KAGAN, 2004) which contours the likelihood surface in 2-D (β, M_c) space to determine both maximum-likelihood estimates and 95%-confidence ranges. In Fig. 3 we display two maps of the likelihood surface for the same tectonic zones as in Fig. 2. For the 'Active continent' the corner moment (M_c) estimate is well constrained, whereas for the 'Slow-spreading ridges' the upper limit is ∞. The results for all zones are shown in Table 1 (as we mentioned above, see Eq. 1, both the moment threshold and the corner magnitude have been calculated with slightly different coefficients from Table 2).

5. Branching Model of Earthquake Occurrence

Our technique for producing short-term hazard estimates is to establish a statistical model to fit the catalogue of earthquake times, locations, and seismic moments, and subsequently to base forecasts on this model. While most of the components of the model have been tested (KAGAN and KNOPOFF, 1987; KAGAN, 1991; JACKSON and KAGAN, 1999; KAGAN and JACKSON, 2000), some require further exploration and may be modified as our research progresses.

The assumptions we make in constructing our initial model have been summarized in KAGAN and KNOPOFF (1987) and KAGAN (1991). A similar model called the ETAS model was proposed by OGATA (1988, 1998), as well as by OGATA and ZHUANG (2006). In both of these models seismicity is approximated by a Poisson cluster process, in which clusters or sequences of earthquakes are statistically independent, although individual earthquakes in the cluster are triggered events. The clusters are assumed to form a Poisson point process with a constant temporal rate. We assumed that the interrelationships between events within a cluster are closely approximated by a stochastic space-time critical branching process. Under this assumption there is a sole trigger for any given dependent event. As shown below, the space-time distribution of interrelated earthquake sources within a sequence is controlled by simple relations justified by analyzing the available statistical data on seismicity.

Below we describe the statistical distributions for the analysis of the CMT earthquake catalog, with the scalar seismic moment M as the measure of earthquake size. The model can be easily reformulated for the PDE and other catalogs (see Sect. 3), where earthquakes are characterized by a magnitude.

We construct an earthquake intensity function, $\Lambda(t, \mathbf{x}, M)$ which indicates the earthquake rate at time t, at location $\mathbf{x}$, with the scalar seismic moment M, given the history of previous seismicity. (For small time/space intervals the probability of occurrence and the occurrence rate are equivalent.) This function is given

$$\Lambda(t, \mathbf{x}, M) = \lambda\phi(\mathbf{x}, M) + \sum_i \psi^{(M_i)}(t - u_i, \mathbf{x} - \mathbf{y}_i, M), \tag{3}$$

where λ (or λ_t) is the rate per time unit of the Poisson occurrence of independent (spontaneous) earthquakes with $M \geq M_t$ in the volume ($\mathbf{Y}$); $\phi(\mathbf{x}, M)$ is their space-seismic moment distribution; $\psi^{(M)}(t - u, \mathbf{x} - \mathbf{y}, M)$

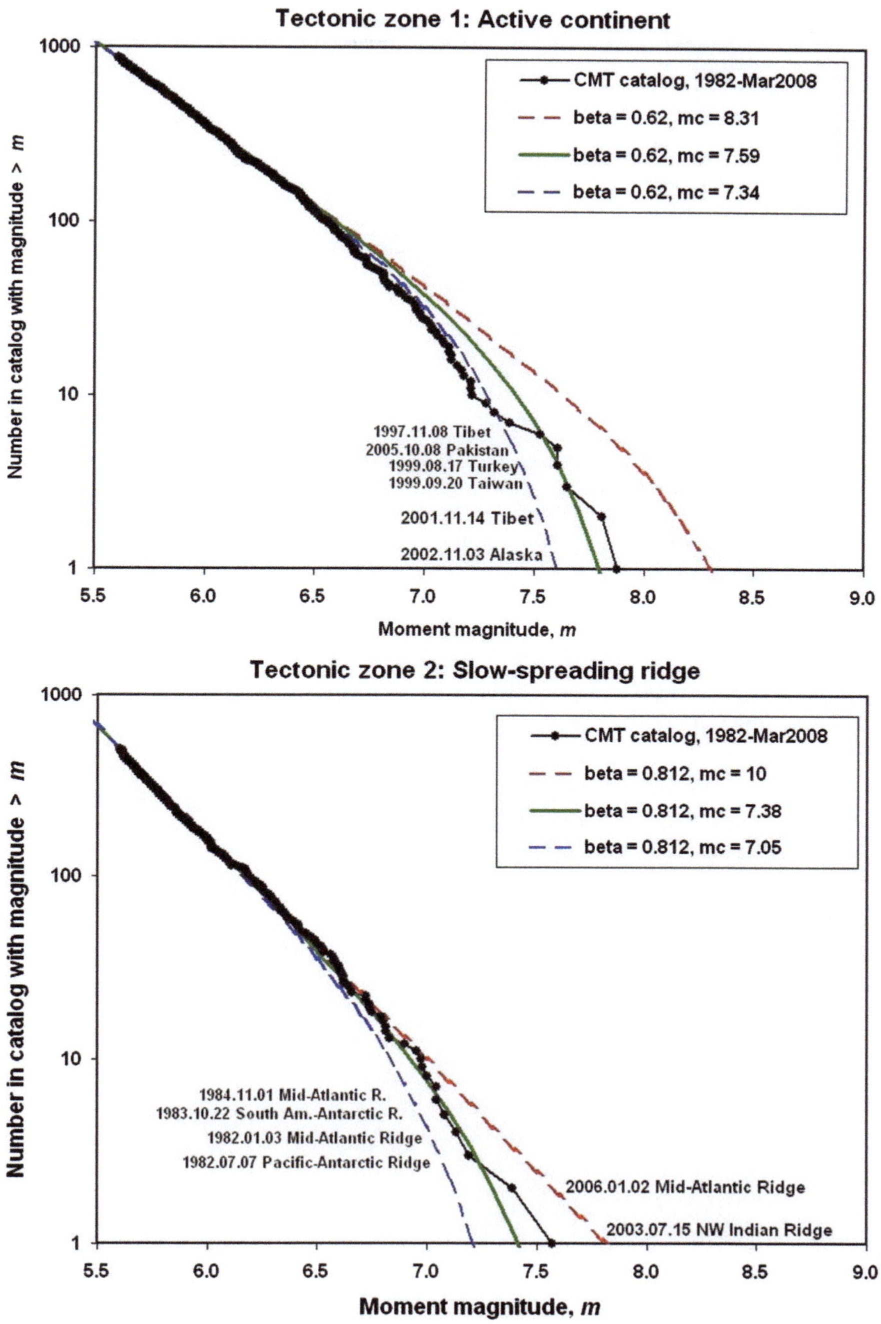

Figure 2
Comparison of tapered Gutenberg–Richter model for the frequency/magnitude distribution to actual cumulative distribution. Earthquakes with $m > 7.5$ are identified. In each frequency/magnitude plot, three model-tapered Gutenberg–Richter distribution curves are shown. Each has the optimal spectral slope β from Fig. 3 below. The three variant curves show the minimum, best-estimate, and maximum corner magnitude (except that a value of m_c = 10 is substituted for an unbounded corner magnitude in the lowest plot). **a** Upper plot, tectonic zone 1: Active continent. **b** Lower plot, tectonic zone 2: Slow-spreading ridges

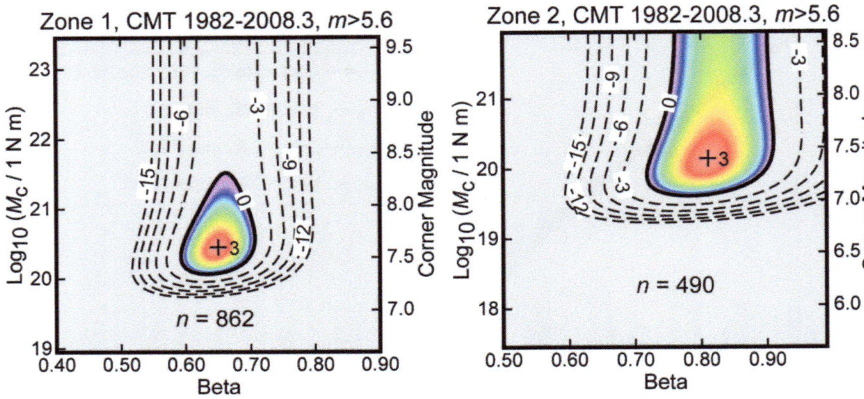

Figure 3

Likelihood surface for the tapered Gutenberg–Richter model in the 2-D parameter space of asymptotic spectral slope β (abscissa) and corner moment M_c or corner magnitude m_c (left and right ordinates). Color scale (*red* high, *blue* low) and cross for optimal value are used inside the contour which is three natural-log-units below the peak; this area corresponds to the 95%-confidence range. *Dashed contours* with interval of three natural-log-units are used outside. **a** Left, tectonic zone 1: Active continent. **b** Right, tectonic zone 2: Slow-spreading ridges. Note that the 95%-confidence area is just barely closed on the high-magnitude side for the "Active continent"; ranges for the "Slow-spreading ridges" and other zones are open on the high side

is a conditional distribution of succeeding events at time t and coordinates $\mathbf{x}$, if preceding earthquakes have occurred at times u_i in the places with coordinates $\mathbf{y}_i$.

We subdivide the spatial coordinates $\mathbf{x}$ into $\mathbf{s} \times z$, where $\mathbf{s}$ are surface coordinates, and z is depth. If the duration of the catalog is T, then λT is the number of independent events, and $\lambda\, T/N$ is the fraction of independent events (N is the total number of earthquakes in a catalog).

5.1. Earthquake Clusters—Independent Events

The first event in a sequence is usually the largest one and is called a main shock. Other dependent events are called aftershocks, though some of them are actually aftershocks of aftershocks. If the first event in a sequence is smaller than subsequent shocks, it is called a foreshock. Retrospectively, it is relatively easy to subdivide an earthquake catalogue into fore-, main-, and aftershocks. However, in real-time forecasting it is uncertain whether the most recent event registered by a seismographic network is a foreshock or a main shock. Although the subsequent triggered events are likely to be smaller and would thus be called aftershocks, there is a significant chance that some of the subsequent earthquakes may be bigger (Kagan, 1991).

5.2. Dependent Events

Similar to the estimation of the long-term seismic hazard, we assume that the distribution of triggered events within a cluster may be broken down into a product of its marginal distributions, i.e., the conditional rate density of the jth shock dependent on the ith shock ($j > i$) with seismic moment M_i is modelled as

$$\psi(\Delta t, \rho, M_j|M_i) = \psi_{\Delta t}(\Delta t) \times \psi_\rho(\rho) \times \psi_M(M_i) \times \phi_M(M_j), \qquad (4)$$

where $\Delta t = t_j - t_i$ and ρ is the horizontal distance between the ith and the jth centroids (or more correctly epicentroids) ($\rho = |\mathbf{x}_j - \mathbf{x}_i|$). $\psi_{\Delta t}$, ψ_ρ, and ψ_M are the marginal temporal, spatial, and moment densities, and are detailed below. The total time-dependent rate density is a sum of effects from all previous earthquakes,

$$\Psi(t_j, \mathbf{x}_j, M_j) = \sum_{i<j} \psi(\Delta t, \rho, M_j|M_i). \qquad (5)$$

The function ψ decays rapidly with time and distance (see below). Thus only neighbouring events substantially contribute to the sum (5), although the range of strong earthquakes is considerably longer than that of weak events.

The first three densities $\psi(x)$ in the right part of (4) depend on M_i. We take a power-law relation for the probability density of time intervals between earthquakes within a cluster:

$$\psi_{\Delta t}(\Delta t) = \theta t_M^{\theta}(\Delta t)^{-1-\theta}, \quad \Delta t \geq t_M, \qquad (6)$$

which is similar to Omori's law. The parameter θ is an 'earthquake memory' factor, and t_M is the coda duration time of an earthquake with seismic moment M_i. We assume for the coda duration

$$t_M = t_r M_i^{1/3}, \qquad (7)$$

where t_r is the coda duration time, taken here as $t_r = 0.0035$ days (about 5 min), of an earthquake with the reference seismic moment $M_r = 10^{15}$ Nm, corresponding to $m_r = 4.0$ (KAGAN, 1991). The probability of the next dependent shock to occur in the time interval (t_1, t_2), for $0 < t_1 < t_2$, given an event of a cluster occurring at time 0, can then be calculated simply as

$$\mathrm{Prob}(t_1 < t < t_2) = t_M^{\theta}(t_1^{-\theta} - t_2^{-\theta}). \qquad (8)$$

The non-normalized function $\psi_M(M_i)$ which corresponds to the number of triggered shocks generated on the average by an earthquake with seismic moment M_i, is assumed to obey (KAGAN, 1991)

$$\psi_M(M_i) = \mu\left(\frac{M_i}{M_t}\right)^{\delta}. \qquad (9)$$

To make the distribution parameter δ of triggered events comparable to the b values, in our Tables we use a parameter

$$a = \delta \times 1.5. \qquad (10)$$

KAGAN (1991, p. 142) discusses the δ-parameter and its relation to other measurements of size distribution of triggered events in more detail.

We approximate by a Rayleigh distribution the probability density of the horizontal distance (ρ) between two earthquake epicentroids in a cluster:

$$\psi_{\rho}(\rho) = \frac{\rho}{\sigma_{\rho}^2} \times \exp\left[-\rho^2/(2\sigma_{\rho}^2)\right] \qquad (11)$$

where σ_{ρ} is a spatial standard deviation. The standard deviation is assumed to depend on the standard errors of epicentroid determination and on the seismic moment M_i of the main event according to the relation (KAGAN,1991):

$$\sigma_{\rho}^2 = \epsilon_{\rho}^2 + s_r^2(M_i/M_r)^{2/3}. \qquad (12)$$

Here ϵ_p is the standard error in epicentroid determination and s_r is a characteristic size of a focal zone of an earthquake with the reference seismic moment M_r (7). Equations 11 and 12 roughly correspond to spatial response functions used in the ETAS model (OGATA, 1998, Eqs. 2.2–2.4).

6. *Likelihood Analysis of Catalogs*

6.1. *Statistical Analysis Results*

Using the branching model of Sect. 5 we can obtain a likelihood function for the earthquake process. The likelihood equations are rather bulky (see KAGAN, 1991). We compare the likelihood function for the branching process with a process based on the spatial inhomogeneous Poisson model of seismicity. By searching for the maximum of the likelihood ratio of two models or of the log-likelihood difference, we obtain the statistical estimates of the model parameters.

Two methods can be used to search for the likelihood function maximum; one is the Newton–Raphson optimization technique (needs derivatives of the likelihood function), and the simplex algorithm (PRESS *et al.*, 1992, p. 402) (no derivatives are needed). The simplex method was applied in HELMSTETTER *et al.* (2006). We used a modification of the Newton–Raphson technique in most of our studies (KAGAN, 1991; KAGAN and JACKSON, 2000) and in the present work. We could not apply available standard packages of the Newton–Raphson technique because many of the parameters of our model have either physical constraints (for example, non-negative) or constraints based on prior geophysical results. We also censor the immediate aftershocks from the catalogs (see Eq. 7). This censoring introduces a Heaviside function in the time history of earthquake occurrence, which significantly complicates the calculation of the derivatives. This also explains why we did not apply a triangular window to remove aftershocks in our computations (KAGAN, 2004; HELMSTETTER *et al.*, 2006). It would be even more difficult to obtain the likelihood function derivatives in such a case. (The above formula (7) defines a rectangular-in-time window for the aftershock removal.) In our calculations below we applied the same likelihood-optimization program to two catalog types: one in its original form and the other with all of the early aftershocks which are closer to the previous shock than t_M (see Eq. 7) being deleted.

We started our search with the initial model of spatial inhomogeneous, temporally constant Poisson process, and iterated the computations until the changes in the log-likelihood difference were smaller than 10^{-5}. For about 90,000 events (see Table 5) one iteration took less than 30 min, using a FORTRAN program on an Alpha HP 600 Mhz workstation; the total iteration number was usually from 90 to 120.

In Tables 2, 3, 4, 5, 6 and 7 we display some of the likelihood search results. The additional results are shown in our Wiki page http://bemlar.ism.ac.jp/wiki/index.php/Bird%27s_Zones. The Tables show the branching model parameter values for four catalogs, with varying magnitude thresholds, and for different tectonic zones. The last Table 7 analyzes only one tectonic zone (Active continent).

Figure 1 shows that the tectonic zones described in Sect. 2 contain many small regions spaced out at large distances throughout the zone. Most of the parameters which depend on the earthquake location separation (like λ, μ, θ) can be expected to be significantly different for these zones compared to the full catalog ('All' column in the Tables). However, such a variation is not observed, most likely because these parameters are defined by a close distance interaction between the earthquakes.

Earthquake number ratios (N/N_{All}) in the tectonic zones for the CMT and the PDE catalogs in Tables 2, 3, 4, 5 and 6 show some differences, that are especially large for the "Fast spreading ridges"—10.8% for the CMT catalog versus 5.4% for $m_t = 4.7$ PDE catalog. Several factors may influence this variation: the change in the β and the corner moment values (Table 1), various systematic effects in the determination of m_b magnitude in the PDE catalog, as well as others.

The b values in Tables 2, 3, 4, 5, 6 and 7 were computed using the Aki/Utsu formula (Utsu, 1999). These values for all the subdivisions of the catalogs are close to 1.0, the β value can be converted to b (see Table 1) as $b = 1.5 \times \beta$. The only significant deviation from this b universality is the values for oceanic ridges. The increased b value for these earthquakes is caused by a relatively low value of their corner magnitude (Table 1) which biases the estimates (cf. Fig. 2) toward higher slope values.

The parameters s_r and ϵ_p (see Eq. 12) characterize the spread of dependent earthquakes (usually aftershocks) in a neighborhood of a 'parent' event. The estimates of the horizontal location accuracy (ϵ_p) are in good agreement with the independent evaluations of these uncertainties (Kagan, 2003): for the CMT catalog the accuracy is considered to be of the order 20 km and for the PDE catalog on average it is close to 10 km. The location uncertainties for the PDE catalog depend on the magnitude threshold, m_t. As one should expect for a larger magnitude ($m_t = 5.3$ in Table 3) the accuracy is higher (from 8 to 12 km) than for the lower magnitude ($m_t = 4.7$ in Table 5) where ϵ_p varies from 12 to 16 km. The lowest ϵ_p values in the PDE catalog are for zone 1 (Active continent), because most of seismographic stations are located on continents.

The vertical error ϵ_h in Tables 2, 3, 4, 5, 6 and 7 is not informative, since many shallow events are assigned 10 or 15 km depth in the CMT catalog and 33 km in the PDE catalog. We kept this parameter in our likelihood procedure since it is useful for the analysis of deep and intermediate

Table 3

Parameters for various subdivisions of PDE catalog, 1968–2007/01/01, $m \geq 5.3$.

		All	Trench (4)	Act-Con. (1)	Inter. (0)	Slow (2)	Fast (3)
1	2	3	4	5	6	7	8
1.	N	22,515	14,827	3,583	960	1,513	1,632
2.	N/N_{All}	1.0	0.659	0.159	0.043	0.067	0.072
3.	$m_{\max}$	8.80	8.80	8.50	8.45	7.60	7.30
4.	I/N	1.71	1.83	2.01	1.08	0.80	0.71
5.	$\lambda T/N$	0.733	0.714	0.720	0.778	0.890	0.917
6.	μ	0.102	0.102	0.127	0.172	0.102	0.067
7.	b	1.07	1.07	1.00	1.17	1.25	1.10
8.	a	0.69	0.71	0.59	0.42	0.10	0.24
9.	θ	0.39	0.47	0.27	0.1*	0.54	*1.0
10.	s_r	0.37	0.36	0.30	0.26	0.15*	0.26
11.	ϵ_p	8.9	8.9	8.1	9.2	11.2	12.0
12.	ϵ_h	4.6	5.8	4.4	3.0*	3.0*	3.0*

This Table is similar in format to Table 2 above.

Table 4

Parameters for various subdivisions of PDE catalog, 1968–2007/01/01, m ≥ 5.0.

		All	Trench (4)	Act-Con. (1)	Inter. (0)	Slow (2)	Fast (3)
1	2	3	4	5	6	7	8
1.	N	45,942	29,980	7,686	2,191	3,296	2,789
2.	N/N_{All}	1.0	0.653	0.167	0.048	0.072	0.061
3.	m_{max}	8.80	8.80	8.50	8.45	7.60	7.30
4.	I/N	1.90	1.99	2.16	1.40	1.28	0.98
5.	$\lambda T/N$	0.680	0.661	0.687	0.716	0.816	0.869
6.	μ	0.141	0.140	0.133	0.234	0.182	0.109
7.	b	1.05	1.04	1.05	1.17	1.17	0.93
8.	a	0.63	0.65	0.64	0.35	0.02	0.18
9.	θ	0.28	0.33	0.23	0.12	0.40	0.55
10.	s_r	0.38	0.37	0.31	0.24	0.15*	0.20
11.	ϵ_p	9.5	9.7	7.9	9.8	10.1	12.6
12.	ϵ_h	3.0*	4.5	4.3	3.0*	3.0*	3.0*

This Table is similar in format to Table 2 above.

Table 5

Parameters for various subdivisions of PDE catalog, 1968–2007/01/01, m ≥ 4.7.

		All	Trench (4)	Act-Con. (1)	Inter. (0)	Slow (2)	Fast (3)
1	2	3	4	5	6	7	8
1.	N	88,920	56,656	16,235	4,492	6,730	4,807
2.	N/N_{All}	1.0	0.637	0.183	0.051	0.076	0.054
3.	m_{max}	8.80	8.80	8.50	8.45	7.60	7.30
4.	I/N	2.01	2.052	2.159	1.526	1.786	1.374
5.	$\lambda T/N$	0.637	0.614	0.660	0.679	0.753	0.813
6.	μ	0.177	0.184	0.152	0.240	0.225	0.158
7.	b	1.00	0.97	1.06	1.09	1.10	0.87
8.	a	0.58	0.57	0.64	0.41	0.12	0.15
9.	θ	0.22	0.23	0.18	0.11	0.40	0.45
10.	s_r	0.36	0.40	0.34	0.21	0.15*	0.15*
11.	ϵ_p	14.0	14.3	11.9	13.5	14.0	16.2
12.	ϵ_h	3.0*	3.0*	3.0*	3.0*	3.0*	3.0*

This Table is similar in format to Table 2 above.

earthquakes, where accuracy of the depth location is higher. This parameter use is also appropriate for those local catalogs which have a low vertical error.

The s_r estimate for the regional catalogs (CIT and ANSS) had to be constrained (Table 7). Most likely it is because the location accuracy of these catalogs significantly varies during the long period of observations, starting at 1932. Table 7 displays four sets of inversions in which the parameter s_r was constrained at two different values: 0.3 and 0.15 km. This was done to see whether such a constraint causes significant modifications in the estimates of other parameters. The changes appear to be minor.

Table 6

Parameters for various subdivisions of PDE catalog, 1968–2007/01/01, close aftershock removed, m ≥ 5.0.

		All	Trench (4)	Act-Con. (1)	Inter. (0)	Slow (2)	Fast (3)
1	2	3	4	5	6	7	8
1.	N	42,925	27,648	7,205	2,127	3,217	2,728
2.	N/N_{All}	1.0	0.644	0.168	0.050	0.075	0.064
3.	m_{max}	8.80	8.80	8.50	8.45	7.60	7.30
4.	I/N	1.25	1.24	1.50	1.17	1.04	0.76
5.	$\lambda T/N$	0.695	0.674	0.709	0.736	0.831	0.881
6.	μ	0.171	0.177	0.146	0.241	0.185	0.121
7.	b	1.04	1.03	1.04	1.17	1.17	0.93
8.	a	0.57	0.56	0.61	0.27	0.0*	0.0*
9.	θ	0.11	0.13	0.10	0.1*	0.32	0.39
10.	s_r	0.37	0.35	0.32	0.15*	0.15*	0.15*
11.	ϵ_p	9.7	9.8	7.8	9.7	10.0	12.5
12.	ϵ_h	3.0*	4.8	4.7	3.0*	3.0*	3.0*

This Table is similar in format to Table 2 above.

Table 7

Parameters for California/Nevada catalogs.

		CIT				ANSS		
		All	All/d	All	All/d	All	All	All/d
	m_t	3.0	3.0	3.0	3.0	4.7	4.0	4.0
1	2	3	4	5	6	7	8	9
1.	N	15,512	13,579	15,512	13,579	774	4,495	3,928
2	m_{max}	7.5	7.5	7.5	7.5	7.5	7.5	7.5
3.	I/N	3.102	2.043	3.105	2.053	1.29	3.263	1.776
4.	$\lambda T/N$	0.475	0.539	0.474	0.539	0.666	0.531	0.586
5.	μ	0.220	0.208	0.213	0.199	0.114	0.172	0.166
6.	b	0.94	0.98	0.94	0.98	0.95	0.95	0.95
7.	a	0.60	0.58	0.61	0.60	0.69	0.66	0.64
8.	θ	0.11	0.10*	0.11	0.10*	0.25	0.22	0.10*
9.	s_r	0.30*	0.30*	0.15*	0.15*	0.30*	0.15*	0.15*
10.	ϵ_p	3.6	3.6	3.9	3.7	7.5	6.1	6.3
11.	ϵ_h	3.0*	3.0*	3.0*	3.0*	3.0*	3.0*	3.0*

This Table is similar in format to Table 2 above.

CIT catalog time limits: 1932–2001/12/31. CIT catalog window: Latitude 32.5°–36.0° N, Longitude 115.0°–121.0° W.

ANSS catalog time limits: 1932–2008/05/31. ANSS catalog window: Latitude 31.5°–43.0° N, Longitude 113.0°–125.0° W.

The global catalogs are considerably more recent and the variations of their location accuracy are not that extensive (Kagan, 2003). The standard deviation s_r values for all the global catalogs (CMT and PDE) are close to 0.30–0.35 km. This would correspond to the size of the focal zone of $m4$ earthquake close to 1 km, the value coinciding with most estimates by other geophysical measurements (ABERCROMBIE, 1995; KAGAN, 2002). The confirmation of the parameters s_r and ϵ_p values, discussed above, makes our results on the statistical analysis of seismicity patterns more credible.

The comparison of the results for the original catalogs with those obtained for the catalogs with immediate aftershocks removed (see Eq. 7 and Tables 2, 3, 4, 5, 6 and 7 where we mark these catalogs by '*/d' note) shows that the values of most of the parameters are not influenced strongly by this procedure. However, three of the tables entries do change. The information score per event (I/N) changes significantly because these aftershocks coming close after a parent event with the high conditional rate of occurrence, strongly increase the log-likelihood ratio. The ratio is used in selecting and comparing the best models for approximating an earthquake occurrence. Therefore, in any such testing it is important to know how the immediate aftershocks have been processed.

The ratio of the number of independent events to total $\lambda T/N$ increases since with the removal of the immediate aftershocks fewer triggered earthquakes are left in a catalog. The branching coefficient μ increases apparently because the immediate aftershocks are incompletely sampled in a catalog, and therefore they influence the μ estimate making it smaller.

These changes are less evident in the CMT catalog (Table 2). This catalog is based on the interpretation of long-period waves and the coda of these waves interferes with the identification of close aftershocks. For that reason fewer such aftershocks (to be deleted) are present in the catalog.

Figure 4 shows the likelihood search iteration procedure for $m \geq 5$ earthquakes in the PDE catalog. It demonstrates that the estimates of parameters λ and μ are negatively correlated. The other parameters look less correlated during the search. In principle, to evaluate parameter uncertainties we need to obtain the Hessian matrix (the second partial derivatives of the likelihood function) of the parameter estimates or at least to plot two-dimensional plots of the parameter estimates similar to what was done in BIRD and

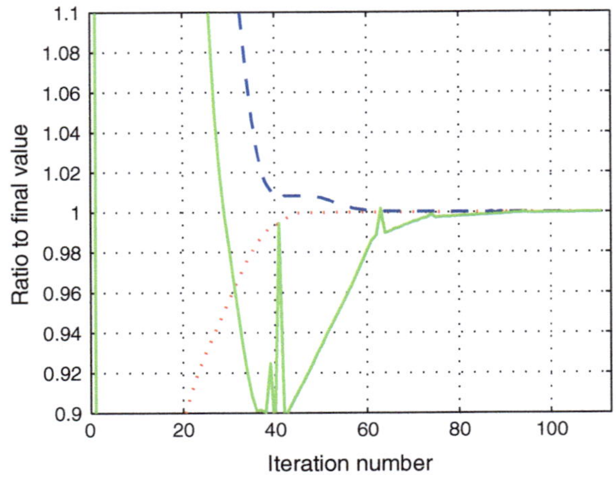

Figure 4
Plot of the log-likelihood function (*dotted line*), independent event rate λ (*dashed line*) and the branching coefficient μ (*solid line*). These variables have been normalized to their final values. All shallow earthquakes with $m \geq 5$ in the PDE catalog are processed

KAGAN (2004) or in Fig. 3. Such plots have an advantage because if the relation is nonlinear or the parameter value needs to be restricted (if for example, it goes into the negative domain or to infinity, etc.) such plots provide us with more accurate information than the second-derivative matrix.

To assess the statistical significance of the differences in the parameter estimates across the tectonic regimes, we need a reliable way to estimate the multidimensional confidence regions. For the overdetermined linear problems with the Gaussian data errors, the confidence regions can be estimated simply from the covariance matrix of the parameter estimates and an estimate of the data uncertainty. For mildly nonlinear overdetermined problems, asymptotic methods can yield approximate confidence regions. These methods assume local linearity of the relationship between the model perturbations and data perturbations, and they use the Hessian matrix to estimate a covariance matrix. In a particular instance the validity of the asymptotic assumption depends on the how well the parameters are constrained by observations and prior information, and on the differentiability of the likelihood function (JACKSON and MATSU'URA, 1985). For catalogs with many earthquakes over a wide magnitude range and with many isolated earthquake clusters, the asymptotic methods might be applicable to branching models. As we explained above, one problem is that positivity and other hard constraints on the parameters, and censoring of immediate aftershocks, introduce discontinuities preventing differentiation of the likelihood function. We have not yet made a serious study of the confidence regions for the parameters, but indirect evidence suggests that the asymptotic method may not be appropriate, even without aftershock censoring. That evidence comes from (1) the shape of the likelihood contours for two relevant parameters in Fig. 3, which would be elliptical under the assumptions in the asymptotic approach, and (2) the behavior of the likelihood during convergence of the parameters while searching for the optimum parameter estimates (Fig. 4). The likelihood converges relatively slowly, and the tradeoffs between parameters clearly vary through the process. We believe that a trustworthy confidence map may require laborious grid search or Monte Carlo techniques.

6.2. Comparison of Models and Results with ETAS

Our branching model which was first proposed and applied to the central California earthquake record by KAGAN and KNOPOFF (1987), is essentially similar in design to the ETAS model (OGATA, 1988, 1998). Both models use the stochastic branching processes theory to approximate the temporal development of earthquake occurrence. Seismicity is approximated by a time-magnitude-space Poisson cluster process (see Sect. 5), with dependent, cluster events concentrating around an earlier 'parent' event (see more detail in KAGAN, 2006, Section 8). The main difference between these two models lies in the parameterization of the influence functions of a dependent event and normalization of these functions.

Our Eq. 6 is similar to the time influence function used in the ETAS model (OGATA, 1998, Eq. 1.1), where the number of aftershocks is expressed as the Omori–Utsu (UTSU, 1961) formula

$$n_i(t) = \frac{1}{(t - t_i + c)^p}, \qquad (13)$$

where the time t is counted from the occurrence of the previous event t_i, p, and c are parameters. The p parameter can be identified with $(1 + \theta)$ in (6), whereas the c-parameter is introduced to account for the lack of aftershocks immediately following a strong earthquake (KAGAN, 2004). In our model we simply excluded such aftershocks (Eq. 7) from the catalog, in a way that is similar to deleting all earthquakes smaller than the moment threshold, M_t (see Eq. 2).

However, many evaluations of the c parameter show that it varies strongly, by orders of magnitude, even for main shocks of similar size. Moreover, the c estimates vary by an order of magnitude even for the same main shock when using different data (catalogs). KAGAN (2004) discusses the reasons for this phenomenon: the frequency range of a network, the distance to the closest station, installation of temporary stations following a strong event, workforce availability, etc. This effect may be of little importance in the consideration of individual sequences, but in the statistical analysis of earthquake catalogs we should expect that the parameters of the model have a property of statistical stability. Otherwise, derived parameter values are an average of various nonpertinent factors and as such may be reproducible. Many researchers interpret the c value as having a physical significance; they should then be able to explain why c estimates are so unstable. This is in contrast to other seismicity parameters, such as the p value of Omori's law or the b value of the Gutenberg–Richter relation, which hardly (if at all) change (KAGAN, 2006).

The other problem is that the coefficient c in the present ETAS implementations (as it is in Eq. 13) does not seem to be dependent on the magnitude of the preceding event. Our Eq. 7 which defines which aftershocks are removed from a catalog due to their incomplete sampling, is scaled with magnitude: for a main shock $m4$ it is about 5 min, whereas for $m8$ earthquake the 'dead-time' extends to more than 8 h. As the above example suggests, a relative temporal decrease of the aftershock numbers modeled by c should vary strongly with the magnitude of a main shock or a preceding event. It can be expected that the model which requires the same short-time behavior for earthquakes of different size may lead to a significant bias in parameter estimation and seismicity forecast.

In calculation with Eq. 13 it is assumed that immediately after a parent event ($t = t_i$), the rate is equal to $1/c$. However, the coda waves of the previous earthquake may hide any aftershocks to be identified for minutes or hours depending on the event magnitude. OGATA (1983) introduced a coefficient 'S' to describe a practical absence of aftershocks in the very initial part of a sequence. His estimate of S is 1/2 day (p.120). Therefore, (13) would cause a bias in the evaluation of the branching coefficient K_0 (see below).

The estimated temporal parameters for the ETAS model (Tables 1–3 in Ogata, 1998; Tables 1–3 in OGATA and ZHUANG, 2006) demonstrate that the c value in (13) depends on the magnitude threshold. It is smaller for a smaller m_t. This is an additional argument against treating this parameter as having a physical significance.

The p value is usually slightly larger than 1.0 (ibid), although for some models the estimate is less than 1.0—the value that can be rejected on general grounds: since it means that each earthquake has an infinite number of aftershocks. Thus, the p value results are in principle consistent, taking into account that $p = 1 + \theta$, with our Tables 2, 3, 4, 5, 6 and 7.

With regards to the spatial distribution (OGATA, 1998, Eqs. 2.2–2.4) there are two problems in the ETAS model: (1) Two effects, earthquake location errors and the dimension of the aftershock zone need to be separated in the modeling (cf., Eq. 12), since as we discussed above (Sect. 6.1), there are good geophysical estimates for these variables. Thus we can test whether the inversion results are trustworthy. (2) The rupture area is used to characterize the source size in the ETAS model. The area is more difficult to measure and is less known than a simple linear size of the focal zone—especially for earthquakes with a near vertical fault plane due to large vertical location errors for shallow earthquakes. Thus, because of the different parameterization in both models, it is difficult to compare our results for the spatial clustering parameters (s_r and $\epsilon_p$in Tables 2, 3, 4, 5, 6 and 7) with the corresponding ETAS values.

Our Eq. 9 corresponds to the right-hand part of the ETAS model dependence (OGATA, 1998, Eq. 1.1),

$$v_i(t) = n_i(t)K_0 \times \exp[\alpha(m_i - m_t)], \qquad (14)$$

where K_0 and α are parameters, m_i is a magnitude of a 'parent' event, and m_t is a magnitude threshold. Then our a (see Eq. 10) should be equal to $\alpha/\log(10) \approx \alpha/2.3$. In the above-mentioned publications the most likely α estimates are between 1.2 and 1.6, that would roughly correspond to our a values in Tables 2, 3, 4, 5, 6 and 7. It is unclear how exactly our branching μ parameter is related to the K_0 parameter, the dimensionality and the normalization of the parameters are different, thus it is difficult to compare their estimates.

Users of the ETAS model do not generally determine the likelihood score for an event. Thus, we cannot compare our information score I/N to the fit of the ETAS model to earthquake occurrence record and we cannot check the model's performance compared to an inhomogeneous spatial Poisson model.

HELMSTETTER *et al.* (2006) performed calculations for the southern California seismicity, using the ANSS catalog and the model they named the ETES model which is largely similar to the ETAS. The parameters' values obtained by the authors (ibid, their Table 2) are close to those shown in Table 7. The p value is around 1.2. Their $\alpha \approx 0.4$-0.8 can be directly compared to our a value, since it is an exponent for 10.0. The focal zone size f_d translates about 0.4 km for an $m4$ earthquake (they fixed the location error at 0.5 km, see their Eq. 19). Their information score $\log_2 G \approx 3.5$ is also close to our values for a $m_t = 3$ subcatalog.

7. *Discussion*

What differences in the zones parameters can be seen in Tables 1, 2, 3, 4, 5, 6 and 7? The β value is essentially the same for all five zones. Perhaps the 'slow-spreading ridges' may have a statistically significant higher β value, although that could be caused by a mixture of different earthquake populations. Our more exhaustive analysis (BIRD and KAGAN, 2004) suggests that the hypothesis of the universality for β value ($\beta \approx 0.63$) cannot be rejected based on the present data. The results of the corner moment evaluation (Table 1) are also consistent with our previous analysis (ibid). As we discussed above (Sect. 6.1), the differences in the b values for various tectonic zones can be most likely attributed to nonlinear tapering of the magnitude–frequency relation due to the corner magnitude influence and to various biases and systematic effects in magnitude determination for the PDE and local catalogs (KAGAN, 2003).

Can we distinguish the differences in the earthquake occurrence patterns and in particular the earthquake clustering in various tectonic zones? The parameters I/N, $\lambda T/N$, μ, a, and θ need to be compared to see the difference. The θ estimates exhibit no consistent pattern. For the remaining parameters, in all the Tables 2, 3, 4, 5, 6 and 7 oceanic ridges seem to stand out compared to continental areas. The ridges exhibit significantly less clustering than the continents: the information score I/N is smaller and the ratio of spontaneous events ($\lambda T/N$) is higher for the oceanic earthquakes.

Two parameters that characterize the degree of branching, i.e., μ and the size distribution of descendent events (a) are also different for these zones, though the dissimilarity is not the same for the CMT (Table 2) and the PDE catalogs (Tables 3, 4, 5, and 6). In the CMT results the ridges a value is similar to the continental (zones 0, 1, and 4) estimates, but the μ values are significantly smaller. The PDE catalog exhibits an opposite pattern: the a values drop for the ridges, whereas the μ estimates essentially stay the same. In the PDE data set the size of smaller earthquakes is characterized by various magnitudes, most often by the m_b magnitude, which exhibit many biases and systematic errors (KAGAN, 2003). Therefore it is likely that the size distribution of triggered events may be biased in the PDE catalog. Unfortunately, the magnitude range of the CMT catalog is too small to see the differences in the branching rate and earthquake clustering due to the variation of the moment threshold. Among continental zones, subduction earthquakes in the CMT catalog exhibit the strongest clustering pattern, however this feature is not prominent in the PDE catalog.

The variations in maximum-likelihood estimates of branching parameters across tectonic zones displayed in Tables 2, 3, 4, 5, 6 and 7 result from a complex interplay of physical variability in the Earth, varying deficiencies of the seismic catalogs, and instability/tradeoff in the maximum-likelihood estimation process. One view is that it is not important to separate these factors, because the branching parameters are expected to have practical value in improved forecasts which can be tested by their performance. However, it is of great interest to know whether we can reject the null hypothesis that the shallow earthquake process (including interevent triggering) can be described as uniform, in some carefully qualified sense, around the planet. A major obstacle to answering this question is that our present software does not provide trade-off analysis between the parameters we solve for, nor does it provide formal uncertainty ranges (either for parameters in isolation, or for combinations of linked parameters). This means that we have to use empirical comparisons to put lower bounds on residual uncertainties, by examining how results change with threshold magnitude, with catalog, and with processing method (i.e., raw vs. windowed or censored catalogs, discussed previously). We first examine results from tectonic zone 1 (Active continent) because there we have the widest range of catalogs and threshold magnitudes to compare. Figure 5a combines productivity coefficients μ from global zone-1 subcatalogs of CMT ($m_t = 5.55$) and PDE ($m_t = 5.3$, or $m_t = 5.0$, or $m_t = 4.7$) and also from regional California-Nevada catalogs CIT and ANSS ($m_t = 4.7$, 4.0 or 3.0). The abscissa chosen for this comparison is the threshold magnitude assumed (and used to truncate the catalog).

We see a strong tendency for apparent productivity coefficient μ to increase with decreasing threshold magnitude, even though existing theory suggests no reason why this should occur. Until we understand this better, it will be important to use equal threshold magnitude in comparing inferred branching parameters across tectonic zones. Another possible signal in Fig. 5a is a possible small offset (by $\sim$0.04) between the trends for global zone-1 subcatalogs and the two California/Nevada catalogs. However, we cannot assign any confidence to this possible difference because our present analysis does not provide confidence ranges for each. A different measure of branching is given by the fraction of independent events ($\lambda T/N$ of Tables 3, 4, 5, and 6). This measure is (negatively) correlated with productivity coefficient μ, but is also influenced by parent productivity exponent a (which we estimate

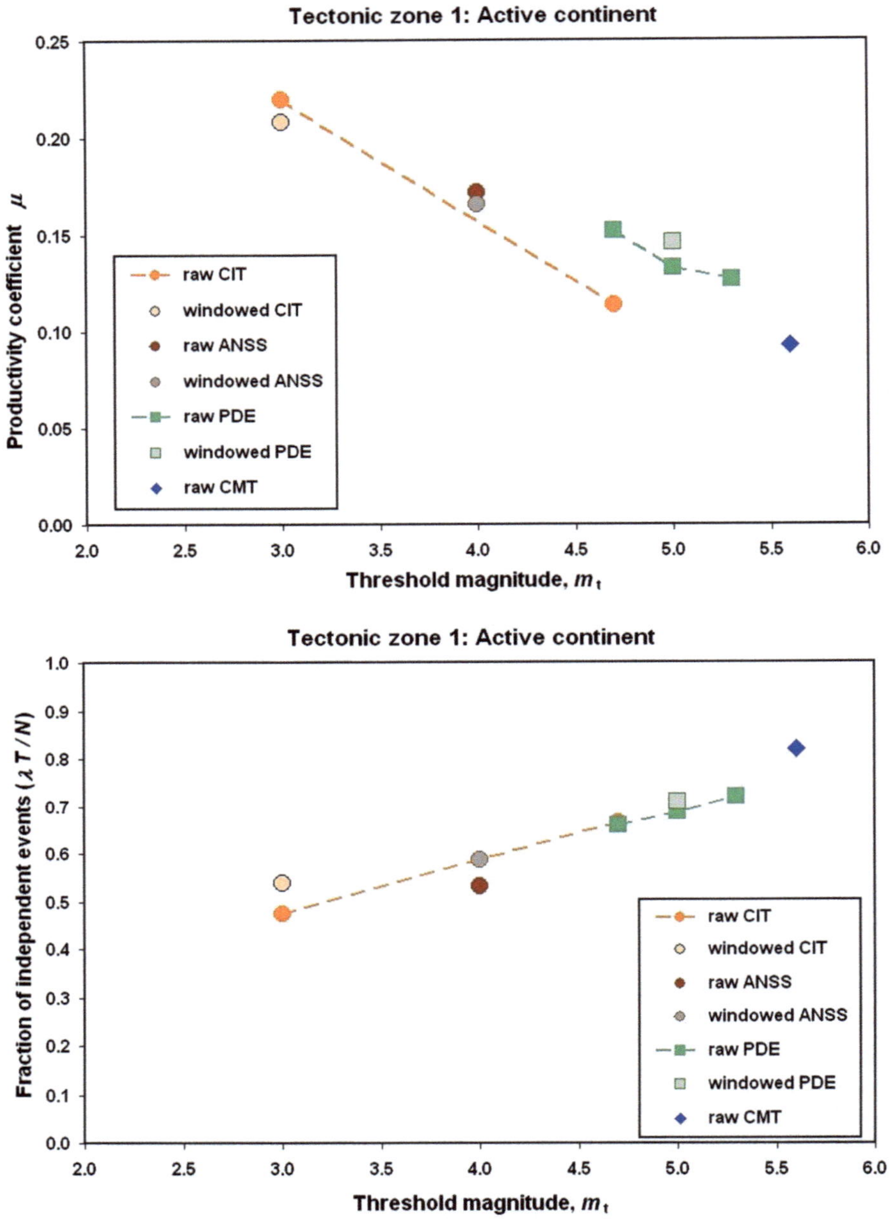

Figure 5

Productivity coefficients and fraction of independent earthquakes for tectonic zone 1: Active continent. See text for details of catalogs, and tables for sources of μ and $\lambda T/N$. **a** Upper plot, branching parameter μ. **b** Lower plot, fraction of spontaneous earthquakes $\lambda T/N$

consistently as $\sim$0.6 in tectonic zone (1)) and by the earthquake geometry and time relations in the catalog. In Fig. 5b, we plot this measure for tectonic zone 1, and see that it also varies consistently with threshold magnitude (either real or assumed) with which we truncate these catalogs. In the results for this measure of branching, the possible offset between global and local catalogs is less clear, as results from raw PDE align with those from raw CIT.

Tectonic zone 4 (Trenches) produces the majority ($\sim$65%) of Earth's moderate and large shallow earthquakes, therefore branching parameters for zone 4 are not very different from global values. In Fig. 6a and 6b we display various inferred values of productivity coefficients and independence estimates $\lambda T/N$ for tectonic zone 4, obtained from the CMT catalog ($m_t = 5.55$) and the PDE catalog ($m_t = 5.3$, or $m_t = 5.0$, or $m_t = 4.7$), and also display two alternative results computed with windowing of the early aftershock times. Figure 6a shows that windowing has a large effect on inferred μ from the PDE catalog, suggesting that PDE is missing numerous early aftershocks. On the other hand, the estimate from CMT is hardly affected by windowing. As we discussed in Sect. 6.1, because the CMT technique uses long-period waves, fewer such early aftershocks are present in the catalog. Using windowing, μ for tectonic zone 4 is fairly consistent at 0.17–0.18 between these two catalogs. Figure 6b shows that mean independence is easier to measure.

Results for tectonic zone 0 (Plate-interior) suggest higher productivity coefficients μ (0.10–0.24) and lower parent productivity exponents a (0.27–0.45) than in the more active tectonic zones 1 and 4. These two anomalies tend to offset each other, so that the fraction of independent events $\lambda T/N$ is in a range (0.68–0.89) that overlaps those of zones 1 and 4. This indicates the need for a multi-dimensional trade-off analysis of the acceptable ranges of triggering parameters before reaching any firm conclusions regarding potential differences between zones. Another potential problem is that plate-interior events more often participate in cross-zone triggering to/from adjacent plate-boundary zones, although such cross-zone triggering is not considered when we analyze this subcatalog in isolation.

Tectonic zones 2 and 3 (Slow-spreading and Fast-spreading ridges) may also have productivity coefficient μ values (0.03–0.23) that have been biased by some failures to resolve parent productivity exponent a. Parameter a has a very low values (0–0.24) inferred from PDE, but the values inferred from CMT (0.52 and 0.59) are more consistent with those of other tectonic zones. Because we cannot trust the lower values of a which we obtain, we also cannot trust associated values of μ for these two tectonic zones; μ will be biased upward whenever a is too low. Therefore, in Fig. 7a and b we plot only the fraction of independent event ($\lambda T/N$) estimates, which we believe to be less affected by these problems.

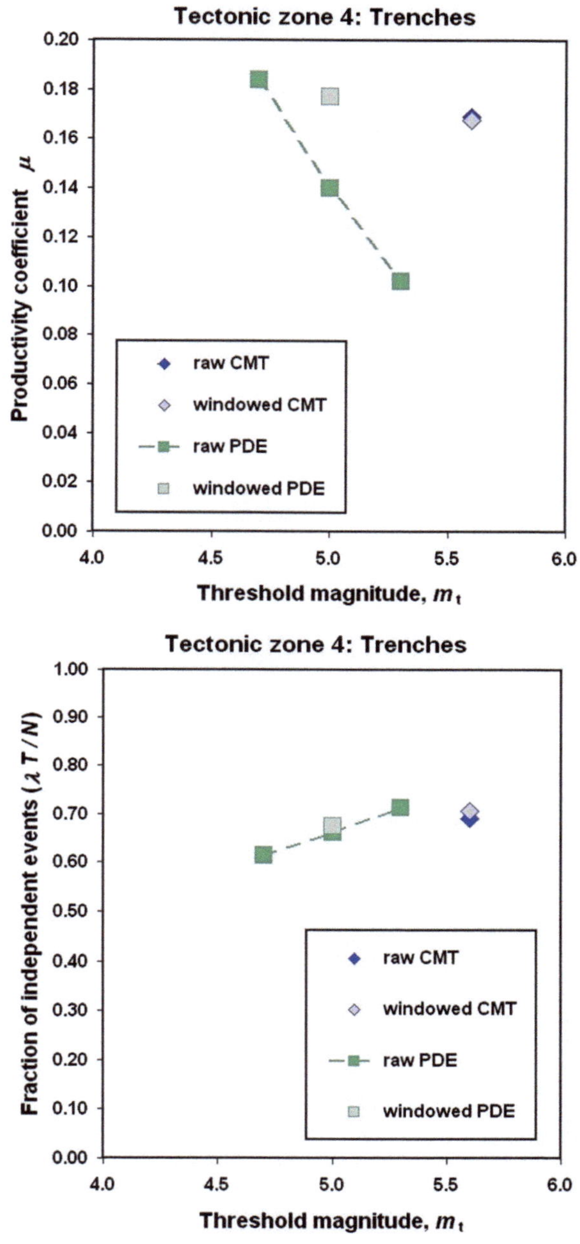

Figure 6
Productivity coefficients and fraction of independent earthquakes for tectonic zone 4: trenches. See text for details of catalogs, and tables for sources of μ and $\lambda T/N$. **a** Upper plot, branching parameter μ. **b** Lower plot, fraction of spontaneous earthquakes $\lambda T/N$

Both Fig. 7a (zone 2: Slow-spreading ridges) and b (zone 3: Fast-spreading ridges) show the apparent fraction of dependent events ($1 - \lambda T/N$) to be lower than in other tectonic zones, nonetheless increasing as

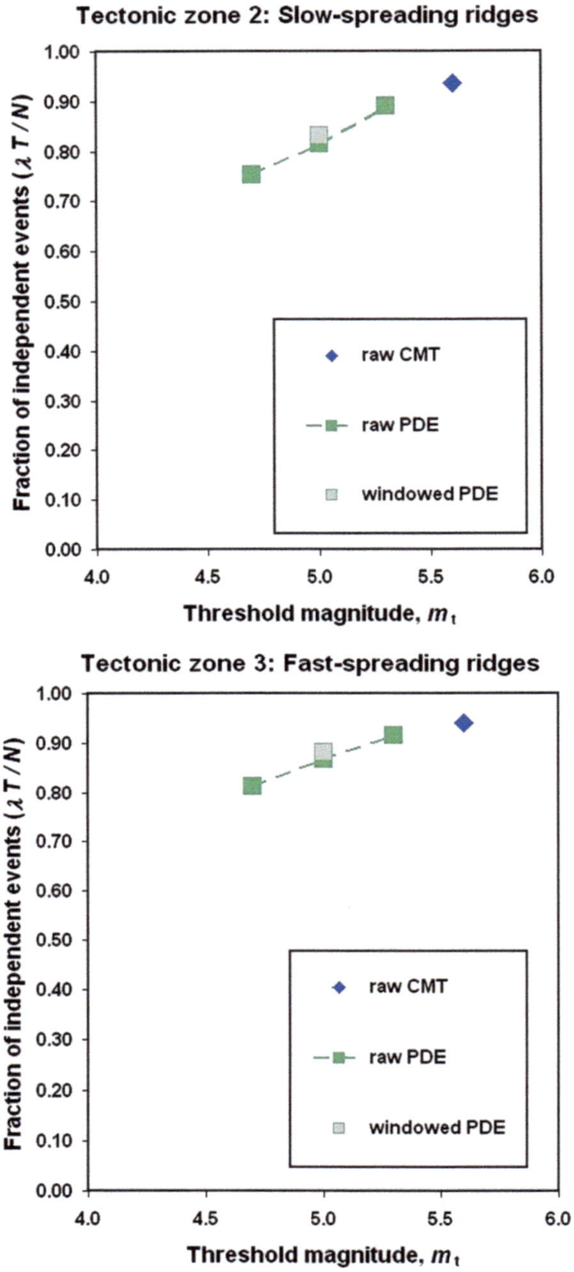

Figure 7
Fraction of independent earthquakes in oceanic ridges. See text for details of catalogs, and tables for sources of $\lambda T/N$. **a** Upper plot, Slow-spreading ridges. **b** Lower plot, Fast-spreading ridges

threshold magnitude is decreased. Based on an analogy to our Fig. 5b, we must consider that additional increases beyond the last values seen (0.25 and 0.19 for $(1 - \lambda T/N)$ of zones 2 and 3 in windowed PDE with $m_t = 4.7$) may occur with even larger subcatalogs that will presumably be obtained from hydrophones and/or ocean-bottom seismometers. At most, the difference in mean fraction of dependent events $(1 - \lambda T/N)$ between spreading-ridge tectonic zones 2 and 3 and zone 4 (Trenches) seems to be a factor-of-two, based on our results.

A series of studies (Boettcher and Jordan, 2004; McGuire *et al.*, 2005) of seismicity of oceanic transform faults (OTFs) has lead to general conclusions that OTFs have more foreshocks than continental transform faults (CTFs), although far fewer aftershocks. The seismicity of our tectonic zones 2 and 3 is clearly dominated by OTF earthquakes (relative to smaller normal-faulting earthquakes on spreading centers), and therefore we expected to see a markedly higher fraction of independent events in these two zones. Instead, we found that this difference is modest, and comparable to differences that appear as assumed threshold magnitude varies. It will be important to resolve this. Boettcher and Jordan (2004) may have exaggerated the contrast between OTFs and CTFs by comparing aftershocks in a teleseismic catalog of OTFs with those in high-quality local catalogs of CTFs. (Compare our Fig.5a, in which a fictitious factor-of-two or greater variation in apparent productivity coefficients μ could be obtained by comparing Global CMT to local CIT catalogs in tectonic zone 1.) The study of McGuire *et al.* (2005) used data from a local hydrophone array on the East Pacific Rise; while this may have lowered the threshold magnitude, it does not approach the global and 26–40-year scope of our analyses of tectonic zones. The prevalence of foreshocks and the resulting enhanced predictability of OTF earthquakes appears robust (McGuire, 2008), however the balance of foreshocks versus aftershocks was not addressed by our study; therefore we only detect the net anomaly in foreshocks plus aftershocks.

In summary, we have observed interesting potential variations in branching parameters between tectonic zones, such as a higher fraction of independent events in the spreading-ridge tectonic zones. Nevertheless we have not been able to document any variations in branching behavior among the tectonic zones with high confidence. At present the variations in apparent branching obtained at different assumed threshold magnitudes (and/or with different catalogs, and/or with different methods of processing) are comparable in size to the effects we seek. More precise and confident discrimination will require some advance in at least one of three areas: (1) Extensions to maximum-likelihood fitting software to provide multi-dimensional confidence bounds on parameters and combinations of related parameters; (2) comparisons of apparent branching parameters obtained from different analysis codes; and/or (3) propagation of apparent differences into earthquake forecasts of global scope which can be scored against future seismicity, to determine whether a forecast that discriminates between zones will outperform one that treats all zones as equivalent.

Acknowledgments

The authors appreciate support from the National Science Foundation (NSF) through grants EAR 04-09890 and EAR-0711515, as well as from the Southern California Earthquake Center (SCEC). SCEC is funded by NSF Cooperative Agreement EAR-0529922 and the U.S. Geologic Survey (USGS) Cooperative Agreement 07HQAG0008. Comments by W. Marzocchi and R. Console have been helpful in revision of the manuscript. Opinions expressed are those of the authors, and do not reflect official positions of the NSF, the USGS, or the SCEC. Publication 1244, SCEC.

Appendix

Flow-chart for Assignment of Tectonic Zone

At every point in the grid (or every epicenter), the computational flow-chart is to:

*Test whether the point lies in any "subduction lane" as defined by KAGAN and BIRD (2004) using SUB (SUBduction zone) plate-boundary steps and Euler poles from the PB2002 model of BIRD (2003). If so: (4) Trench.
*Otherwise, test whether the point is adjacent to any OCB (Oceanic Convergent Boundary) of PB2002. If so: (4) Trench.
*Otherwise, test whether the point lies within any of the 13 orogens of PB2002. If so:

*Test whether crust is "continental" at that point? Yes: (1) Active continent.
*Non-continental crust is "oceanic". Assign tectonic zone according to orogen:

* "Alps" orogen: (4) Trench.
* "Laptev Sea" orogen: (2) Slow-spreading ridge.
* "Persia-Tibet-Burma" orogen: (4) Trench.
* "Ninety East-Sumatra" orogen: (4) Trench.
* "Philippines" orogen: (4) Trench.
* "New Hebrides-Fiji" orogen: (3) Fast ridge.
* "western Aleutians" orogen: (4) Trench.
* "Alaska-Yukon" orogen: (4) Trench.
* "Gorda-California-Nevada" orogen: (3) Fast ridge.
* "Rivera-Cocos" orogen: (3) Fast-spreading ridge.
* "Peru" orogen: (4) Trench.
* "Puna-Sierras Pampeanas" orogen: (4) Trench.
* "west central Atlantic" orogen: (2) Slow-spreading ridge.

*For points NOT within any orogen of PB2002:

*Test whether crust is "continental" at that point?

*Yes, continental: Test for proximity to any non-SUB/OCB plate boundary:

*Yes: (1) Active continent.
*No: (0) Plate interior.

*No, not continental; therefore, oceanic crust:

*Adjacent to any slow OTF and/or OSR? (2) Slow-spreading ridge.
*Else, adjacent to any fast OTF and/or OSR? (3) Fast-spreading ridge.
*Else: (0) Plate interior.

The terms "lane" and "adjacent" and "proximity" in the lines above refer to the finite-width seismicity zones around each of the seven classes of plate boundaries whose widths were defined by Bird and Kagan (2004). The definition of "continental" crust is based on elevation and/or lithosphere age, as specified in Bird (2003).

The result of this classification algorithm is shown in Fig. 1. We note that there are a few regions (Andaman Sea, Lau-Havre Trough) where our algorithm produces "speckled" zonation with many small polygons intermixed. However, these are rare, and for most active plate boundaries the map is divided into strips and/or blocks of large size, appropriate for the study of branching parameters.

References

Abercrombie, R. E. (1995) *Earthquake source scaling relationships from − 1 to 5 M_L using seismograms recorded at 2.5-km depth*, J Geophys Res *100*, 24015–24036.

Anonymous (1988) *Digital Relief of the Earth, Data Announcement*, NOAA/NGC, 88-MGG-02, CD-ROM.

Advanced National Seismic System (ANSS) Catalog Search (2008) Northern California Earthquake Data Center (NCEDC) http://www.ncedc.org/anss/catalog-search.html

Bird, P. (2003) *An updated digital model of plate boundaries*, Geochem Geophys Geosys *4*(3), 1027. doi:10.1029/2001GC000252.

Bird, P., Kagan, Y. Y., and Jackson, D. D. (2002) *Plate tectonics and earthquake potential of spreading ridges and oceanic transform faults*. In Plate Boundary Zones, AGU Geodynamics Series Monograph Volume 30, (eds. S. Stein and J. T. Freymueller), pp. 203-218. doi:10/1029/030GD12.

Bird, P. and Kagan, Y. Y. (2004) *Plate-tectonic analysis of shallow seismicity: apparent boundary width, beta, corner magnitude, coupled lithosphere thickness, and coupling in seven tectonic settings*, Bull Seismol Soc Am *94*(6), 2380–2399 (plus electronic supplement).

Boettcher, M. and Jordan, T. H. (2004) Earthquake scaling relations for mid-ocean ridge transform faults, J Geophys Res *109*(B12), Art. No. B12302. doi:10.1029/2004JB003110.

Console, R. and Murru, M. (2001). *A simple and testable model for earthquake clustering*, J Geophys Res *106*, 8699–8711.

Console, R., Murru, M., and Lombardi, A. M. (2003) *Refining earthquake clustering models*, J Geophys Res *108*(B10), Art. No. 2468.

Ekström, G., Dziewonski, A. M., Maternovskaya, N. N., and Nettles, M. (2005) *Global seismicity of 2003: centroid-moment-tensor solutions for 1087 earthquakes*, Phys Earth Planet Inter *148*(2–4), 327–351.

Flinn, E. A., Engdahl, E. R., and Hill, A. R. (1974) *Seismic and geographical regionalization*, Bull Seismol Soc Am *64*, 771–992.

Hanks, T. C. (1992) *Small earthquakes, tectonic forces*, Science *256*, 1430–1432.

Hanks, T. C. and Kanamori, H. (1979) *A moment magnitude scale*, J Geophys Res *84*, 2348–2350.

Helmstetter, A., Kagan, Y. Y., and Jackson, D. D. (2006) *Comparison of short-term and time-independent earthquake forecast models for southern California*, Bull Seismol Soc Am *96*(1), 90–106.

Hileman, J. A., Allen, C. R., and Nordquist, J. M., *Seismicity of the Southern California Region, 1 January 1932 to 31 December 1972* (Cal. Inst. Technology, Pasadena 1973).

Hutton, L. K. and Jones, L. M. (1993) *Local magnitudes and apparent variations in seismicity rates in Southern California*, Bull Seismol Soc Am *83*, 313–329.

Jackson, D. D. and Kagan, Y. Y. (1999) *Testable earthquake forecasts for 1999*, Seismol Res Lett *70*(4), 393–403.

Jackson, D. D. and Matsu'ura, M. (1985) *A Bayesian approach to nonlinear inversion*, J Geophys Res *90*, 581–591.

Kagan, Y. Y., (1991) *Likelihood analysis of earthquake catalogues*, Geophys J Int *106*(1), 135–148.

Kagan, Y. Y., (2002) *Aftershock zone scaling*, Bull Seismol Soc Am *92*(2), 641–655.

Kagan, Y. Y., (2003) *Accuracy of modern global earthquake catalogs*, Phys Earth Planet Inter (PEPI), *135*(2–3), 173–209.

Kagan, Y. Y., (2004) *Short-term properties of earthquake catalogs and models of earthquake source*, Bull Seismol Soc Am *94*(4), 1207–1228.

Kagan, Y. Y., *Why does theoretical physics fail to explain and predict earthquake occurrence?*, In *Modelling Critical and Catastrophic Phenomena in Geoscience: A Statistical Physics Approach, Lecture Notes in Physics*, **705**, pp.303–359, (eds P. Bhattacharyya and B. K. Chakrabarti) (Springer Verlag, Berlin–Heidelberg 2006).

Kagan, Y. Y. and Jackson, D. D. (2000) *Probabilistic forecasting of earthquakes*, Geophys J Int *143*(2), 438–453.

Kagan, Y. Y. and Jackson, D. D. (2010) *Earthquake forecasting in diverse tectonic zones of the Globe*, Pure Appl. Geophys., this issue.

Kagan, Y. Y. and Knopoff, L. (1987) *Statistical short-term earthquake prediction*, Science *236*(4808), 1563–1567.

Lombardi, A. M. and Marzocchi, W. (2007) *Evidence of clustering and nonstationarity in the time distribution of large worldwide earthquakes*, J Geophys Res *112*, B02303. doi:10.1029/2006JB004568.

Macdonald, K. C. (1986) *The crest of the Mid-Atlantic Ridge: Models for crustal generation processes and tectonics*. In The Western North Atlantic Region, The Geology of North America, vol. M, Geological Society of America pp. 512–68 (eds P. Vogt and B. Tucholke)

Marzocchi, W. and Lombardi, A. M. (2008) *A double branching model for earthquake occurrence*, J. Geophys. Res., *113*, B08317. doi:10.1029/2007JB005472.

McGuire, J.J. (2008) *Seismic cycles and earthquake predictability on East Pacific Rise transform faults*, Bull Seismol Soc Am *98*(3), 1067–1084.

McGuire, J. J., Boettcher, M. S., and Jordan, T. H. (2005) *Foreshock sequences and short-term earthquake predictability on East Pacific Rise transform faults*, Nature *434*(7032), 457–461; Correction—Nature *435*(7041), 528.

Mueller, D., Roest, W. R., Royer, J.-Y., Gahagan, L. M., and Sclater, J. G. (1997) *Digital isochrons of the world's ocean floor*, J Geophys Res *102*(B2), 3211–3214.

Ogata, Y. (1983) *Estimation of the parameters in the modified Omori formula for aftershock frequencies by the maximum likelihood procedure*, J Phys Earth *31*, 115–124.

Ogata, Y. (1988) *Statistical models for earthquake occurrence and residual analysis for point processes*, J Am Stat Assoc *83*, 9–27.

OGATA, Y. (1998) *Space-time point-process models for earthquake occurrences*, Ann Inst Statist Math *50*(2), 379–402.

OGATA, Y., JONES, L.M., and TODA S. (2003) *When and where the aftershock activity was depressed: contrasting decay patterns of the proximate large earthquakes in southern California*, J Geophys Res *108*(B6), 2318.

OGATA, Y. and ZHUANG, J. C. (2006) *Space-time ETAS models and an improved extension*, Tectonophysics *413*(1–2), 13–23.

PRESS, W. H., TEUKOLSKY, S. A., VETTERLING, W. T., and FLANNERY, B. P. (1992) *Numerical Recipes in FORTRAN*, 2-nd ed. (Cambridge University Press, Cambridge 1992) 63 pp.

U.S. GEOLOGICAL SURVEY, *Preliminary Determination of Epicenters (PDE)*, 2008. U.S. Dep. of Inter., Natl. Earthquake Inf. Cent., http://neic.usgs.gov/neis/epic/epic.htm.

UTSU, T. (1961) *A statistical study on the occurrence of aftershocks*, Geophys Magazine *30*, 521–605.

UTSU, T. (1999) *Representation and analysis of the earthquake size distribution: A historical review and some new approaches*, Pure Appl Geophys *155*, 509–535.

YOUNG, J. B., PRESGRAVE, B. W., AICHELE, H., WIENS, D. A., and FLINN, E. A. (1996) *The Flinn-Engdahl regionalisation scheme: The 1995 revision*, Phys Earth Planet Inter *96*, 223–297.

ZHUANG, J. C., CHANG, C.-P., OGATA, Y., and CHEN, Y.-I. (2005) *A study on the background and clustering seismicity in the Taiwan region by using point process models*, J Geophys Res *110*, B05S18. doi:10.1029/2004JB003157.

(Received August 3, 2008, revised February 1, 2009, accepted March 7, 2009, Published online March 13, 2010)

Pure Appl. Geophys. 167 (2010), 743–749

DOI 10.1007/s00024-010-0069-1

Pure and Applied Geophysics

Forecasting the Locations of Future Large Earthquakes: An Analysis and Verification

ROBERT SHCHERBAKOV,[1] DONALD L. TURCOTTE,[2] JOHN B. RUNDLE,[3] KRISTY F. TIAMPO,[1] and JAMES R. HOLLIDAY[3]

Abstract—The objective of this paper is to quantify the use of past seismicity to forecast the locations of future large earthquakes and introduce optimization methods for the model parameters. To achieve this the binary forecast approach is used where the surface of the Earth is divided into $l^\circ \times l^\circ$ cells. The cumulative Benioff strain of $m \geq m_c$ earthquakes that occurred during the training period, ΔT_{tr}, is used to retrospectively forecast the locations of large target earthquakes with magnitudes $\geq m_T$ during the forecast period, ΔT_{for}. The success of a forecast is measured in terms of hit rates (fraction of earthquakes forecast) and false alarm rates (fraction of alarms that do not forecast earthquakes). This binary forecast approach is quantified using a receiver operating characteristic diagram and an error diagram. An optimal forecast can be obtained by taking the maximum value of Pierce's skill score.

Key words: Earthquake forecasting, ROC diagram, error diagram, forecast verification.

1. *Introduction*

Forecasts of the locations of future major earthquakes play an important role in earthquake preparedness and determining earthquake insurance costs. Many such forecasts have been carried out, one example is the National Seismic Hazard Map for the United States (FRANKEL *et al.*, 1996). This is a probabilistic estimate of the ground-shaking risk. A major input into this assessment is the observed rate of occurrence of past earthquakes. KOSSOBOKOV *et al.* (2000) utilized the rates of occurrence of $m \geq 4.0$ earthquakes globally to forecast the expected rates of occurrence of larger earthquakes. Recently, the Regional Earthquake Likelihood Models (RELM) forecasts were published (FIELD, 2007). This represents the first-generation of forecasts based on a variety of approaches and methods. It also introduces a testing methodology to verify and compare these forecasts (SCHORLEMMER *et al.*, 2007).

In this paper we have presented a method to construct a binary forecast of where future large earthquakes are likely to occur globally. We divide the Earth's surface into $l^\circ \times l^\circ$ cells in order to construct the forecast and evaluate its success. The input for our forecast is a relative intensity (RI) map of $m \geq m_c$ earthquakes that occur in the cells during a specified training period, ΔT_{tr}. As a measure of RI we have utilized the cumulative Benioff strain in each cell during that period. The cumulative Benioff strain has been computed using the data from the CMT (Harvard) catalog (http://www.globalcmt.org). This catalog is complete from 1/1/1976 to 1/1/2008 to magnitude $m_c = 5.5$. The RI map is then converted to a binary forecast by introducing a threshold cumulative Benioff strain. Cells with Benioff strains greater than this threshold constitute *alarm cells* where future earthquakes are forecast to occur. A high threshold reduces the forecast areas but leads to more events that are not predicted. A low threshold reduces the failures to predict but increases the false alarms. We have introduced a standard optimization procedure for binary forecasts in order to select the optimal threshold.

In order to test our method we have performed a retrospective analysis. To define training and forecasting time periods we subdivide the catalog using a base year T_b, a forecasting year T_f, and an end year T_e.

[1] Department of Earth Sciences, University of Western Ontario, London, ON N6A 5B7, Canada.
E-mail: rshcherb@uwo.ca; ktiampo@seis.es.uwo.ca
[2] Department of Geology, University of California, Davis, CA 95616, USA. E-mail: turcotte@geology.ucdavis.edu
[3] Department of Physics, University of California, Davis, CA 95616, USA. E-mail: rundle@physics.ucdavis.edu

We have utilized seismicity during the training period from $T_b = 1/1/1976$ to $T_f = 1/1/2004$ ($\Delta T_{tr} \equiv T_f - T_b = 28$ years) to make a forecast for the locations of future target earthquakes of a specified magnitude, $m_T \geq 7.2$. The locations of earthquakes during the forecasting period $T_f = 1/1/2004$ to $T_e = 1/1/2008$ ($\Delta T_{for} \equiv T_e - T_f = 4$ years) are then used to test the validity of the method. The success of the binary forecast has been evaluated using techniques developed in the atmospheric sciences (MASON, 2003). The evaluation is in terms of a hit rate (fraction of events that are successfully forecast) versus a false alarm rate (the fraction of forecast cells in which earthquakes do not occur). Both hit rate and false alarm rate are functions of the tuning parameter (in our case the threshold Benioff strain). Their parametric plot produces a receiver operating characteristic (ROC) diagram. This approach has been used by HOLLIDAY *et al.* (2005) to test the success of forecasting algorithms in California. Similar analysis was also applied to the verification of the intermediate-term earthquake prediction algorithm M 8 by MINSTER AND WILLIAMS (1999).

We have also quantified our approach using the variation of the error diagram introduced by MOLCHAN (1991), MOLCHAN AND KEILIS-BOROK (2008). The comparison of the error diagram approach and the maximum likelihood method for evaluation of earthquake forecasting methods was given by KAGAN (2007). The connection between the entropy score and log-likelihood estimates with applications to several seismicity models is given in HARTE AND VERE-JONES (2005). The analysis of the error diagram approach applied to California seismicity was performed by ZECHAR AND JORDAN (2008). KOSSOBOKOV AND SHEBALIN (2003) and ZALIAPIN *et al.* (2003) have applied the error diagram to test a number of earthquake predictions algorithms.

2. Relative Intensity of Global Seismicity

The proposed analysis is based on the RI of past seismicity where large target earthquakes with magnitudes greater than $m_T \geq 7.2$ are forecasted. The moment magnitude m of earthquakes has been calculated from the scalar moment according to the formula $m = \frac{2}{3}\log_{10}(M_0) - 10.7$, where M_0 is a scalar moment reported in the CMT catalog. The grid size plays an important role in this analysis. In this study the Earth's surface has been subdivided into $2° \times 2°$ cells

As a measure of the RI the cumulative Benioff strain has been computed in each cell for a specified training time period ΔT_{tr}. The cumulative Benioff strain B up to time t is defined as

$$B_{xy}(t) = \sum_{i=1}^{N_{xy}(t)} \sqrt{E_{xy}^{(i)}}, \tag{1}$$

where $E_{xy}^{(i)}$ is the seismic energy release in the ith earthquake and $N_{xy}(t)$ is the number of earthquakes considered up to time t, and (xy) is the coordinates of a given cell. The seismic energy E is related to the seismic moment M_0 as $E \sim M_0$.

The normalized RI in a cell is given by B_{xy}/B_{max} where B_{xy} is the cumulative Benioff strain in the active cell under consideration and B_{max} is the Benioff strain for the most active cell. The cumulative Benioff strain is also normalized by the actual area of the corresponding cell measured in square kilometers.

In order to construct a binary map a decision threshold of RI is applied. All cells with intensities greater than the specified threshold are alarm cells and constitute the binary forecast. To illustrate our approach we consider a specific example. The analysis is based on the seismicity during a training period from $T_b = 1/1/1976$ to $T_f = 1/1/2004$ ($\Delta T_{tr} = 28$ years). The threshold for the normalized Benioff strain $B/B_{max} = 0.0417$ is used. The reason for selecting this particular threshold will be discussed in the next section. The alarm cells in which earthquakes are forecast to occur are illustrated in Fig. 1. During the entire time period ($T_b = 1/1/1976$ to $T_e = 1/1/2008$) there were 1,503 active cells (cells with at least one earthquake with magnitudes $m \geq 5.5$) out of total of 16,200 cells. Note that the area of the active cells depends on the chosen cell size and their geographic location. Also shown in Fig. 1 are the 34 earthquakes with $m_T \geq 7.2$ that occurred during the forecast period $T_f = 1/1/2004$ to $T_e = 1/1/2008$ ($\Delta T_{for} = 4$ years). The next step is to quantify the success of our binary forecast of the locations of these earthquakes.

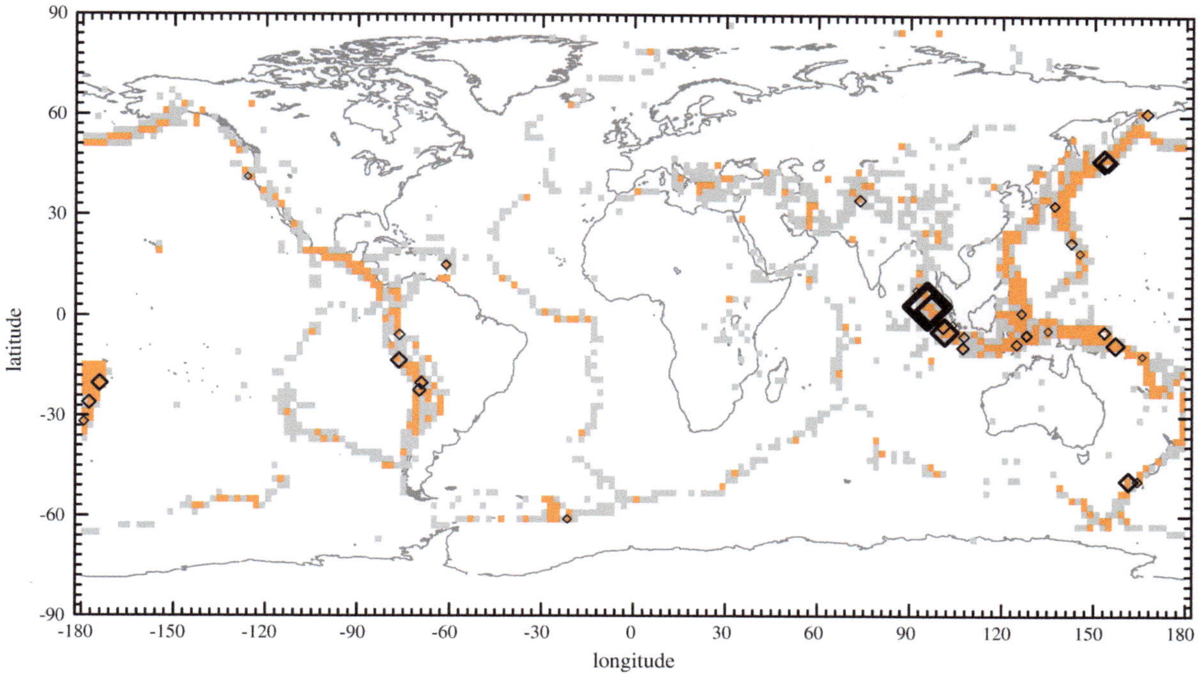

Figure 1
Spatial distribution of $2° \times 2°$ alarm cells. Earthquakes with magnitudes greater than $m \geq 5.5$ are considered during the time period 1/1/1976 until 1/1/2004 ($\Delta T_{tr} = 28$ years). Also shown as *squares* are the locations of earthquakes with $m_T \geq 7.2$ that occurred during the forecast period $T_f = 1/1/2004$ to $T_e = 1/1/2008$ ($\Delta T_{for} = 4$ years). The threshold value corresponds to the maximum Pierce's skill score given in Fig. 3a

3. *Forecast Verification*

The success of a forecast is based on maximizing the fraction of the earthquakes that occur in alarm cells and minimizing the fraction of the alarm cells in which no earthquake occurs. This analysis is carried out by utilizing a *contingency table* approach. Each cell in which an earthquake is forecast to occur constitutes an alarm cell. Earthquakes occur in some of these alarm cells but also in cells where alarms have not been issued. As a result of this test each cell falls in one of four categories: a, alarm and earthquake (successful forecast); b, alarm but not earthquake (false alarm); c, no alarm but an earthquake (failure to predict); d, no alarm and no earthquake (successful forecast of non-occurrence). In addition a is the number of target earthquakes that occurred in alarm cells, b is the number of alarm cells with no earthquakes in them, c is the number of target earthquakes that occurred outside alarm cells, and d is the number of cells with no target earthquakes and which are not alarm cells.

As a specific example we consider the forecast given in Fig. 1. We retrospectively test this forecast in terms of the locations of $m_T \geq 7.2$ target earthquakes that occurred during the forecast period $T_f = 1/1/2004$ to $T_e = 1/1/2008$. The total number of $2° \times 2°$ cells in which a $m \geq 5.5$ occurred during the entire time period (active cells) was 1,503. During the forecast period $a + c = 34$ earthquakes with $m_T \geq 7.2$ occurred. Of these $a = 23$ occurred in alarm cells (were forecast) and $c = 11$ occurred in the other cells (were not forecast).

The success of a forecast can be quantified using a hit rate, H, and a false alarm rate, F:

$$H = \frac{a}{a+c}, \qquad F = \frac{b}{b+d}. \tag{2}$$

The hit rate H is the fraction of earthquakes that are successfully forecast. For this example we have

$H = 23/34 = 0.676$. The false alarm rate F is the fraction of cells with no earthquake for which an alarm was issued. The hit rate H and the false alarm rate F define conditional probabilities $H = P(A|E)$ and $F = P(A|\text{not } E)$ respectively, where A denotes an event of having an alarm in given cells and E denotes an event of having a target earthquake in a given cell. For this example we have $F = 371/1469 = 0.253$. The result of a forecast is given in a contingency table (Table 1).

The above example has given the values of H and F for the specific value of the tuning parameter $B/B_{\max} = 0.0417$. The plot of H and F for a range of $B/B_{\max}$ values is known as a ROC diagram. An ROC diagram for our forecast is shown in Fig. 2a. Results are given for large earthquakes with magnitudes $m_T \geq 7.2$. The diagonal line from (0.0) to (1.1) would be the long-run expectation for alarms that are declared randomly. A perfect forecast would have $F = 0$ (no false alarms) and $H = 1$ (all earthquakes are forecast).

The choice of the optimal forecast (preferred value of $B/B_{\max}$) is the subject of extensive discussions in the atmospheric science literature (MASON, 2003). Several quantitative measures find wide use. One of these is *Pierce's skill score*:

$$\mathrm{PSS_{ROC}} = H - F = \frac{ad - bc}{(a+c)(b+d)}, \tag{3}$$

which is the difference between the observed hit rate H and the value on the diagonal line corresponding to the random forecast. This score is also known as *Hansen-Kuiper's score* or *Kuiper's performance index* (MASON, 2003; HARTE AND VERE-JONES, 2005). The history and applications of this score along with its properties can be found in MASON (2003).

Table 1

Contingency table for earthquakes with $m_T \geq 7.2$ that occurred during the forecast period $T_f = 1/1/2004$ to $T_e = 1/1/2008$ ($\Delta T_{for} = 4$ years)

Forecast	Observed		
	Yes	No	Total
Yes	$a = 23$	$b = 371$	$a + b = 394$
No	$c = 11$	$d = 1098$	$c + d = 1109$
Total	$a + c = 34$	$b + d = 1469$	$a + b + c + d = 1503$

A perfect forecast would be $\mathrm{PSS_{ROC}} = 1.0$. The values of $\mathrm{PSS_{ROC}}$ corresponding to the results in Fig. 2a are given in Fig. 3a. They are plotted as a function of the tuning parameter values $B/B_{\max}$ (Fig. 2a).

The maximum value of $\mathrm{PSS^{max}_{ROC}} = \mathrm{PSS_{ROC}}$ from an ROC diagram can be used to find an optimizing value of the tuning parameter. For large earthquakes $m_T \geq 7.2$ the maximum value of Pierce's skill score is $\mathrm{PSS^{max}_{ROC}} = 0.424$ corresponding to the threshold value $B/B_{\max} = 0.0417$. This corresponds to $H = 0.676$ and $F = 0.253$ and was the basis for selecting this threshold value for the example given in Fig. 1. The value of the hit rate $H = 0.676$ (67.6%) gives the probability of having a large earthquake in the alarm cells during the forecast period. The threshold values obtained from retrospective forecasts can be applied a priori in subsequent prospective forecasts. This can be done by analyzing a particular region of study where the magnitude of completeness of a catalog is obtained and the optimal threshold and cell size values were determined from the retrospective analysis.

Another method of measuring the success of a forecast is the *error* or *Molchan's diagram* approach (MOLCHAN, 1991; MOLCHAN and KAGAN, 1992; MOLCHAN and KEILIS-BOROK, 2008). In order to construct the error diagram the miss and alarm rates

$$\nu = \frac{c}{a+c} = 1 - H, \qquad \tau = \frac{a+b}{a+b+c+d} \tag{4}$$

are introduced. The sum of the hit rate and miss rate is unity, i.e. $H + \nu = 1$. The alarm rate τ is the fraction of all cells that are alarm cells and defines the probability of a forecast of occurrence in weather forecast verification (MASON, 2003). The parametric plot of ν and τ for a range of threshold parameters $B/B_{\max}$ is known as an *error diagram*. The error diagram for the above example is given in Fig. 2b. An analog of the Pierce's skill score for the error diagram is

$$\mathrm{SS_{ED}} = 1 - \nu - \tau = \frac{ad - bc}{(a+c)(a+b+c+d)}. \tag{5}$$

From the analysis of the error diagram curve the maximum value of this skill score is $\mathrm{SS^{max}_{ED}} = 1 - 0.324 - 0.262 = 0.414$ corresponding to the threshold value $B/B_{\max} = 0.0417$. It is worthwhile to note

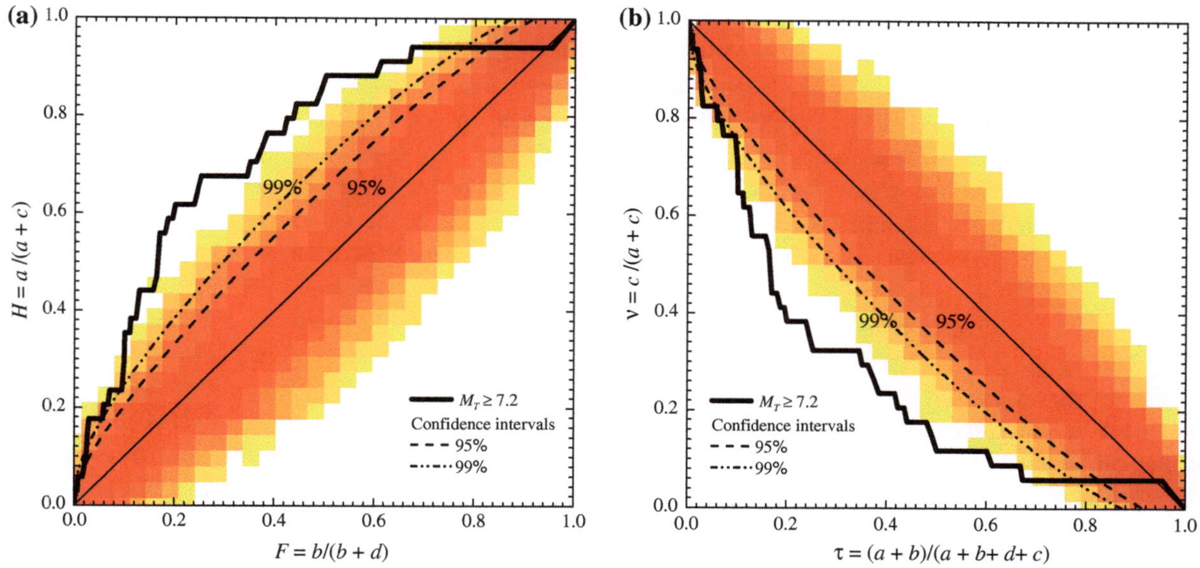

Figure 2

a The ROC diagram for our forecast of global seismicity. **b** The error diagram for our forecast. Also included the results of 150,300 Monte-Carlo simulations (*shaded area*) in which alarm cells have been redistributed randomly and the corresponding confidence intervals

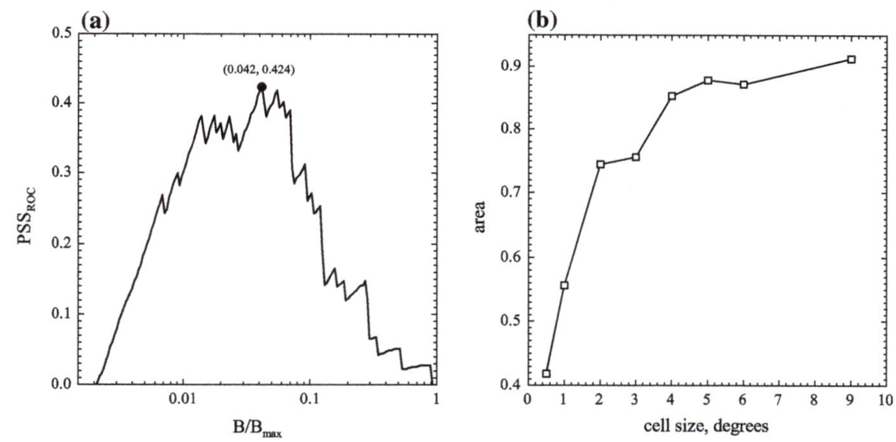

Figure 3

a Pierce's skill scores PSS_{ROC} from the ROC diagram given in Fig. 2a as a function of the threshold values of the normalized cumulative Benioff strain B/B_{max}. The maximum value of the Pierce's skill score is $PSS_{ROC}^{max} = 0.424$ corresponding to the threshold value $B/B_{max} = 0.042$. **b** Dependence of the area under ROC diagram on different cell sizes

that the threshold values corresponding to the maximum values of the skill scores for both ROC and error diagrams are the same. This is due to the fact that the values of F and τ are very close to each other for corresponding threshold values as they are both dominated by large $b + d$ values compared to small $a + c$. As a result the error diagram curve is almost a mirror image of the ROC curve due to the fact that $\nu = 1 - H$. In general this is not always the case.

In the above analysis we have used a uniform prior distribution for the seismically active cells and assumed a null hypothesis where alarm cells are redistributed randomly. To reject this null hypothesis in favor of the method presented here we have also included in Fig. 2a and b the results of 150,300 Monte-Carlo simulations in which alarm cells have been redistributed randomly and the same target earthquakes were used to construct ROC and error diagrams. The confidence intervals were evaluated using the exact analytical result for the binomial distribution given in Zechar and Jordan (2008).

4. Discussion

The primary purpose of this paper is to quantify the extent to which locations of past smaller earthquakes can be used to forecast the locations of future large earthquakes. Our study has been carried out globally. The surface of the Earth has been divided into a grid of $l° \times l°$ cells with $l = 2°$. The objective is to forecast the locations of future large target earthquakes with magnitudes greater than m_T based on the past seismicity. Another important aspect of this work is to introduce different optimization procedures to find the best forecast and optimal values for model parameters. It is also important to note that in the presented analysis the measure of seismicity based on cumulative Benioff strain does not perform better than the predictor based on a long-run rate of occurrence of earthquakes.

The sensitivity of the method to the model parameters is an important issue. The cell size $l = 2°$ has been chosen based on the analysis of the areas under the ROC curves constructed for different cell sizes and different forecasting magnitudes. The analysis shows that the areas tend to increase (better forecast) with increasing cell sizes (Fig. 3b). Bigger cell sizes cover a larger Earth's surface and increase the total success rate. However, bigger cell sizes are less practical in forecasting as they include larger geographical regions and decrease the value of the forecast. In this analysis we assume all earthquakes are point events. Cell size can be an important factor if forecast earthquakes are defined through their rupture areas which can span several cell sizes.

Other important parameters are the durations of the training and forecasting intervals and the lower magnitude cutoff. The analysis of the areas under the ROC curves shows that increasing the training time interval (ΔT_{tr}) usually improves the forecast. As to the forecasting time interval (ΔT_{for}) changes in the duration of the interval result in different numbers for forecasting earthquakes which, as was shown by Zechar and Jordan (2008), affects the ROC and error diagrams. To investigate the performance of the method for different forecasting time intervals a contour plot of the values of the areas between the ROC curve and 95% confidence interval for different forecasting years, T_f, and base years, T_b, is given in Fig. 4. Darker areas indicate a better performance with respect to the 95% confidence interval of the random forecast. It shows a prominent increase in the values for intervals less than 4 years and another increase for 15-year interval. This was a reason to use a 4-year forecasting interval in our analysis.

Finally, the dependence of the method on the lower magnitude cutoff, m_c, shows a strong trend in the performance increase of the method with decreasing m_c. However, the completeness of the catalog prevents the use of events smaller than $m < 5.5$. The full analysis of the dependence of the method on the above parameters will be presented elsewhere. By varying the threshold B/B_{max} the method generates a spectrum of forecasts. The use of the Pierce's skill score is probably the most accepted optimization tool, but several others have been proposed (Mason, 2003; Harte and Vere-Jones, 2005). An actual optimization should also be based on other

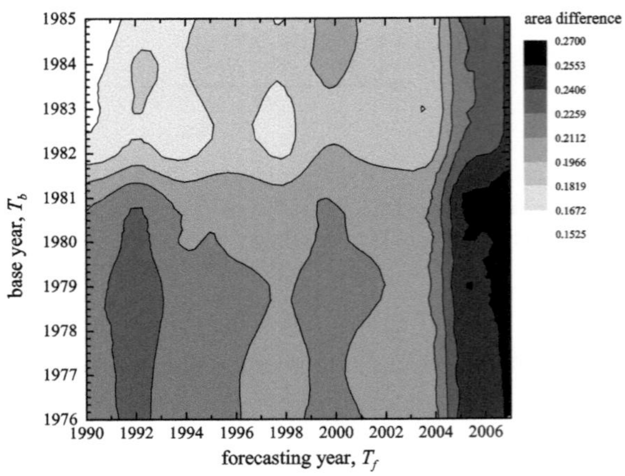

Figure 4
A contour plot of the values of the areas between the ROC curve and the 95% confidence interval for the random forecast for different forecasting (T_f) and base (T_b) years

factors such as available resources, population densities, quality of construction, and other risk factors.

Acknowledgments

This work has been supported by NSERC Discovery grant (RS), by NSF grant ATM-0327558 (DLT), by NASA grant NGT5 (JRH), by a US Department of Energy grant to UC Davis DEFG03-14499 (JRH, JBR), and by the NSERC and Benfield/ICLR Chair in Earthquake Hazard Assessment (KFT).

References

Field, E.H. (2007), *Overview of the working group for the development of Regional Earthquake Likelihood Models (RELM)*, Seismol. Res. Lett. *78*, 7–16.

Frankel, A., Mueller, C., Barnhard, T., Perkins, D., Leyendecker, E., Dickman, N., Hanson, S., and Hopper, M. (1996), *National seismic-hazard maps*, Tech. Rep. Open-File Report 96–532, USGS.

Harte, D. and Vere-Jones, D. (2005), *The entropy score and its uses in earthquake forecasting*, Pure Appl. Geophys. *162*, 1229–1253.

Holliday, J.R., Nanjo, K.Z., Tiampo, K.F., Rundle, J.B., and Turcotte, D.L. (2005), *Earthquake forecasting and its verification*, Nonlinear Proc. Geophys. *12*, 965–977.

Kagan, Y.Y. (2007), *On earthquake predictability measurement: Information score and error diagram*, Pure Appl. Geophys. *164*, 1947–1962.

Kossobokov, V.G., Keilis-Borok, V.I., Turcotte, D.L., and Malamud, B.D. (2000), *Implications of a statistical physics approach for earthquake hazard assessment and forecasting*, Pure Appl. Geophys. *157*, 2323–2349.

Kossobokov, V.G. and Shebalin, P., *Earthquake prediction.* In *Nonlinear Dynamics of the Lithosphere and Earthquake Prediction* (eds. Keilis-Borok, V.I., and Soloviev, A.A.) (Springer-Verlag, Berlin 2003), pp. 141–207.

Mason, I.B., *Binary events.* In *Forecast Verification* (eds. Jolliffe, I.T. and Stephenson, D.B.) (Wiley, Chichester 2003), pp. 37–76.

Minster, J.-B. and Williams, N., *Systematic global testing of intermediate-term earthquake prediction algorithm.* In *1-st ACES Workshop Proceedings* (ed. Mora, P.) (GOPRINT, Brisbane 1999).

Molchan, G. and Keilis-Borok, V. (2008), *Earthquake prediction: Probabilistic aspect*, Geophys. J. Int. *173*, 1012–1017.

Molchan, G.M. (1991), *Structure of optimal strategies in earthquake prediction*, Tectonophysics *193*, 267–276.

Molchan, G.M. and Kagan, Y.Y. (1992), *Earthquake prediction and its optimization*, J. Geophys. Res. *97*, 4823–4838.

Schorlemmer, D., Gerstenberger, M.C., Wiemer, S., Jackson, D.D., and Rhoades, D.A. (2007), *Earthquake likelihood model testing*, Seismol. Res. Lett. *78*, 17–29.

Zaliapin, I., Keilis-Borok, V., and Ghil, M. (2003), *A Boolean delay equation model of colliding cascades. Part II: Prediction of critical transitions*, J. Stat. Phys. *111*, 839–861.

Zechar, J.D. and Jordan, T.H. (2008), *Testing alarm-based earthquake predictions*, Geophys. J. Int. *172*, 715–724.

(Received September 12, 2008, revised November 1, 2008, accepted March 6, 2009, Published online February 27, 2010)

Pure Appl. Geophys. 167 (2010), 751–761

DOI 10.1007/s00024-010-0070-8

Pure and Applied Geophysics

PI Forecast for the Sichuan-Yunnan Region: Retrospective Test after the May 12, 2008, Wenchuan Earthquake

CHANGSHENG JIANG[1] and ZHONGLIANG WU[1]

Abstract—The May 12, 2008, Wenchuan M_S 8.0/M_w 7.9 earthquake occurred in the middle part of the north–south seismic zone in central west China, being one of the greatest thrust events on land in recent years. To explore whether there were some indications of the increase of strong earthquake probabilities before the Wenchuan earthquake, we conducted a retrospective forecast test applying the Pattern Informatics (PI) algorithm to the earthquakes in the Sichuan-Yunnan region since 1992. A regional earthquake catalogue complete to M_L 3.0 from 01/01/1977 to 15/06/2008 was used. A 15-year long 'sliding time window' was used in the PI calculation, with 'anomaly training time window' and 'forecast time window' both set to 5 years. With a forecast target magnitude of M_S 5.5, the ROC test shows that the PI forecast outperforms not only random guess but also the simple number-counting approach based on the clustering hypothesis of earthquakes (the RI forecast). 'Hotspots' can be seen in the region of the northern Longmenshan fault which is responsible for the Wenchuan earthquake. However, when considering bigger grid size and higher cutoff magnitude, such 'hotspots' disappear and there is very little indication of an impending great earthquake.

Key words: Pattern informatics, statistical physics, intermediate-term earthquake forecast, ROC test, Wenchuan earthquake.

1. Introduction

In the study of earthquake predictability, the testing of forecast algorithms, even retrospective testing, plays an important role in model development and improvement (FIELD, 2007). The unexpected occurrence of the May 12, 2008, Wenchuan M_S 8.0/M_w 7.9 earthquake reminds us of the importance of the retrospective evaluation of predictive models for time-dependent earthquake hazard. In such retrospective investigation, one of the candidates is the Pattern Informatics (PI) algorithm, based on the statistical physics of complex systems. The algorithm has been successfully applied to California (RUNDLE *et al.*, 2000, 2003; TIAMPO *et al.*, 2002), central Japan (NANJO *et al.*, 2006) and Taiwan (CHEN *et al.*, 2005). Our analysis of PI includes two parts. First, a general retrospective forecast test was conducted to explore the performance of the PI forecast. Then we performed a retrospective forecast test that focused on the Wenchuan Earthquake, to investigate whether there was any indication of this great earthquake's approach.

2. Region under Study and Data Used for Analysis

The region investigated in this paper is located between latitudes 20.8°–34.0°N and longitudes 97.2°–107.0°E, including Sichuan and Yunnan Provinces as well as their surrounding regions. The Sichuan-Yunnan diamond and its boundary fault zones form one of the seismically active regions in continental China, with the Xianshuihe, Anninghe, Zemuhe, Xiaojiang, Honghe, Lancangjiang, and Longmenshan fault systems located in this region. These fault systems as well as other neighboring fault systems cut the entire region into different 'tectonic blocks' and control the distribution of earthquakes (XU and DENG, 1996). YI *et al.* (2002) and XU *et al.* (2005) observed that strong earthquakes in this region have a significant component of stochastic clustering, with a recurrence process that does not demonstrate simple periodicity and does not seem to be well described by the time-predictable model or magnitude-predictable model. Intense seismicity, spatial–temporal

[1] Institute of Geophysics, China Earthquake Administration, Beijing 100081, China. E-mail: wuzl@cea-igp.ac.cn

heterogeneity, and complicated recurrence behavior motivate the application of the PI algorithm which seeks to identify both seismic activation and quiescence. Figure 1 shows the location of this region, together with the epicenter of the Wenchuan main shock and its aftershocks. JIANG and WU (2008) conducted a retrospective test of PI algorithm for a smaller region than used in this study. In their study only part of the Longmenshan fault was included (see Fig. 2), and this ironic result indicated that the great Wenchuan earthquake had not been predicted.

For the present study we used the Monthly Earthquake Catalogue from 01/01/1970 to 15/06/2008 provided by the China Earthquake Networks Center (CENC), a composite of catalogues provided by regional/local seismic networks, with magnitude unified as M_L. SU *et al.* (2003) studied the regional earthquake catalogues of Sichuan and Yunnan from 1970 to 2001 and concluded that since the early time of earthquake monitoring in the 1970s, the earthquake catalogues of the Sichuan-Yunnan region have been complete to magnitude M_L 3.0. The PI algorithm requires that the catalogue magnitude of completeness be at least two magnitude units smaller than the 'target earthquake' (HOLLIDAY *et al.*, 2006). In this study, based on previous results (SU *et al.*, 2003) and the Gutenberg-Richter frequency–magnitude statistics as shown in Fig. 1b, the cutoff magnitude M_c is set as M_L 3.0. This means that anomalies related to earthquakes with magnitudes larger than M_S 5.5 should be detected and can be reflected by the PI algorithm. Surface-wave magnitude is used for the definition of 'target earthquake' to avoid the bias caused by the difference between different local magnitudes. For this purpose, the 'target earthquakes' are selected from the catalogue of the China National Seismograph Network (http://www.csndmc.ac.cn/newweb/data.htm#). Figure 1 also displays the spatial–temporal distribution of these earthquakes. In this figure, the 'anomaly training window' and the 'forecast window' are also indicated.

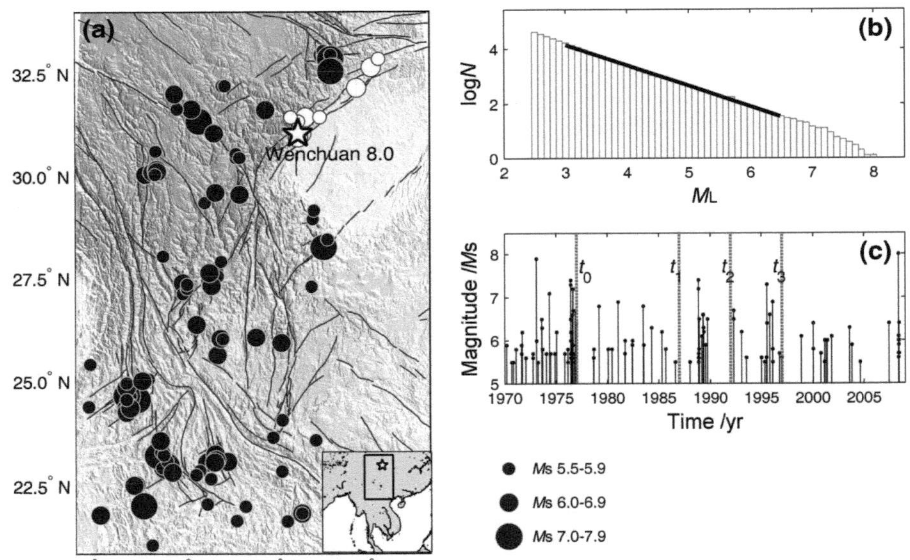

Figure 1

a Earthquakes larger than M_S 5.5 since 1970 in the Sichuan-Yunnan region, with tectonic faults shown by *gray lines*. *White star* and *white dots* indicate the epicenters of the Wenchuan main shock and its aftershocks, respectively. The region under study is shown in the indexing figure to the *bottom right*, in which the Wenchuan earthquake is shown by the *white star*. **b** Frequency–magnitude distribution showing the selection of the magnitude of completeness. **c** Temporal distribution, with the *three vertical dash lines to the right* representing t_1 the starting time of the 'anomaly training window', t_2 the ending time of the 'anomaly training window' and the starting time of the 'forecast window', and t_3 the ending time of the 'forecast window'. For the sliding window considered in this example, the catalogue is selected to start from t_0. See text for details

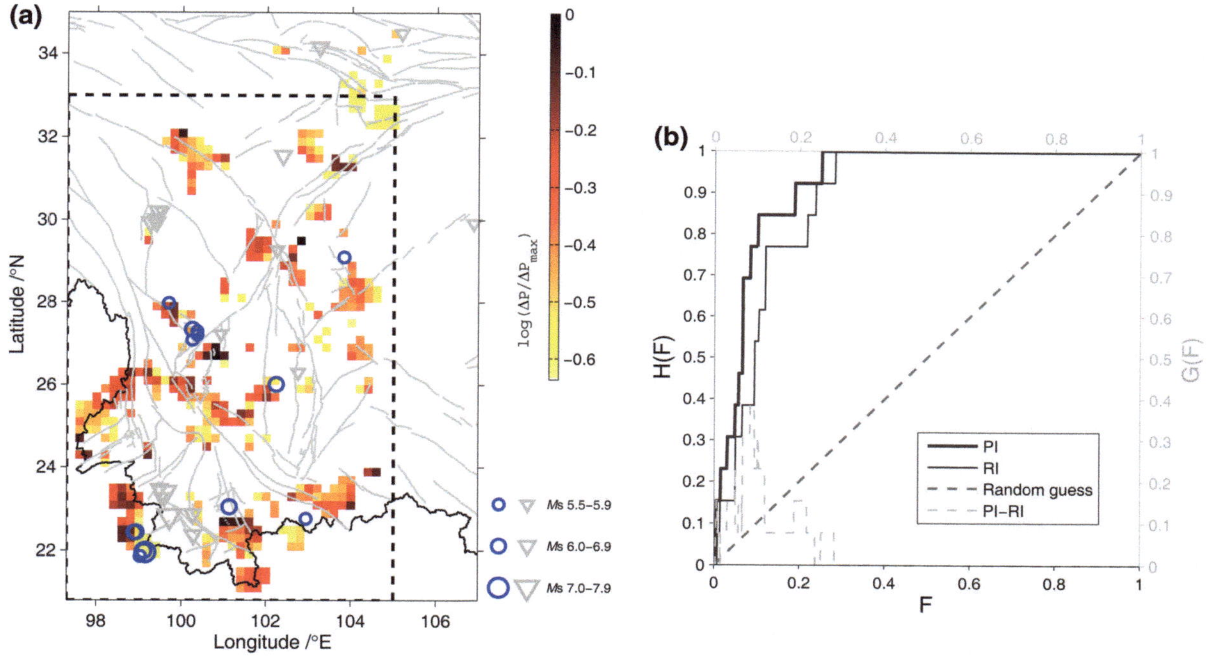

Figure 2
Retrospective test of the PI algorithm: Sichuan-Yunnan region, period of 01/01/1992 to 01/01/1997. **a** Hotspot map of the PI algorithm. *Color-coded hotspots* highlight the relative probability increase for earthquakes above M_S 5.5, with spatial grid size 0.2°. *Blue circles* stand for the earthquakes above M_S 5.5 occurring within the 'forecast window', while *gray reverse triangles* show the earthquakes above M_S 5.5 occurring within the 'anomaly training window'. *Dashed lines* delimitate the region studied by Jiang and Wu (2008). **b** ROC diagram. *Thick solid line* represents the ROC result for the PI forecast, *thin solid line* the ROC result for the RI forecast, and *black dashed line* the result for random forecast. *Gray dashed line* shows the difference between the 'hit rate' of the PI algorithm and that of the RI algorithm. *H* hit rate, *F* false alarm rate, *G* difference between the hit rate of the PI and that of the RI

3. The PI Algorithm

The PI algorithm applies the concepts of statistical physics of complex systems to the analysis of seismic activity, assuming that seismogenetic dynamics can be regarded as a 'threshold system' driven by persistent forces or currents. By analyzing the fluctuations of seismicity, the PI algorithm estimates the increase of the probability of earthquakes at an intermediate-term time scale. Details of the PI algorithm are given by Rundle *et al.* (2002). The code used for the calculations in this test is the modified Matlab version of the PI algorithm provided by the Rundle group. Simply speaking, the input data for the PI calculation are regional earthquake catalogues. The output of the PI calculation is the relative increase of the probability of earthquakes, or the distribution of 'hotspots', described by the top 30% of the normalized probability increase. The PI algorithm is applicable mainly for seismically active regions such as California and Japan. The Sichuan-Yunnan region also possesses the characteristics of high seismicity, and thus is appropriate for using the PI algorithm.

To select proper parameters of predictive models is one of the critical issues in statistical seismology. The PI algorithm suffers from the same difficulty and complexity. Previous works provided clues for the selection of PI parameters. Nanjo *et al.* (2006) used a 35-year catalogue, 27-year 'anomaly training window' and more than 9-year 'forecasting window' to forecast the $M \geq 5$ target events by setting the box size as 0.1° × 0.1°. For earthquake forecasting verification of the PI method, Holliday *et al.* (2005) used a 68-year catalogue and took the 'anomaly training window' and the 'forecast window' to be both 10 years, forecasting the $M \geq 5$ target events of southern California and central Japan by setting box size as 0.1° × 0.1°, and for worldwide application with magnitudes <7.0 by setting the box size as

1.0° × 1.0°. To detect the precursory seismic activity preceding the 1999 Chi-Chi, Taiwan, earthquake by using PI analysis, CHEN *et al.* (2005) used a near 12-year catalogue, and an approximately 6-year 'anomaly training window'. HOLLIDAY *et al.* (2006) applied the modified PI method to a 5-year term forecast, and partitioned the area by using square bins with edge length 0.1°, corresponding to the linear size of a magnitude 6 earthquake. As a contributed forecasting model for RELM testing of Southern California $M \geq 5$ target events forecasting, HOLLIDAY *et al.* (2007) used a catalogue of more than 55 years, took the 'anomaly training window' and the 'forecast window' to be more than 20 and 5 years, respectively, and the research area was divided into 0.1° × 0.1° pixels. WU *et al.* (2008) discuss the migration of PI hotspots of the December 26, 2006, Pingtung doublets by setting the box size as 0.1° × 0.1°, and using a 13-year catalogue. In their work t_2 was fixed, with t_1 shifting from 5 to 2 years before t_2 using the sliding step of 3 months. JIANG and WU (2008) conducted retrospective forecast tests of the PI algorithm for earthquakes in the Sichuan-Yunnan region, to investigate the stability of the algorithm against the selection of model parameters. They adjusted the parameters systematically and investigated the effect of such parameter variation. As a result, optimizing parameters were selected for the 'target magnitude' of M_S 5.5: A 15-year long 'sliding time window'; the 'anomaly training time window' and 'forecast time window' both being 5 years, and the spatial grid taken as $D = 0.2°$. Considering these previous works, in this paper, the spatial grid is taken as $D = 0.2°$. Due to the uncertainty of hypocenter location and also following previous work (KEILIS-BOROK and ROTWAIN, 1990; HOLLIDAY *et al.*, 2007; JIANG and WU, 2008), in this study only shallow earthquakes with depths ranging from 0 to 70 km are considered.

In Fig. 1c, the period of 01/01/1992 to 01/01/1997 is a seismically active period for earthquakes larger than M_S 5.5. As an example of the retrospective test, the 'forecast window' is taken as 5 years. Accordingly, the 'anomaly training window' is selected as 01/01/1987 to 01/01/1992, with the starting time of the catalogue $t_0 = 01/01/1977$, i.e., 15 years before the starting time of the 'forecast window'. In the 'anomaly training window', 17 earthquakes occurred larger than M_S 5.5, with the November 6, 1988 Lancang-Gengma earthquakes with M_S 7.4 and M_S 7.2 being the largest. In the 'forecast window', there were 15 earthquakes with magnitude larger than M_S 5.5, including the July 12, 1995 China-Burma border earthquake with M_S 7.3 and the February 3, 1996 Lijiang, Yunnan earthquake with M_S 7.0. Figure 1c shows the positions of t_1, t_2 and t_3. Figure 2a depicts the PI 'hotspots', using logarithm amplitudes $\log(\Delta P/\Delta P_{max})$ to represent the relative probability increase for strong earthquakes. Following CHEN *et al.* (2005), only the top 30% values are considered as the 'hotspots' and shown in the PI map.

4. *The ROC Test*

As RUNDLE *et al.* (2000, 2003) and TIAMPO *et al.* (2002) did in the evaluation of the performance of forecasts, a receiver operating characteristic (ROC) test (SWETS, 1973; MOLCHAN, 1997) was conducted by systematically changing the 'alarm threshold' of the 'forecast region' and counting the 'hit rate' and 'false alarm rate' relative to real earthquake activity. Here 'hit rate' means the number of 'forecasted' events divided by the total number of 'target' events. 'Hit' or 'forecasted' is defined as the case in which the 'target event' occurs within any 'alarmed cell' or one of the nearest neighbors. In Fig. 2b, the PI forecast is compared with a random guess. It is well known that the larger the area under the ROC curve when compared to the area of the triangle for random forecast, the better the performance of the forecast. From the ROC diagram in Fig. 2b, it may be seen that for the testing period 01/01/1992 to 01/01/1997, the PI forecast significantly outperforms the random forecast.

Similar to CHEN *et al.* (2005) and HOLLIDAY *et al.* (2006), we also compared the PI forecast with the 'relative intensity' (RI) forecast. Here the 'relative intensity' forecast is used as a null hypothesis to test the necessity of using the PI algorithm. In the RI algorithm, it is assumed that strong earthquakes will occur in the regions where earthquakes occurred before, being similar to the 'clustering' argument (KAGAN and JACKSON, 2000). Here, the RI index is computed simply by the box-counting of the number of events. Similar to the PI algorithm, the RI statistics were normalized, and the 'alarm threshold' was determined by the relative value of the RI statistics ranging from 0 to 1. For visual comparison with the PI forecast, only the top 30% of the RI increases is regarded as 'hotspots' and is shown on the map. Figure 2b also shows the ROC curve of the RI forecast. It may be seen that the PI algorithm outperforms not only the random guess but also the RI forecast. The reason why the PI algorithm has a better performance than the RI can be understood, to a great extent, from Fig. 2a: In the period under consideration, some 'target earthquakes', as shown by the blue circles, did not occur in the regions near the earthquakes occurring in the 'anomaly training time window', as shown by the gray reverse triangles. That is, the PI algorithm describes not only seismic activation but also seismic quiescence, while the RI algorithm describes only the former.

5. Retrospective Test for the Sichuan-Yunnan Region: 1992–2007

Figure 3 shows the ergodicity test results for the Sichuan-Yunnan region, indicating that the PI algorithm is valid for this region and for the time period under consideration. The ergodicity test is conducted following TIAMPO *et al.* (2004, 2007) by plotting the Thirumalai-Mountain (TM) metric (THIRUMALAI *et al.*, 1989) versus time, as shown in Fig. 3. For different grid sizes and different cutoff magnitude values, the TM metric plots show that since 1978 seismicity in Sichuan-Yunnan region has had a strong ergodicity, and the PI algorithm has been able to be used for the estimation of time-dependent earthquake rates.

To systematically investigate the performance of the PI algorithm, we consider the 'anomaly training window' and the 'forecast window' to be 5 years, with a 'sliding window' of 15 years, and sliding t_2 from 01/01/1992 to 01/01/2003, with the sliding step

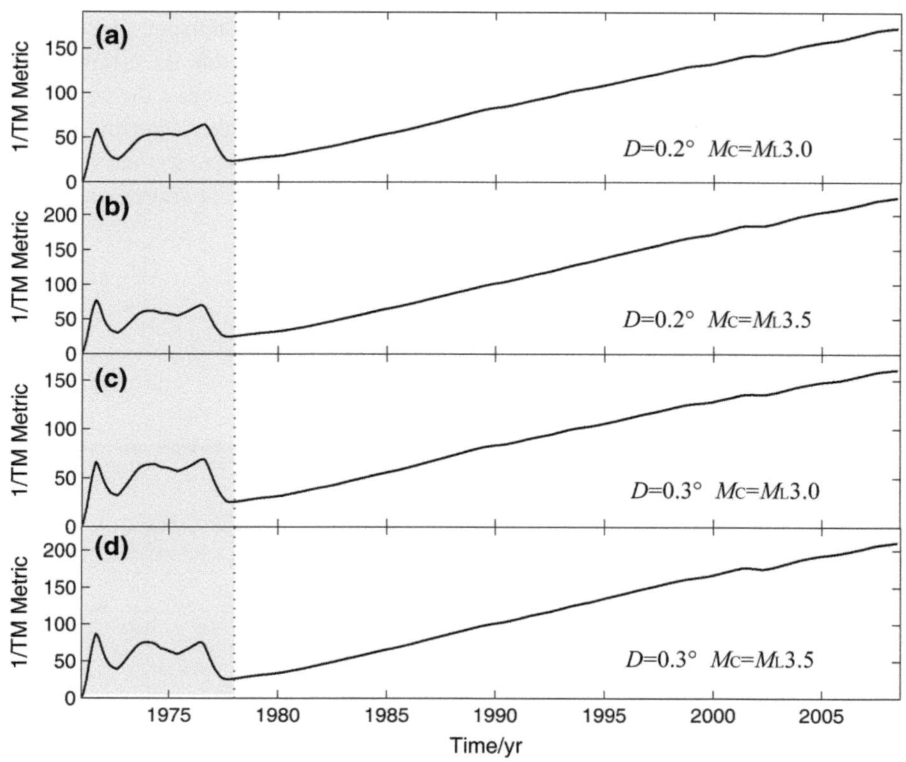

Figure 3
Plots of the inverse TM metric for the Sichuan-Yunnan region from 01/01/1971 to 15/06/2008. In *each subplot D* is the spatial grid size, and M_c is the cutoff magnitude

being 0.5 year. Figure 4a gives the ROC test result, in which the gray zone delimits the range of all the ROC curves, showing the overall performance of the PI forecast. From the figure, it can be seen that despite the variation of the performance with time, generally the PI forecast seems considerably better than a random guess. Figure 4b compares the forecasts of the PI algorithm and the RI algorithm. It can be seen that PI also outperforms RI. For a smaller region, Jiang and Wu (2008) systematically changed the parameter settings to test the robustness of the forecast against the variation of parameters, which shows that the parameter setup as shown in Fig. 2 is stable under the variation of parameter values, and meanwhile seems to be optimal for the Sichuan-Yunnan region.

6. *'Forecast' Performance for the Wenchuan Earthquake*

The Wenchuan earthquake occurred on May 12, 2008 at 14:28 p.m. local time, with magnitudes M_S 8.0 and M_w 7.9, having a thrust-dominated focal mechanism with its aftershock belt being 330 km long, and causing fatalities that reached 69,227 and approaching 18,000 missing (by September 25, 2008, from the Ministry of Civil Affairs of China). The earthquake was one of the largest in China in recent decades and caused considerable attention not only within the scientific community but also in the public (Zhang *et al.*, 2008). After the earthquake, there was a debate on whether the earthquake could have been 'forecast' if observational data were carefully examined. This question is important not only for this earthquake but also for the study of future earthquake predictability.

To explore the answer to this tricky question we investigated whether the PI algorithm, which performances well for the Sichuan-Yunnan region, could have provided clues to the approach of the Wenchuan earthquake. Figure 5 shows the 'forward forecast' for different time ranges from $t_2 = 01/01/2004$ to $t_2 = 01/01/2008$ with a 1-year step, covering the time of the Wenchuan earthquake. In the figure we mark the segment of the Longmenshan fault which accommodated the main rupture and epicenter (or, initiation point) of the Wenchuan earthquake. This area also covers the aftershock zone. From the figure it may be seen that hotspots exist in the 'forecast time window', and the 'hotspot' cluster spanned the entire rupture area. Note that the original objective of the PI algorithm was to forecast the position of the epicenter of earthquakes, rather than the rupture area. On the other hand, however, this earthquake is too large to be treated as a 'point source', and such a correlation

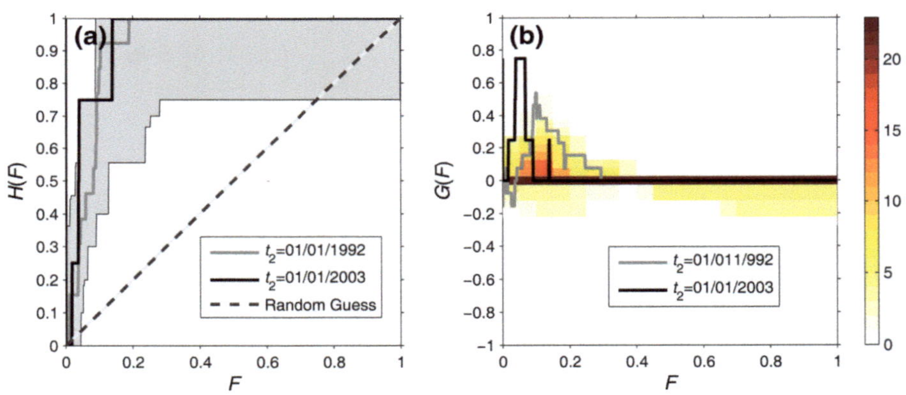

Figure 4
ROC test of the PI algorithm. 'Anomaly training window' and 'forecast window' are taken both as 5 years. The sliding window is taken as 15 years. 'Forecast window' slides from $t_2 = 01/01/1992$ to $t_2 = 01/01/2003$, with sliding step being 0.5 year. **a** ROC curve for the PI forecasts. *Gray zone* delimits the range of all the ROC curves, with the *gray line* and *black line* representing the results of the first and the last sliding, respectively. **b** Difference between the hit rate of the PI algorithm and the RI algorithm. The *gray line* and *black line* represent the results of the first and the last sliding, respectively. The *color bar* indicates the stacking of the areas encompassed by the G curves and the horizontal zero-line

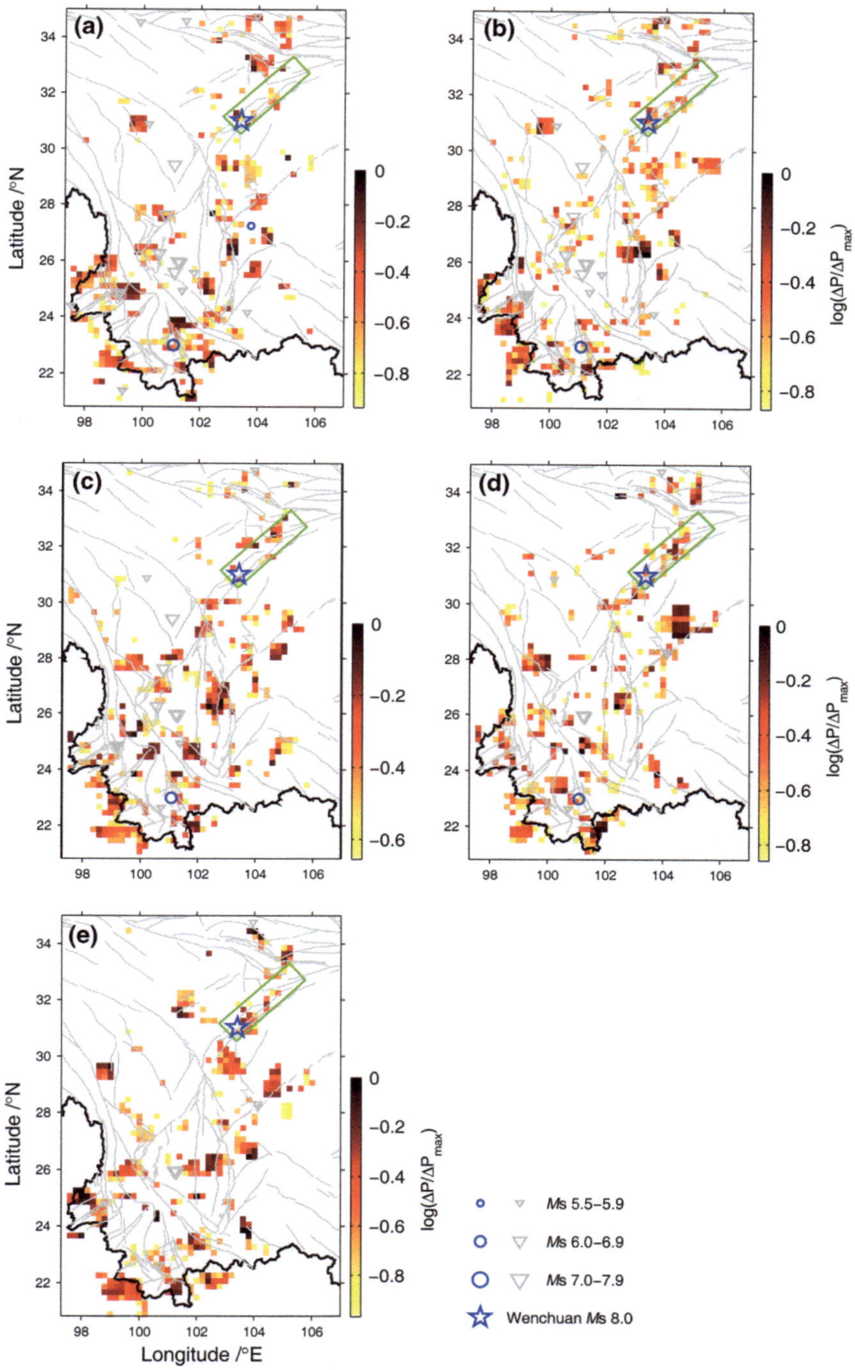

Figure 5
Hotspot maps of the PI algorithm for different 'forecasting time windows'. **a** $t_2 = 01/01/2004$; **b** $t_2 = 01/01/2005$; **c** $t_2 = 01/01/2006$; **d** $t_2 = 01/01/2007$; **e** $t_2 = 01/01/2008$. 'Forecast time window' t_3–t_2 = 5 years. *Green box* delimits the northern segment of the Longmenshan fault which accommodated the Wenchuan earthquake. *Star* shows the epicenter (or, nucleation point) of the great earthquake

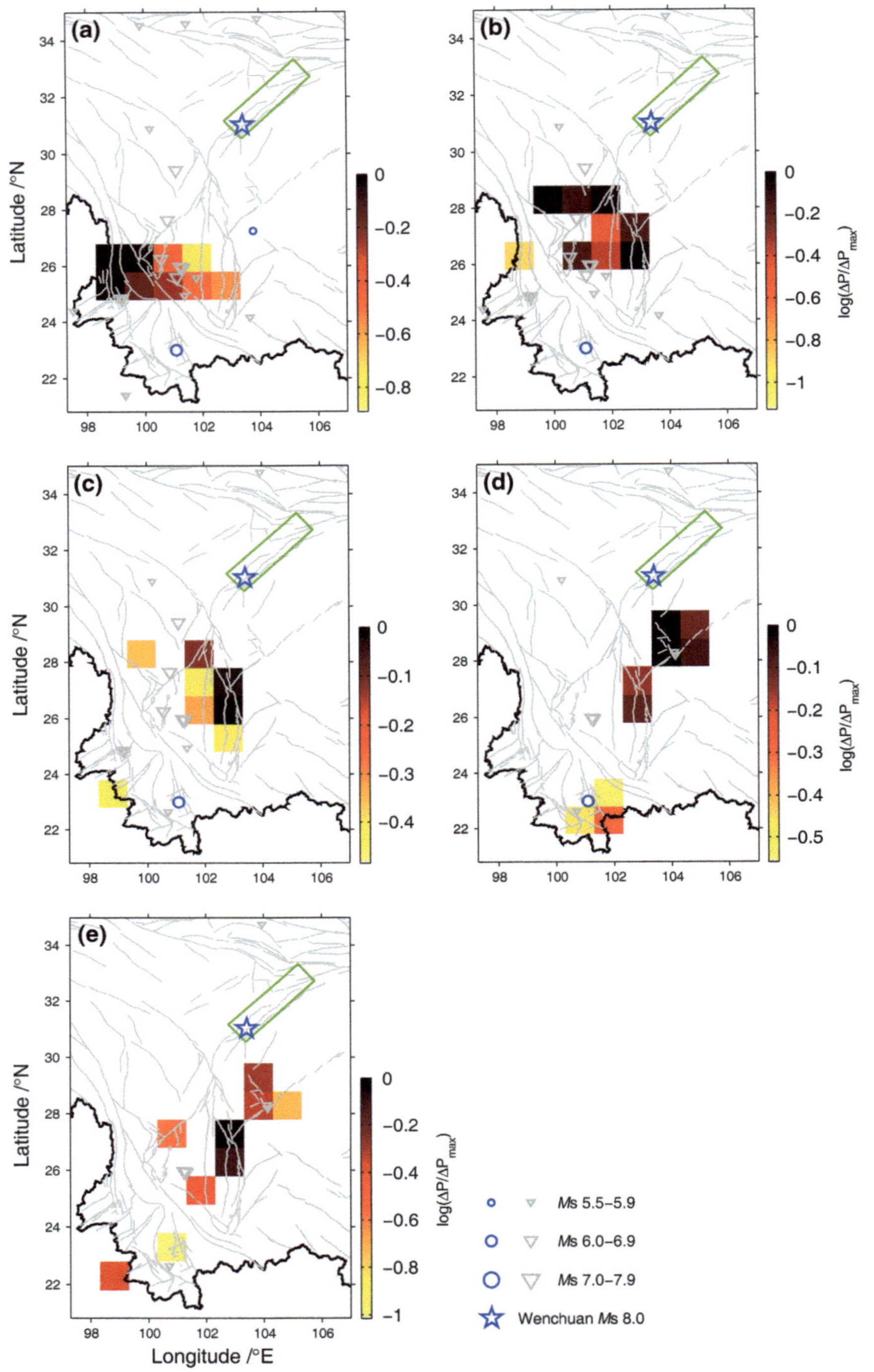

Figure 6
Hotspot maps of the PI algorithm for different 'forecasting time windows' with grid size 1.0° and cutoff magnitude M_L 3.0. **a** t_2 = 01/01/2004; **b** t_2 = 01/01/2005; **c** t_2 = 01/01/2006; **d** t_2 = 01/01/2007; **e** t_2 = 01/01/2008. 'Forecast time window' t_3–t_2 = 5 years. *Green box* delimits the northern segment of the Longmenshan fault which accommodated the Wenchuan earthquake. *Star* shows the epicenter (or, nucleation point) of the great earthquake

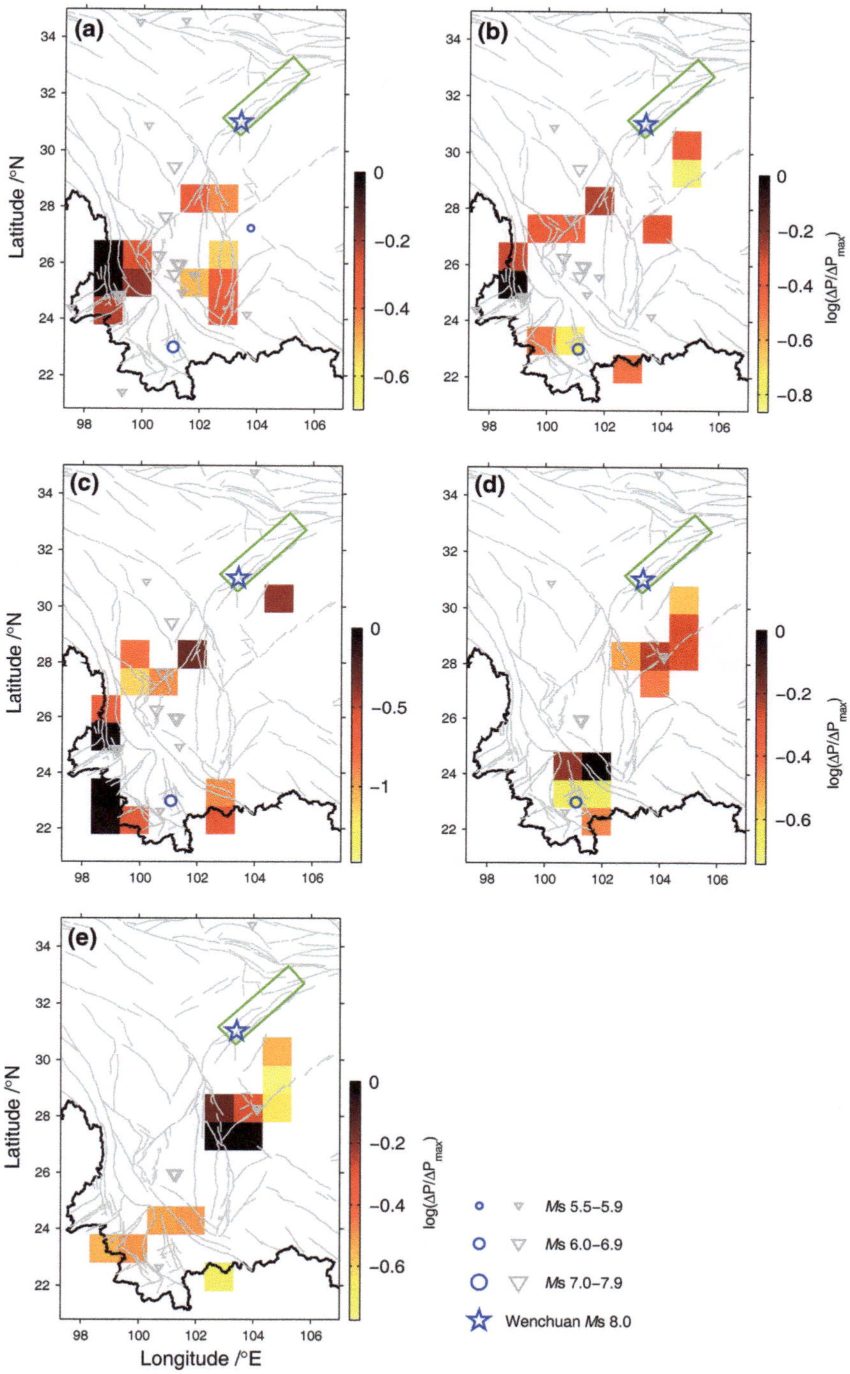

Figure 7
Hotspot maps of the PI algorithm for different 'forecasting time windows' with grid size 1.0° and cutoff magnitude M_L 4.0. **a** $t_2 = 01/01/2004$; **b** $t_2 = 01/01/2005$; **c** $t_2 = 01/01/2006$; **d** $t_2 = 01/01/2007$; **e** $t_2 = 01/01/2008$. 'Forecast time window' t_3–t_2 = 5 years. *Green box* delimits the northern segment of the Longmenshan fault which accommodated the Wenchuan earthquake. *Star* shows the epicenter (or, nucleation point) of the great earthquake

between 'hotspots' and the rupture area may make sense in the physics of earthquakes.

7. Conclusions and Discussion

We tested PI for the Sichuan-Yunnan region, which is to a great extent based on our previous work (Jiang and Wu, 2008) and has been reactivated by the recent Wenchuan earthquake. In the retrospective test, we used the PI algorithm for forecasting the earthquakes with magnitudes larger than M_S 5.5. The result shows that the identical algorithm and parameter setup of Holliday *et al.* (2006) leads to the forecast results outperforming both random guess and RI forecast.

For the Wenchuan earthquake, we examined the PI 'hotspot' map in a different 'forecast time window' covering the time of this great thrust event. Because the epicenter of this earthquake occurs in one of 'hotspot clusters', the PI forecast likely contains information pertaining to about the approaching Wenchuan earthquake. Moreover, the 'hotspots' span the entire rupture area, forming a cluster striking north-eastward along the northern Longmenshan fault zone. One problem is, relative to the hotspot distribution in the whole region, it is difficult to make a clear forecast of the Wenchuan earthquake based on the present mapping. In the same figure, there are many 'hotspots' existing in this region, with most of them substantially more 'intense' than the Longmenshan 'hotspot cluster'. Another problem might be the selection of the grid size: A larger grid size may be more appropriate for forecasting larger events, as pointed out by Kossobokov and Soloviev (2008) for the 2004 Sumatra earthquake. To explore the possibility of this case, Figs. 6 and 7 show the pictures in which the increases of grid size and cutoff magnitude cause the 'hotspots' near the northern Longmenshan fault to disappear. Anomalies are seen 150 km from the rupture zone, but are not seen in the epicentral region. These results are consistent with those of Holliday *et al.* (2005), which were published prospectively (i.e., before the earthquake occurred) using a 1°-grid size. In that paper, a prominent PI anomaly is located at about 101°E, 30°N, roughly about 200 km distant. This may be indicative that, in dealing with inland earthquakes, the convention that the cutoff magnitude be two units less than that of the 'target earthquake' is problematic. Physically, it implies that the population of small events may contain information about the increase in probability of strong earthquakes. If only larger events and larger grids are used, then other patterns, such as the migration of PI 'hotspots' (Wu *et al.*, 2008), must be considered. For example, it seems that there is a 'migration pattern' of PI hotspots with the approach of the earthquake in Figs. 6 and 7, which is similar to the observation of Wu *et al.* (2008). Related to this problem, using longer time windows may lead to the 'cleaning up' of the 'hotspot' picture. However, in the debate regarding whether the Wenchuan earthquake could be forecast, the more important time scales are basically 1 to 5 years. Last but not least, it is important to point out that this work is only a retrospective case study, and has limited significance in the statistical analysis of the PI method.

Acknowledgments

Appreciation to MOST for financial support (contract No.2004CB418406), and also thanks to Professor J. B. Rundle for guidance in PI algorithm, to Dr. C. Chen for providing the Matlab code, and Dr. Peng Hanshu for assistance during the work. Authors are also grateful to Dr. Jeremy Zechar, Dr. Matt Gerstenberger and an anonymous referee for their constructive comments and suggestions, which enhanced our work.

References

Chen, C. C., Rundle, J. B., Holliday, J. R., Nanjo, K. Z., Turcotte, D. L., Li, S. C., and Tiampo, K. F. (2005), *The 1999 Chi-Chi, Taiwan, earthquake as a typical example of seismic activation and quiescence*, Geophys. Res. Lett. *32*, L22315, doi: 10.1029/2005GL023991.

Field, E. D. ed. (2007), *Special Issue—Regional Earthquake Likelihood Models*, Seism. Res. Lett. *78*, 1–140.

Holliday, J. R., Chen, C. C., Tiampo, K. F., Rundle, J. B., and Turcotte, D. L. (2007), *A RELM earthquake forecast based on pattern informatics*, Seism. Res. Lett. *78*, 87–93.

Holliday, J. R., Nanjo, K. Z., Tiampo, K. F., Rundle, J. B., and Turcotte, D. L. (2005). *Earthquake forecasting and its verification*, Nonlinear Process. Geophys. *12*, 965–977.

Holliday, J. R., Rundle, J. B., Tiampo, K. F., Klein, W., and Donnellan, A. (2006), *Modification of the pattern informatics method for forecasting large earthquake events using complex eigenfactors*, Tectonophysics *413*, 87–91.

JIANG, C. S. and WU, Z. L. (2008), *Retrospective forecasting test of a statistical physics model for earthquakes in Sichuan-Yunnan region*, Sci. China Ser. D Earth Sci. *51*, 1401–1410.

KAGAN, Y. Y. and JACKSON, D. D. (2000), *Probabilistic forecasting of earthquakes*, Geophys. J. Int. *143*, 438–453.

KEILIS-BOROK, V. I. and ROTWAIN, I. M. (1990), *Diagnosis of time of increased probability of strong earthquakes in diferent regions of the world*, Phys. Earth Planet Inter. *61*, 57–72.

KOSSOBOKOV V. G. and SOLOVIEV, A. A. (2008), *Prediction of extreme events: Fundamentals and prerequisites of verification*, Russ. J. Earth Sci. *10*, ES2005, doi:10.2205/2007ES000251.

MOLCHAN, G. M. (1997), *Earthquake prediction as a decision-making problem*, Pure Appl. Geophys. *149*, 233–247.

NANJO, K. Z., RUNDLE, J. B., HOLLIDAY, J. R., and TURCOTTE, D. L. (2006), *Pattern informatics and its application for optimal forecasting of large earthquakes in Japan*, Pure Appl. Geophys. *163*, 2417–2432.

RUNDLE, J. B., KLEIN, W., TURCOTTE, D. L., and MALAMUD, B. D. (2000). *Precursory seismic activation and critical-point phenomena*, Pure Appl. Geophys. *157*, 2165–2182.

RUNDLE, J. B., TIAMPO, K. F., KLEIN, W., and MARTINS, J. S. A. (2002), *Self-organization in leaky threshold systems: The influence of near mean field dynamics and its implications for earthquakes, neurobiology, and forecasting*, Proc. Natl. Acad. Sci. USA *99*(Suppl. 1), 2514–2521.

RUNDLE, J. B., TURCOTTE, D. L., SHCHERBAKOV, R., KLEIN, W., and SAMMIS, C. (2003), *Statistical physics approach to understanding the multiscale dynamics of earthquake fault systems*, Rev. Geophys. *41*(4), 1019, doi:10.1029/2003RG000135.

SU, Y. J., LI, Y. L., and LI, Z. H. (2003). *Analysis of minimum complete magnitude of earthquake catalogue in Sichuan-Yunnan region*, J. Seismol. Res. (in Chinese with English Abstract) *26*(Suppl), 10–16.

SWETS, J. A. (1973), *The relative operating characteristic in psychology*, Science *182*, 990–1000.

THIRUMALAI, D., MOUNTAIN, R. D. and KIRKPATRICK, T. R. (1989). *Ergodic behavior in supercooled liquids and in glasses*, Phys. Rev. A. *39*, 3563–3574.

TIAMPO, K. F., RUNDLE, J. B., KLEIN, W. and MARTINS, J. S. Sá. (2004). *Ergodicity in natural fault systems*, Pure Appl. Geophys. *161*, 1957–1968, doi:10.1007/s00024-004-2542-1.

TIAMPO, K. F., RUNDLE, J. B., KLEIN, W., HOLLIDAY, J., MARTINS, J. S. Sá, and FERGUSON, C. D. (2007), *Ergodicity in natural earthquake fault networks*, Phys. Rev. E. *75*, 066107, doi:10.1103/PhysRevE.75.066107.

TIAMPO, K. F., RUNDLE, J. B., MCGINNIS, S., GROSS, S. J., and KLEIN, W. (2002), *Mean-field threshold systems and phase dynamics: An application to earthquake fault systems*, Europhys. Lett. *60*, 481–487.

XU, X. W. and DENG, Q. D. (1996), *Nonlinear characteristics of paleoseismicity in China*, J. Geophys. Res. *101*, 6209–6231.

XU, X. W., ZHANG, P. Z., WEN, X. Z. (2005), *Features of active tectonics and recurrence behaviors of strong earthquake in the western Sichuan Province and its adjacent regions*, Seismol. Geol. (in Chinese with English abstract) *27*, 446–461.

YI, G. X., WEN, X. Z., and XU, X. W. (2002), *Study on recurrence behaviors of strong earthquakes for several entireties of active fault zones in Sichuan-Yunnan region*, Earthq. Res. China (in Chinese with English abstract) *18*, 267–276.

ZHANG, P. Z., XU, X, W., WEN, X. Z., and RAN, Y. K. (2008), *Slip rate and recurrence intervals of the Longmen Shan active fault zone, and tectonic implications for the mechanism of the May 12 Wenchuan earthquake, 2008, Sichuan, China*, Chin. J. Geophys. (in Chinese with English abstract) *51*, 1066–1073.

WU, Y.-H., CHEN, C., and RUNDLE, J. B. (2008), *Detecting precursory earthquake migration patterns using the pattern informatics method*, Geophys. Res. Lett. *35*, L19304, doi: 10.1029/2008GL035215.

(Received August 15, 2008, revised February 4, 2009, accepted September 7, 2009, Published online March 4, 2010)

Pure Appl. Geophys. 167 (2010), 763–782
© 2010 Birkhäuser / Springer Basel AG
DOI 10.1007/s00024-010-0076-2

Pure and Applied Geophysics

Ergodicity and Earthquake Catalogs: Forecast Testing and Resulting Implications

K. F. Tiampo,[1] W. Klein,[2] H.-C. Li,[3] A. Mignan,[4] Y. Toya,[1] S. Z. L. Kohen-Kadosh,[1] J. B. Rundle,[5] and C.-C. Chen[3]

Abstract—Recently the equilibrium property of ergodicity was identified in an earthquake fault system (Tiampo *et al.*, Phys. Rev. Lett. *91,* 238501, 2003; Phys. Rev. E *75,* 066107, 2007). Ergodicity in this context not only requires that the system is stationary for these networks at the applicable spatial and temporal scales, but also implies that they are in a state of metastable equilibrium, one in which the ensemble averages can be substituted for temporal averages when studying their behavior in space and time. In this work we show that this property can be used to identify those regions of parameter space which are stationary when applied to the seismicity of two naturally-occurring earthquake fault networks. We apply this measure to one particular seismicity-based forecasting tool, the Pattern Informatics index (Tiampo *et al.*, Europhys. Lett. *60,* 481–487, 2002; Rundle *et al.*, Proc. National Acad. Sci., U.S.A., Suppl. 1, *99,* 2463, 2002), in order to test the hypothesis that the identification of ergodic regions can be used to improve and optimize forecasts that rely on historic seismicity catalogs. We also apply the same measure to synthetic catalogs in order to better understand the physical process that affects this accuracy. We show that, in particular, ergodic regions defined by magnitude and time period provide more reliable forecasts of future events in both natural and synthetic catalogs, and that these improvements can be directly related to specific features or properties of the catalogs that impact the behavior of their spatial and temporal statistics.

Key words: Earthquake forecasting, ergodic behavior, PI method, Thirumalai–Mountain metric, Canadian seismicity, Taiwanese seismicity.

[1] Department of Earth Sciences, University of Western Ontario, London, ON N6A 5B7, Canada. E-mail: ktiampo@uwo.ca
[2] Department of Physics, Boston University, Boston, MA 02215, USA.
[3] Institute of Geophysics, National Central University, Jhongli 320, Taiwan, ROC.
[4] Risk Management Solutions, Science and Technology Research, London, UK.
[5] Center for Computational Science and Engineering, University of California, Davis, CA 95616, USA.

1. Introduction

In recent years, several different forecasting methods based upon the quantification of patterns in seismicity data but relying on different approaches, have been proposed with varying degrees of success (Mogi, 1969; Ogata, 1992; Bufe and Varnes, 1993; Dodge *et al.*, 1996; Bowman *et al.*, 1998; Keilis-Borok, 1999; Orihara *et al.*, 2002; Tiampo *et al.*, 2002; Evison and Rhoades, 2004; Holliday *et al.*, 2007; Papazachos *et al.*, 2007; Mignan, 2008b). This approach tends to result in forecasting techniques which either are related to an inferred or hypothesized underlying physical process, or are purely statistical in nature. The behavior and origin of the precursory seismicity patterns determined by this approach therefore often remain obscure and difficult to comprehend. This lack of understanding presents a significant barrier to understanding the accuracy of these methods, the best route to improving their forecasting capability, and the possible limits to their accuracy as imposed by the physics of the process.

Over the past several years we have pursued the idea that at least some earthquake fault systems behave in the same way as a thermal system in equilibrium; or more precisely in punctuated equilibrium. Integral to this theory is the hypothesis that small to moderate earthquakes can be treated as fluctuations of a system in thermal equilibrium but large events represent a different class. Large events drive the system out of equilibrium, which after some time is re-established until the next large event again forces the system out of equilibrium. This scenario was developed from the study of models of earthquake faults (Ferguson *et al.*, 1999; Klein *et al.*, 1997) although the picture of punctuated equilibrium was recently verified in some real fault systems (Tiampo *et al.*, 2003, 2007). In this work, it was demonstrated that the application of a measure of ergodicity commonly used in materials science, the Thirumalai–Mountain (TM) fluctuation metric (Thirumalai *et al.*, 1989), could be used to define the ergodic ranges of the seismic catalog, during which the statistics of the natural catalogs were stationary. The intriguing aspect of this discovery is that the property of punctuated equilibrium appears to be associated with the effectiveness of the Pattern Informatics (PI) method for forecasting large events (Tiampo *et al.*, 2002; Holliday *et al.*, 2006a; Toya *et al.*, 2009). Specifically, those fault systems that display punctuated equilibrium, in which periods of ergodic behavior define equilibrium behavior, have large ($M \geq 5$) events that are forecast by the PI method with considerable accuracy, while the accuracy of the PI method on systems that do not display punctuated equilibrium is severely diminished.

The supposition that we wish to test here is that the accurate forecasting of large events using techniques based on isolating fluctuations in the seismicity rate, either quiescence or activation, are dependent on an accurate assessment of the spatial and temporal averages. Although the impact on the stationary behavior of these averages may vary depending on the particular technique and the significance of those averages to a particular technique, the PI method is particularly relevant for this study. As will be shown in Section 2.2, the calculation of the PI index includes both the mean and variance of the regional, or background, seismicity in its functional form and therefore is particularly sensitive to the behavior of these quantities.

Here we present results from two different tectonic regions, eastern Canada and Taiwan, and synthetic catalogs. These analyses provide evidence that the statistics of natural fault systems are ergodic over particular spatial and temporal scales. This property is used to define the extent and nature of specific parameter ranges and spatial and temporal regions in which long-term spatial and temporal averages are both stationary and equivalent. We then employ the PI index to test the hypothesis that earthquake forecasting methodologies based upon linearized analyses of seismic catalog data require that certain equilibrium properties, such as stationarity, are necessary to produce accurate, effective results, and that ergodicity measures can be used to quantify the spatial and temporal scales for which the fault network behaves as an equilibrium system in order to improve these forecasts.

This paper is organized as follows: In Section 2 we present the two methodologies employed in this study, the TM metric and the PI index. In Section 3 we present analysis and results for two select tectonic regions, eastern Canada and Taiwan. In the last subsection of Section 3 we introduce the development and analysis of synthetic catalogs. In Section 4 we summarize our results and conclusions.

2. Method

2.1. Thirumalai–Mountain Metric

The TM fluctuation metric, as originally proposed by Thirumalai and Mountain, measures effective ergodicity, or the difference between the time average of a quantity, generally related to the energy, E_i, at each site, and its ensemble average over the entire system (Thirumalai *et al.*, 1989; Mountain and Thirumalai, 1992; Thirumalai and Mountain, 1993). A necessary condition for ergodicity is stationarity, so that regions of phase space identified as effectively ergodic are maintaining stationary statistics over a given period of time. During those time periods, the temporal and spatial averages are equal, and can be substituted for one another. In addition, it is a behavior generally limited to equilibrium states. A system in equilibrium is said to have statistical symmetry, meaning that it is impossible to distinguish individual particles in terms of the averaged properties associated with those particles (Thirumalai and Mountain, 1993). Practically, if the actual measurement time is finite, but long enough, most of the accessible phase space of a system will be equally likely sampled, and the system is effectively ergodic.

E_i is a given quantity which characterizes the energy, or a proxy for that energy, at a discrete particle, i. The time average of an individual particle over the interval is

$$\varepsilon_i(t) = \frac{1}{t}\int_0^t E_i\left(t'\right) \mathrm{d}t'. \qquad (1)$$

For this particular application of the TM metric, we use $E_i(t) \equiv \psi_i(t)$, the number of events greater than some minimum magnitude cutoff, M_{cutoff}, calculated for each year. Specifically, we specify for each region or subregion under analysis in each tectonic zone a grid, or set of boxes, that we will use to separate our seismic events by location, designated by i. For each box we count the number of seismic events to occur during a specific time period which, for all subsequent analyses discussed here, is set at one year. This is our seismicity rate, $\psi_i(t)$, or number of events for each year at each location i. As detailed in Ferguson *et al.* (1999) and Tiampo *et al.* (2007), for our purposes in calculating the TM metric and its inverse, $\psi_i(t)$ is a proxy for energy, and substituted for $E_i(t)$in equation (1), above.

The ensemble average of those temporal averages over N particles of a system in the same time interval is

$$\bar{\varepsilon}(t) = \frac{1}{N}\sum_{i=1}^{N} \varepsilon_i(t). \qquad (2)$$

The Thirumalai–Mountain fluctuation metric, $\Omega_e(t)$, is defined as

$$\Omega_e(t) = \frac{1}{N}\sum_{i=1}^{N} [\varepsilon_i(t) - \bar{\varepsilon}(t)]^2. \qquad (3)$$

From Equation (3), the TM metric is the spatial variance of the temporal averages over all particles in the same time interval. Thirumalai *et al.* (1989) showed that, for an effectively ergodic system, the inverse of the fluctuation metric will grow linearly with time. This behavior is related to the central limit theorem, where the variance becomes a constant controlled by the large sample size N and divided by the increasing time t. Therefore

$$\Omega_e(t) = \frac{D_e}{t}, \qquad (4)$$

where D_e is related to the rate at which the phase space is explored, and can be used to calculate the time to reach effective ergodicity, τ_e, where $\tau_e \approx 1/D_e$ (Thirumalai *et al.*, 1989; Mountain and Thirumalai, 1992). As noted above, in this research, the number of earthquakes greater than a given magnitude, m, is adopted as a proxy for energy release (Tiampo *et al.*, 2007). The expected linearly increasing relationship between the inverse TM metric and time was obtained in both simulated and natural earthquake fault systems (Ferguson *et al.*, 1999; Tiampo *et al.*, 2003, 2007). Specifically, Tiampo *et al.* (2003, 2007), showed that, for California seismicity between 1932 and the present, the inverse TM metric is piecewise linear for particular realizations of boxsize discretization and minimum magnitude cutoff, allowing for the identification of regions of punctuated ergodicity in which the seismicity rate, for those particular parameter values, was stationary.

2.2. *Pattern Informatics Index*

The PI index is an analytical method for quantifying the spatiotemporal seismicity rate changes in historic seismicity (Rundle *et al.*, 2002; Tiampo *et al.*, 2002; Holliday *et al.*, 2006a). The quantitative basis for this technique lies in phase dynamics analysis, where the temporal evolution of a dynamical quantity is described by changes in the amplitude and phase angle of a state vector (Holmes *et al.*, 1996; Mori and Kuramoto, 1998; Rundle *et al.*, 2000). By integrating these rotations over time and differencing them at various observation times, space–time patterns of anomalous activity or quiescence emerge that seem to precede large events. Analysis supports the interpretation that the evolution of these seismicity patterns is the surface expression of changes in the underlying stress level, where smaller seismicity is acting as a sensor, responding to and illuminating locations of stress change prior to large events (Tiampo *et al.*, 2006a). The spatial locations of these coherent structures can be related to the locations of large earthquakes that occur a few years after their formation, and appear to have some forecasting capability (Tiampo *et al.*, 2002, 2006a; Holliday *et al.*, 2006a).

To calculate the PI index in natural catalogs, seismicity data are taken from existing observations in seismically active areas. In order to ensure that all the data are consistent with the completeness of the catalog, a minimum magnitude cutoff, M_{cutoff}, is imposed on the natural catalogue. The time period for analysis and magnitude cutoff is related to the fact that the PI analysis uses numbers of events, and relies on changes in the relatively large numbers of smaller earthquakes as a proxy for stress change (Dieterich, 1994; Dieterich *et al.*, 2000; Toda *et al.*, 2002; Tiampo *et al.*, 2006a). The data are discretized by location into coarse-grained grid boxes. A gridding boxsize of 0.1 degrees, latitude and longitude, has proven successful in California, but may not be applicable in all tectonic areas. Time series are created for each of these gridded locations, where an individual time bin represents the total number of events at each location that occurred during that time step. Each location is denoted x_i, where i equals 1 to N locations.

We define the observed seismic activity rate $\psi_{obs}(x_i,t)$ as the number of earthquakes per unit time, $M \geq M_{cutoff}$, within the box centered at x_i at time t. In the following analysis the time period will be one year, so that $\psi_{obs}(x_i,t)$ is the number of events per year. We then define the time-averaged seismicity function $\mathbf{S}(x_i,t_0,t)$ over the interval $(t - t_0)$:

$$\mathbf{S}(x_i, t_0, t) = \frac{1}{(t - t_0)} \int_{t_0}^{t} \psi_{obs}(x_i, t)\, dt. \tag{5}$$

There are N locations for $\mathbf{S}(x_i,t_0,t)$. Denoting spatial averages over the N boxes by $\langle\rangle$, the phase function $\mathbf{S}'(x_i,t_0,t)$ is then defined to be the mean-zero, unit-norm function obtained from $\mathbf{S}(x_i,t_0,t)$:

$$\mathbf{S}'(x_i, t_0, t) = \frac{\mathbf{S}(x_i, t_0, t) - \langle \mathbf{S}(x_i, t_0, t) \rangle}{\|\mathbf{S}(x_i, t_0, t)\|}, \tag{6}$$

where $\|\mathbf{S}(x_i,t_0,t)\|$ is the square root of the variance (the standard deviation) calculated for all spatial boxes.

Dividing by the constant variance normalizes the seismicity to its background and illuminates small, local variations in seismicity. Changes in seismicity are given by $\Delta\mathbf{S}'(x_i,t_1,t_2) = \mathbf{S}'(x_i,t_0,t_2) - \mathbf{S}'(x_i,t_0,t_1)$. The physical picture is that $\Delta\mathbf{S}'(x_i,t_1,t_2)$, the local change in seismicity rate, identified by the relative change in spatial and temporal seismic activity, is related to the underlying stress changes in the system. $\Delta\mathbf{S}'(x_i,t_1,t_2)$ will be positive or negative, depending on whether it is identifying seismic activation or quiescence (Tiampo *et al.*, 2002, 2006b). In the final calculation, $\Delta\mathbf{S}'(x_i,t_1,t_2)$ is averaged over all possible base years, t_0, in order to reduce the effects of noise in the seismic catalog. Note that the calculation of $\mathbf{S}'(x_i,t_0,t)$ and, therefore, $\Delta\mathbf{S}'(x_i,t_1,t_2)$, is a function of both the long-term mean and spatial variance of the background seismicity rate. As a result, accurate calculation of $\Delta\mathbf{S}'(x_i,t_1,t_2)$ is dependent on constant long-term averages, or stationarity.

The PI index, $\Delta\mathbf{P}$, is related to the power in $\Delta\mathbf{S}'(x_i,t_1,t_2)$. $\Delta\mathbf{P} = \Delta\mathbf{P}(x_i,t_1,t_2) = \{\Delta\mathbf{S}'(x_i,t_1,t_2)\}^2 - \mu_{p}$.. μ_p is the spatial mean of $\{\Delta\mathbf{S}'(x_i,t_1,t_2)\}^2$ and represents the time-dependent background. For any given catalog, it is possible to compute the power in the relative change in seismicity over any given time period using the PI index.

In recent years, the PI technique has been applied to California seismicity for the purpose of identifying systematic space–time variations in seismicity that are a measure of current and future seismic moment release (Tiampo *et al.*, 2002). These variations, or anomalies, objectively identify regions where the local stress variations result in anomalous seismicity changes at a location x, during a time interval, $t_2 - t_1$ (Tiampo *et al.*, 2006a). For example, the PI index has been used to identify stress shadows associated with the small but persistent changes in seismicity associated with large events in southern California (Tiampo *et al.*, 2006b). In addition, given that the spatiotemporal variations in seismicity related to underlying stress change serve to identify precursory activity prior to large earthquakes, the PI method has been tested and shown to be a reliable method for forecasting these upcoming events (Holliday *et al.*, 2006a). Holliday et al. (2006a) also demonstrated that ongoing studies of this forecast methodology suggest that the PI index represents a threshold probability, in which a value may be chosen above which there is 100% expectation of a large event.

3. Analysis

The results in 2.1, above, suggest that for particular ranges of the important spatial and temporal parameters the statistical averages of the seismic fault system are stationary, and the spatial averages can be substituted for the temporal averages, and vice versa. In addition, it implies that, for some measurable and significant period of time, the system is in equilibrium. This has potential implications for earthquake forecasting, as many seismicity-based forecasting methods, such as the PI index, rely on identifying variations in the seismicity relative to some quantifiable measure of what is considered the longer-term, averaged behavior of the system. In addition, many of these methods rely on some type of linear behavior of the system in their forward projections. Our hypothesis is that the TM metric can be used to identify those spatial and temporal regions of the seismic catalog that will provide better seismicity-based forecasts.

The PI index, in particular, relies on both of the above assumptions (RUNDLE *et al.*, 2000, 2002; TIAMPO *et al.*, 2002). Specifically, as can be seen in equation (6), above, the PI index uses both the spatial mean and variance in its calculation. If either the mean or the variance fluctuates too much over time, the estimate of the relative fluctuation in the seismicity will be inaccurate. Therefore, it should be possible to optimize, or at the very least improve, PI forecasts for various natural historic catalogs using the TM metric to identify the ergodic limits of the spatial and temporal parameters that factor into its calculations and to test the above hypothesis. Specific parameters affecting the ergodic behavior are the time period and spatial range of the analysis, the grid boxsize, and the minimum magnitude cutoff. In the following sections, we will present results from eastern Canada and Taiwan.

3.1. Natural Catalogs

3.1.1 Eastern Canada

The eastern Canadian shield is an intraplate tectonic region in which most of the earthquake activity is associated with lithospheric-scale geological structures arising from past orogenic and rifting episodes, and these appear to control the spatial distribution of its seismicity (ADAMS and BASHAM, 1989). Figure 1 shows the regional seismicity between 1903 and the

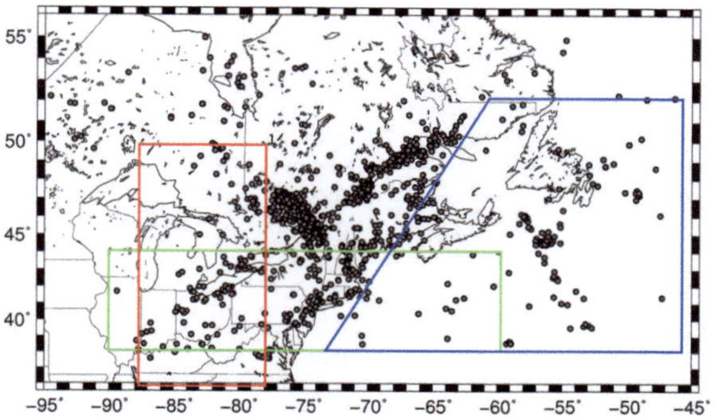

Figure 1
Seismicity in northeastern U.S. and southeastern Canada from 1903 to 2002, $M \geq 2.5$. The three subregions used for testing the spatial heterogeneity of the catalog completeness are outlined by the *blue* (subcatalog 1), *red* (subcatalog 2), and *green* (subcatalog 3) *polygons*. SHEEF catalog courtesy of S. Halchuck and J. Adams, GSC

present. Three subregions used for testing the spatial heterogeneity of the catalog completeness are outlined by the blue (subcatalog 1), red (subcatalog 2), and green (subcatalog 3) polygons. Activity varies from low-level background seismicity to moderate ($4 \leq M \leq 6$) and rare large (magnitude $M \sim 7$) earthquakes. The bulk of the seismicity occurs within the top 25 km of the crust (ADAMS and BASHAM, 1989).

The data used here come from the Seismic Hazard Earthquake Epicentre File (SHEEF) catalog, compiled by the Geological Survey of Canada. The original catalog includes all located events of magnitude 2.5 or greater reverting to 1627. The magnitudes given in the catalog are Nuttli magnitudes (M_N), a body-wave magnitude scale. We tested the completeness of the catalog by calculating the Gutenberg-Richter frequency–magnitude statistics for approximately 20-year-long catalog intervals from 1900 through 2002. For each time window, we apply a least-squares fitting routine to the frequency data for a range of minimum magnitudes. We then calculate the root-mean-square (RMS) error of the fit associated with each minimum magnitude. Finally, we assign a minimum magnitude of completeness (MC) to the time window based on a combination of the lowest RMS error and visual inspection of the data (Table 1). The results indicate that the catalog is complete above magnitude 3.0 back to 1950; however the seismicity is clustered

Table 1

Frequency–magnitude statistics for the SHEEF catalog

Years	*a*-value	*b*-value	RMS	Mc
Full catalog				
1900–1920	2.874	0.435	0.031	4.5
1930–1950	5.467	0.903	0.049	4.3
1950–1970	5.531	0.997	0.076	3.0
1960–1980	5.581	1.021	0.058	2.9
1970–1990	5.206	0.868	0.064	2.5
1980–2002	5.543	0.920	0.072	2.5
Subcatalog 1				
1950–1970	2.180	0.356	0.044	3.6
1960–1980	3.791	0.736	0.064	3.5
1970–1990	5.207	1.048	0.077	3.2
1980–2002	5.573	1.130	0.084	3.2
Subcatalog 2				
1950–1970	4.993	1.038	0.061	2.9[a]
1960–1980	7.119	1.642	0.125	3.1[a]
1970–1990	5.575	1.197	0.074	2.9[a]
1980–2002	5.503	1.208	0.066	2.9[a]
Subcatalog 3				
1950–1970	3.985	0.795	0.038	2.9
1960–1980	4.612	1.010	0.087	2.9
1970–1990				2.9[b]
1980–2002	5.832	1.355	0.066	2.9[a]

Least-squares fitting is used to determine the minimum magnitude of completeness (Mc) for each time period. Three spatial subcatalogs are also tested

[a] Least-squares fitting was only applied to the events of $M \leq 4.0$ due to the paucity of larger events

[b] Based on visual inspection. The fitting algorithm could not be applied due to a lack of $M > 3.5$ events

centrally in the region, suggesting that the spatial coverage of the network is heterogeneous. We repeated the above analysis for the Atlantic off-shore seismicity (Fig. 1, subcatalog 1), the western edge of the catalog (Fig. 1, subcatalog 2) and the southern portion containing the U.S. activity (Fig. 1, subcatalog 3). The results, summarized in Table 1, suggest that the off-shore coverage is very poor, even in recent years. However, the southern and western subcatalogs appear to be complete above magnitude 2.9 from 1950 onwards, with only a slight, questionable decrease in the completeness for the western boundary for the period 1960–1980. To further confirm these results, we calculated the cumulative number of events of $M \geq 3.0$ region-wide as a function of time (not shown here). The data are distinctly linear, with no noticeable changes in slope. As a result, for this analysis, calculations are limited to the area between 42° to 52° latitude and −60° to −85° longitude. However, it should be noted that the a-value does increase over time in all subregions, as seen in Table 1, which is a result of the increase in number and quality of the networks over time (personal communication, Dr. R. Mereu). This is confirmed with a calculation of the average number of events per year of $M \geq 3$. While in the first half of the century there are significant jumps in rate, followed by periods of constant production, a sudden increase in the early 1980s is followed by a steady rise in the rate of events per year. This effect is almost nonexistent for $M \geq 4$.

Figure 2 shows the normalized inverse TM metric for eastern Canada, 1903–2001 and boxsize equal to 0.2 degree. Other boxsizes were tested (not shown), however, there was no significant improvement at smaller sizes. In Fig. 2a, calculated for $M \geq 3.0$, there are no periods in which the inverse TM metric appears linear, and after the mid-1970s it begins to curve downward and continues to do so throughout the 1990s, until the present. Figure 2b shows a plot for $M \geq 3.0$, 1970–2001. The inverse TM metric in eastern Canada, $M \geq 3.0$, for 1970–2001 is not linear, or piecewise linear, but continues to maintain its downward curve after 1985. Our interpretation here is that this is associated with the increase in number of events of $M \geq 3.0$ per year that is seen over that same period.

In Fig. 2c, now calculated for $M \geq 4.0$ and 1903–2001, the inverse TM metric displays linear behavior for periods prior to 1980, followed by a change in the average slope and a subsequent piecewise linear behavior. Figure 2d shows the same metric, $M \geq 4.0$, but for the time period 1970–2001. Here the behavior is primarily linear, with a constant slope.

The PI index, $\mathbf{\Delta P}$, calculated for eastern Canada for the period 1995 to 2000, inclusive, is shown in Fig. 3. Here the gridded boxsize is again 0.2 degrees. Using the interpretation of Holliday et al. (2006a), we assign a threshold probability of zero to the PI index, in which any location with a value of $\mathbf{\Delta P} > 0$ is chosen such that there is 100% expectation of a large event, and is subsequently colored red. Figures 3a and 3b are $\mathbf{\Delta P}$ for $M \geq 3.0$ and $M \geq 4.0$, respectively. Inverted triangles represent events of $M \geq 4.0$ during the period 1995–2000

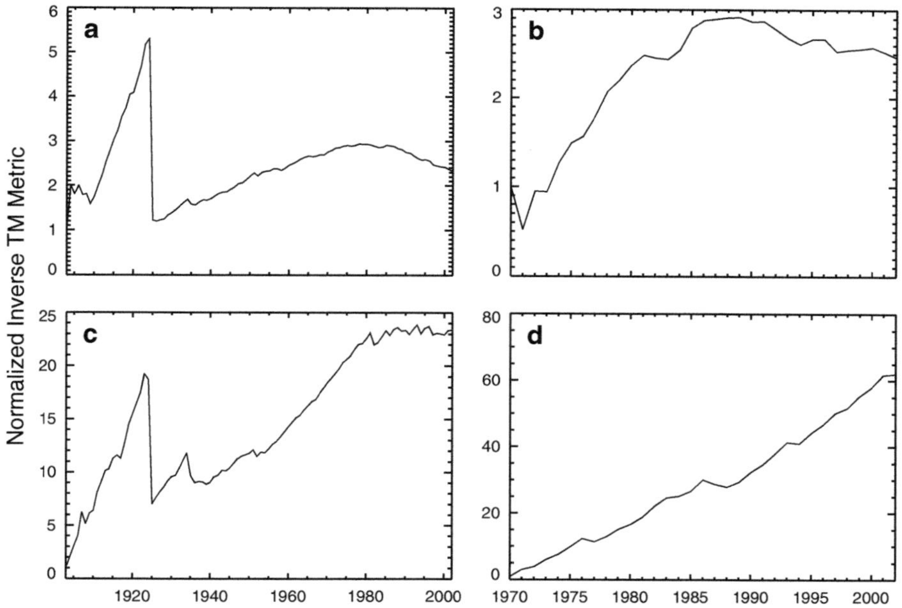

Figure 2
Normalized inverse TM metric for eastern Canada, 1903–2001 and boxsize = 0.2 degree: (**a**) $M \geq 3.0$, 1903–2001; (**b**) $M \geq 3.0$, 1970–2001; (**c**) $M \geq 4.0$, 1903–2001 and (**d**) $M \geq 4.0$, 1970–2001

(inclusive), while circles represent those events of $M \geq 4.0$ that occurred between 2001 and the present.

Relative operating characteristic (ROC) diagrams were computed in order to compare the performance of the results shown in Fig. 3. In an ROC diagram, the hit rate, H, is plotted versus the false alarm rate, F. Details of the ROC diagram computation can be found elsewhere (Jolliffe and Stephenson, 2003), but the calculation is designed to ensure that if the area under the ROC curve is larger, the forecast is better (Holliday *et al.*, 2005). A random forecast will generate a straight line along the diagonal of an ROC plot. A more stringent null hypothesis for comparison of forecast results is the Relative Intensity (RI) index, as described in Holliday *et al.* (2006b). Relative intensity is defined as the long-term average seismic rate for each location during the time period from t_0 to t_2, and is a measure of historical seismicity. As such, it provides a reasonable null hypothesis to test against the PI index. Here, a forecast is successful (i.e., a hit) if the epicenter of the earthquake lies within a hotspot box or in one of the eight adjoining boxes (Moore, 1962).

Figure 4 shows ROC diagrams for the PI maps shown in Fig. 3. The PI calculation is shown in blue, the RI calculation is shown in red. Figure 4a shows the ROC diagram for PI and RI calculations for $M \geq 3$, a forecast time period of 2001 through 2006 (inclusive) and events of $M \geq 4.5$. Figure 4b shows the same calculation, but for $M \geq 4$. Here the ROC plots in Fig. 4 only include values for $F = [0,0.5]$. For $F > 0.5$, the PI and RI plots are virtually indistinguishable, as the PI forecast begins to approach the RI forecast. In addition, the implication of effective binary forecasting implies that any forecast with a high false alarm rate will result in too many false positives for practical application.

Comparing these figures, it is clear that the forecast in Fig. 3a, $M \geq 3.0$, has a significantly higher false positive rate than Fig. 3b, although with approximately the same number of successful forecast events. In contrast, the forecast for $M \geq 4.0$, Fig. 3b, has a much lower false positive rate and the same amount of false negatives. The ROC diagrams in Fig. 4 support this result. In Fig. 4a, the RI index performs better than the PI index for the forecast of Fig. 3a, while the PI index is significantly improved in Fig. 3b. The ROC diagram of Fig. 4b shows that the PI index is above the RI index, because the PI index has a lower false positive rate, as seen in

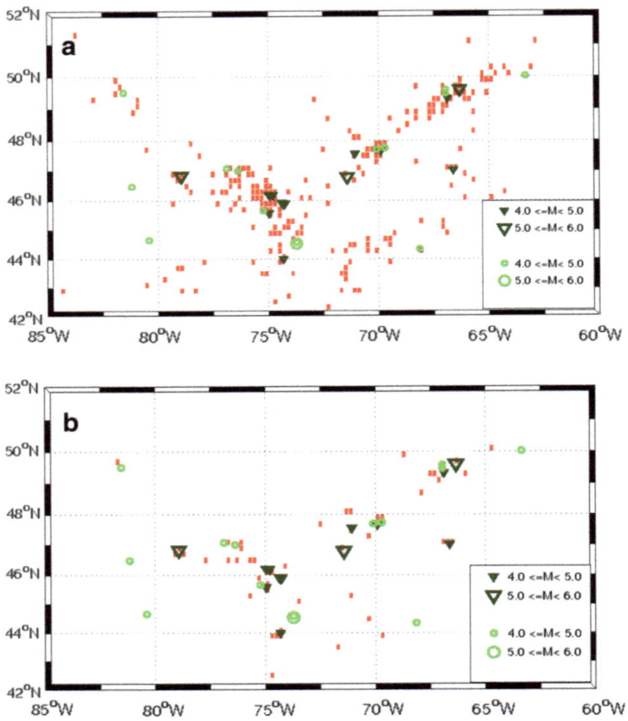

Figure 3

PI index, **ΔP,** for eastern Canada for the period 2000–1991 (inclusive), and a boxsize = 0.2 degrees. Here *red boxes* indicate all locations where $\Delta\mathbf{P} > 0$. **ΔP** is shown for (**a**) $M \geq 3.0$ and (**b**) $M \geq 4.0$. *Inverted triangles* represent events of $M \geq 4.0$, scaled with increasing magnitude, during the period 1995 through the end of 2000, while *circles* represent those events that occurred between 2001 and 2006 (inclusive)

Fig. 3b. The Pierce Skill Score (PSS) (Jolliffe and Stephenson, 2003), the integrated area under the complete ROC curves, supports this result, where the PSS for the PI index is greater than that of the RI index in Fig. 4b.

This analysis supports the hypothesis that configuration of the PI index calculation in accordance with the effectively ergodic regimes as delineated by the TM metric can improve its forecasting accuracy. It should be pointed out that this particular result is not surprising, in light of what is known about this particular network and the details of the PI calculation. If the apparent seismicity rate increases over time (in this case $M \geq 3.0$), then the PI index will calculate that as an increase in activation over any time difference during that period. PI anomalies will be interpreted incorrectly as false positives, as seen in Fig. 3. In this case, ensuring that the system is ergodic over a particular time period ensures that the background rate is stationary; an assumption implicit in the formulation of the PI method and a requirement for an accurate evaluation.

3.1.2 Taiwan

Taiwan is situated on the western portion of the Pacific Rim, along the Ryukyu subduction zone. East of the island the leading edge of the Pacific plate subducts under the Philippine Sea plate, while on the western side the Philippine Sea plate subducts below the Eurasian plate. South of the island, the South China subplate subducts under the Philippine Sea plate. Finally, on the southeast portion of Taiwan, the Eurasian and Philippine Sea plate suture zone separates the region into two separate tectonic provinces (Wu *et al.*, 2008). As a result of significant compression across the island, Taiwan has a relatively high seismic activity rate with a significant depth component.

The seismicity data employed in this research was the Taiwanese Central Weather Bureau (CWB)

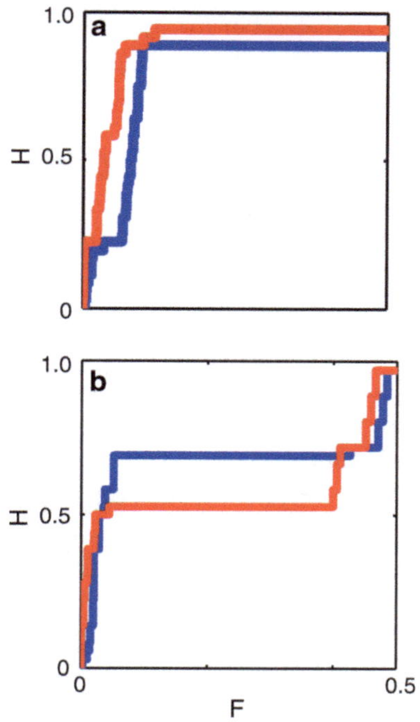

Figure 4
ROC diagrams for the PI maps shown in Fig. 3. The PI calculation is shown in *blue*, the RI calculation is shown in *red*. In **a** is shown the ROC diagram for PI and RI calculations for $M \geq 3$, a forecast time period of 2001 through 2006 (inclusive), and a forecast for events of $M \geq 4.5$. In **b** is shown the same calculation, but for PI and RI calculations of $M \geq 4$

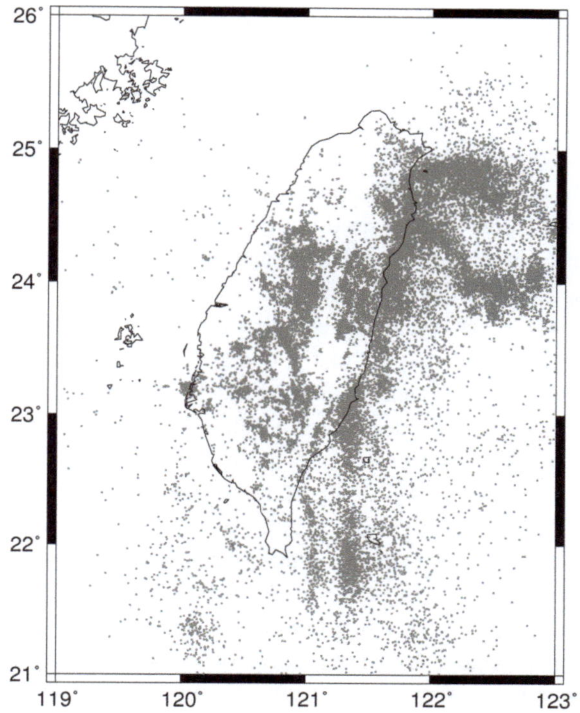

Figure 5
Seismicity map, $M \geq 3$, for Taiwan, from the CWB seismic network catalog

catalog from 1 January, 1973 to 31 December, 2006, shown in Fig. 5. The research region was 21° to 26° latitude, 119° to 123° longitude. Spatial variation of Mc for the entire catalog was computed using the maximum curvature method (WIEMER and WYSS, 2000; WIEMER, 2005). The catalog generally appears to be complete for $M \sim 3$ for the entire region, with the exception of some off-shore areas to the south and east with relatively little seismic activity or poor network coverage. Nevertheless, temporal catalog homogeneity, rather than a static spatial variation of M_{cutoff}, is particularly important for PI studies. While time series of M_{cutoff} maps can be computed, this is a cumbersome, time-consuming process that is strongly dependent on the choice of time step. Therefore, an evaluation of the catalog's homogeneity was performed using the method of HABERMANN (1983).

Results from the application of Habermann's test of catalog homogeneity (HABERMANN, 1983; WIEMER, 2005) for the Taiwanese catalog suggest that times of apparent seismicity rate increases (for magnitude bands with negative Z values) and rate decreases (for magnitude bands with positive Z values) are identified at the 0.01 significance level based on the assumption that all samples derive from asymptotically probable standard Gaussian distributions. To offset limitations in this method, such as a lack of normality for short-time windows and sensitivity to variations in statistical properties (MARSAN 2003; MATTHEWS and REASENBERG, 1987), a catalog is often declustered with Reasenberg's algorithm (REASENBERG, 1985) before further analysis. Results after declustering appear to be affected by the artificial noise introduced in the declustering procedure, however. While there are no obvious magnitude shifts (HABERMANN, 1987), significant rate increases (likely due to detection level improvements) early in the catalog (pre-1980) and rate anomalies corresponding to large earthquake sequences do exist in the data. Again, while declustering by Reasenberg's algorithm is not appropriate for the Taiwanese subduction zone, whose tectonics and seismicity patterns vary significantly in

3-D, it is often included in the application of Habermann's test for catalog homogeneity (above) (e.g., Wyss and Wiemer, 1997), making the results for that analysis difficult to interpret.

In order to better understand the applicability of the declustering procedure, we performed a separate evaluation of the variation in energy, magnitude and number of events with depth and time, which we also applied to the declustered catalog (Reasenberg, 1985). This result suggests that there is an artificial clustering in depth, likely due to assigned default depths for inadequately located events, which remained after declustering. The declustering tool also did not clean the circa 1986 clusters, likely due to the inappropriate use of the windowing scheme. Catalog inhomogeneity is apparent for pre-1980s and for deep events over the entire catalog, likely due to the detection-level improvement of the observation network in both space and time. There is also an apparent decrease in the number of reported shallow events (< 3 km) since 1990, although this could be due to the improvement of event location accuracy.

Based on our previous research, we applied the TM method to the original data set, with no declustering applied, because an evaluation of stationarity will locate those regions in which both the spatial and temporal averages converge to constant values, making this methodology simpler to interpret than those attempted above. In addition, these stationary averages must approach the same value in order for the system to be ergodic (Tiampo *et al.*, 2007).

Figure 6 shows a number of inverse TM metric calculations for the Taiwanese subduction zone. As a result of the evaluation of the variation in energy, magnitude and number of events with depth and time detailed above, only events of $M \geq 3$ were used. The focal depth threshold adopted was 30 km to limit the effect of decreasing accuracy with depth. We implemented both a two-dimensional (2-D) and a three-dimensional (3-D) calculation of both the TM metric and PI calculations, in order to study the effect of seismic variations with depth. Figure 6a is a 2-D analysis of events of $M \geq 3$ and a gridding boxsize of 0.1°. Figures 6b and 6c are 3-D analyses, $M \geq 3$, but with a discretization of $0.1° \times 0.1° \times 1$ km and $0.1° \times 0.1° \times 0.25$ km, respectively. Figure 6d is another 2-D analysis, but for $M \geq 4$ and a boxsize of 0.02°. Figures 6e and 6f are again 3-D analyses and a discretization of $0.1° \times 0.1° \times 1$ km and $0.1° \times 0.1° \times 0.25$ km, $M \geq 4$.

Here all of the plots except for Fig. 6a have periods of linear behavior after the mid-1980s. However, it is also clear that ergodicity improves with increasing magnitude and decreasing boxsize. It is also apparent that the 3-D analysis provides significantly better ergodic behavior than the 2-D analysis. Finally, Figs. 6a through 6f all show breaks in ergodicity related to events of $M \geq 5$, including the occurrence of the $M = 7.3$ Chi–Chi earthquake in 1999. One interesting point is that ergodic behavior is evident at small depth increments, where the accuracy is on the order of 4 km. This likely is a result of the central limit theorem, again, as the variance, or error, in location is reduced by averaging over many realizations. However, additional testing using synthetic catalogs with various noise levels will be necessary to confirm that this hypothesis is correct.

The TM method successfully evaluates the general homogeneity of the catalog both in space and time, while Habermann's test of catalog homogeneity was too sensitive and impractical for this catalog data, in part because of the necessity of declustering the catalog first. Note that this particular Habermann's test required almost two weeks for a full analysis (2-D only) on an Intel processor with 2.4 GHz CPU and 3 GB Memory, while the TM analysis takes minutes to hours, depending on whether a 2-D or 3-D computation is performed. For the specific purpose of determining which regions are stationary for forecasting applications, the TM method provides a clearer and simpler interpretation, and is also an efficient and accurate method for locating these regions.

A 3-D calculation of the PI index is shown in Fig. 7 for the ergodic time period, 1994–1998 and the parameters of Fig. 6c, i.e., $M \geq 3$, a unit time step of 1 year and a discretization of $0.1° \times 0.1° \times 0.25$ km. In addition, we have included plots for the RI index (Holliday *et al.*, 2006b). In these figures, green circles indicate the large events of $M \geq 6$ that occur after the forecast period, 1999–2003. Figures 6a and 6b are for a depth range of 7.75–8.00 km, while Figs. 6c and 6d are for a depth range of 13.75–14.00 km. The color

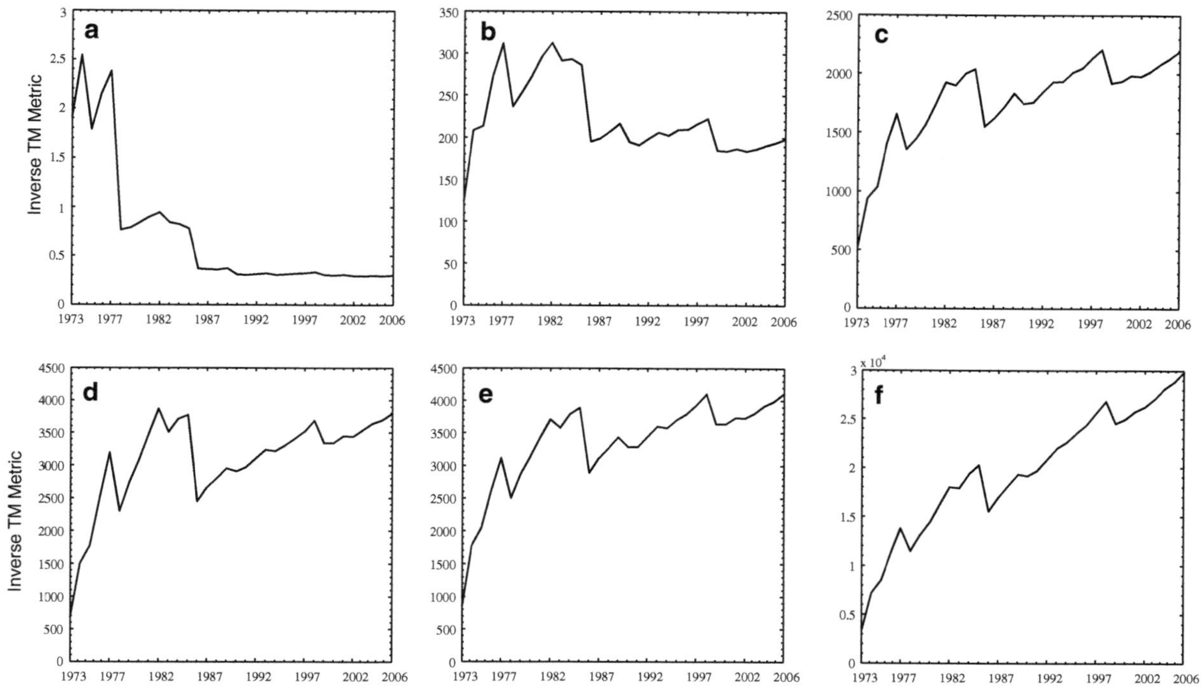

Figure 6
Plots of the inverse TM metric, unit time step of 1 year, for the Taiwanese subduction zone. **a** $M \geq 3$, 2-D analysis, and a gridding boxsize (coarse-graining discretization) of 0.1°; **b** $M \geq 3$, 3-D analysis, and a discretization of 0.1° × 0.1° × 1 km; **c** $M \geq 3$, 3-D analysis, and a discretization of 0.1° × 0.1° × 0.25 km; **d** $M \geq 4$, 2-D analysis, and a boxsize discretization of 0.02°; **e** $M \geq 4$, 3-D analysis, coarse-grained box is 0.1° × 0.1° × 1 km; and **f** $M \geq 4$, 3-D analysis, discretization of 0.1° × 0.1° × 0.25 km

scale is logarithmic, where the number represents the exponent. All values are normalized by the maximum values of each plot. Note that the PI index forecasts the location of all future events (green circles) in those depth ranges, including the Chi–Chi earthquake in Fig. 6a, as does the RI index, although with fewer false positives.

Figure 8 is also a 3-D calculation of the PI and RI indices, similar to that shown in Fig. 7, but for the *nonergodic* time period 1988–1997. Here the inverse TM metric shows a significant break in 1992 (Fig. 6c). Here, while the RI index continues to forecast the events post-1998, in addition to a numerous false positives, the PI index fails to forecast the same events, although it still limits false positives to a small percentage.

Figure 9 quantifies the difference in performance of these two forecasts (1994–1998 and 1988–1997). ROC diagrams were computed in order to compare the performance of the 3-D PI calculation for the entire depth range for the two different time periods. Here, again, the F axis is shortened to $F < 0.3$. In Fig. 9a, for the ergodic time period 1994–1998, the curve of the PI map remains significantly above that of the RI map until their intersection at approximately 0.25 on the horizontal axis, illustrating that the PI map provides a better forecast than the RI map for this case. However, in Fig. 9b, we see that while the performance of the RI index is remarkably stable for long-time periods, and varies little whether the system is ergodic or nonergodic, the performance of the PI index is significantly poorer for the nonergodic time period. For those ergodic cases tested to date, the PI index outperforms the RI index, primarily because the RI index does not identify quiescence. These results again support the hypothesis that PI index forecasts can be improved significantly by restricting the important parameters of the calculation to those associated with ergodic behavior.

However, the above test is different from that of eastern Canada. In eastern Canada, the ergodic calculation is an improvement over the nonergodic

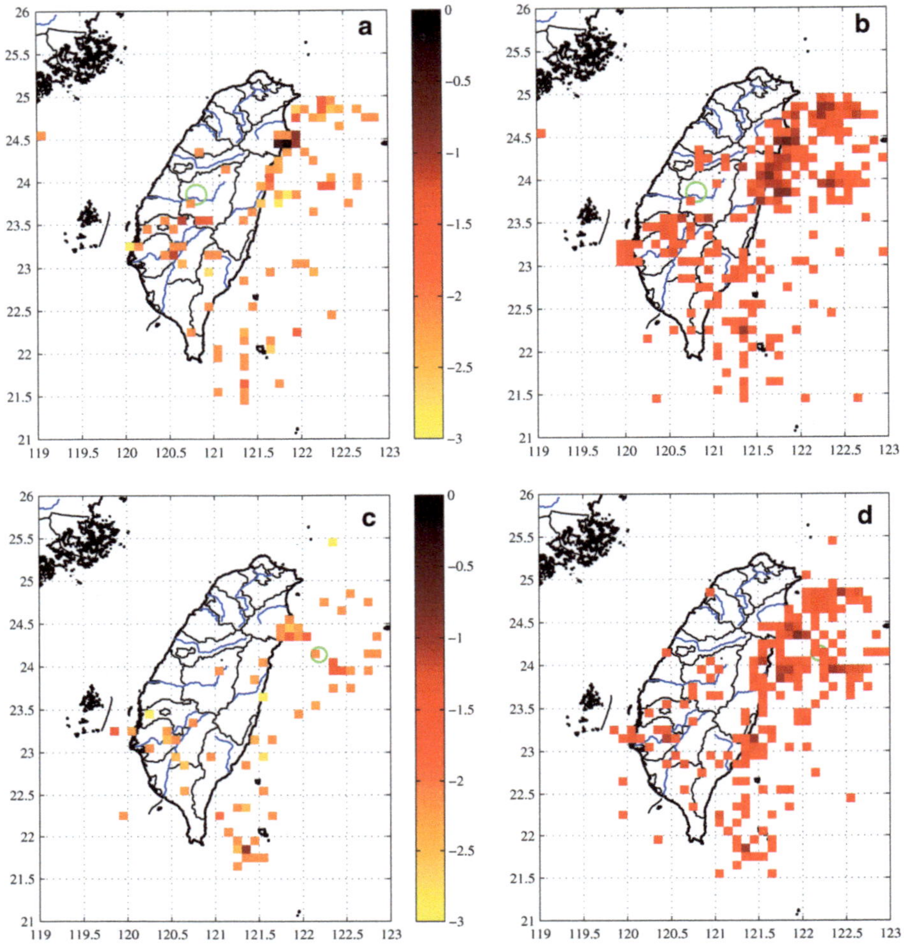

Figure 7
Plots of 3-D PI and RI maps the Taiwanese subduction zone for varying depth ranges. The analysis corresponds to those parameters used for Fig. 6c; i.e., $M \geq 3$, a unit time step of 1 year, and a discretization of $0.1° \times 0.1° \times 0.25$ km. The forecast period is 1994–1998. *Green circles* indicate the large events which $M \geq 6$ that occur in the period 1999–2003. Depth ranges of **a** and **b** 7.75 to 8.00 km, **c** and **d** 13.75 to 14.00 km. The *color scale* is logarithmic, where the number represents the exponent. All values are normalized by the maximum values of each plot

calculation because it reduces the number of false positives but still forecasts the same number of events; i.e., there are no additional misses. In the case of Taiwan, the false positive rate is the same for the nonergodic and ergodic time periods; it is an increase in misses that apparently degrades the forecast for the nonergodic time period. This suggests that the cause of the nonergodic behavior may be different in these examples.

The above analysis for natural catalogs provides evidence that the occurrence of large events and their aftershock sequences associated with these breaks in ergodicity are affecting the ability of the PI index to detect precursory quiescence patterns. This possibility, and potential mechanisms for its occurrence, is investigated using synthetic catalogs in Section 3.2.

3.2. Synthetic Seismicity Catalogs

In a preliminary attempt to better understand the behavior of the PI index and potentially other linear forecasting algorithms based on historic seismicity, we studied synthetic seismicity catalogs that were formulated to match the statistics of natural

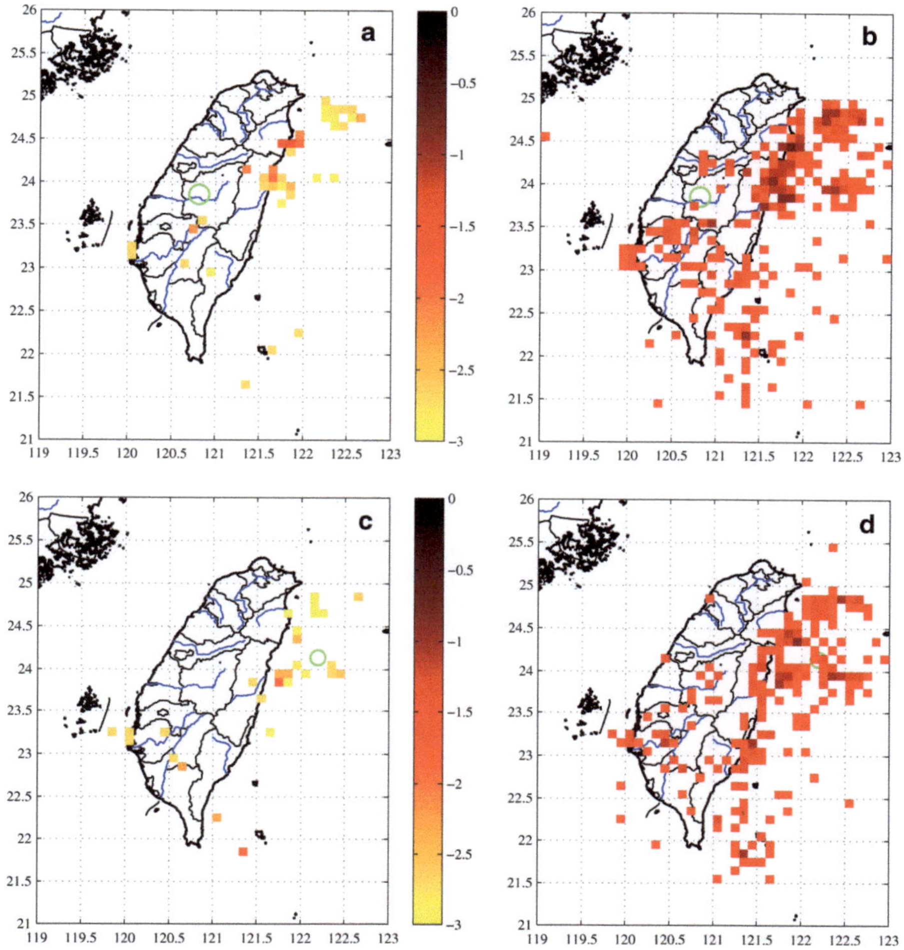

Figure 8
Plots of 3-D PI and RI maps with parameters as shown in Fig. 6c, but for the forecast period of 1988–1997. *Green circles* indicate the large events which $M \geq 6$ that occur in the period 1998–2006. Depth ranges of **a** and **b** 7.75 to 8.00 km, **c** and **d** 13.75 to 14.00 km. The *color scale* is as seen in Fig. 7, above

seismicity catalogs. While these synthetic catalogs are only a simplification of the natural earthquake process, a first step toward a more detailed set of synthetic catalogs, in this case we have included an additional signal reminiscent of those hypothesized to act as precursors to larger events in order to study one potentially quantifiable precursory signal, anomalous quiescence. This particular precursory seismic pattern, based upon Non-Critical Precursory Accelerating Seismicity Theory (or Non-Critical PAST), recently was proposed to explain the formation of accelerating seismicity that may be observed before large earthquakes (MIGNAN, 2008a). In this particular formulation, physically realistic catalogs with activation and quiescence are formulated that depend on only two input parameters: the regional events rate and the noise level in the theoretical precursory signal. In recent work (MIGNAN and TIAMPO, 2010), it was shown that the PI index is successful at identifying the anomalous quiescence associated with the Non-Critical PAST phenomena.

Non-Critical PAST is based upon the hypothesis that Accelerating Moment Release (AMR) occurs, in a significant number of cases, prior to the occurrence of large events. AMR is typically modeled by a simple power-law time-to-failure equation (BUFE and VARNES, 1993), defined by the relation is

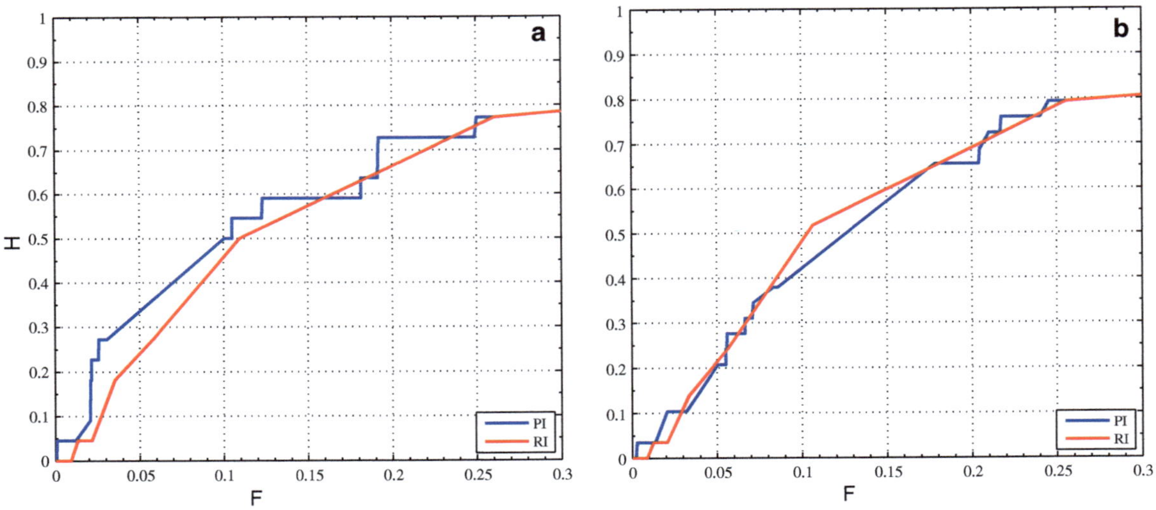

Figure 9
ROC diagrams for the PI and RI maps shown in Figs. 7 and 8. On the *left* is shown the ROC diagram calculated for the ergodic time period, 1994–1998, of Fig. 7. On the *right* is shown the ROC diagram corresponding to the nonergodic time period, 1988–1997, of Fig. 8. *Key* is as shown on plots

$$\varepsilon(t) = A + B(t_f - t)^m, \quad (7)$$

where t_f is the time of the main shock, and A, B and m are constants. Originally, and in subsequent work, $\varepsilon(t)$ was chosen to be the cumulative Benioff strain (Bowman *et al.*, 1998; Brehm and Braile, 1998; Zöller *et al.*, 2000; Papazachos *et al.*, 2007) but recent work suggests that the cumulative number of events is preferable (Jiménez *et al.*, 2005; Mignan *et al.*, 2007). Because the cumulative number of events is a quantity directly input into the PI index, the relative effects of quiescence and activation should be quantifiable using this particular measure.

In Non-Critical PAST, the earthquake cycle is formulated using simple geometrical and stress transfer concepts, and physically-based synthetic catalogues can be constructed in order to test existing forecasting techniques and to understand how they react to different seismicity characteristics (Mignan, 2008a). A spatiotemporal distribution of background seismicity exists, surrounding and incorporating the stress field of a loading fault. At $t = 0$, a large earthquake occurs on the fault and creates a stress shadow which, for some period of time, appears as anomalous quiescence. The size of the stress shadow decreases through time, due to loading at depth, until it finally disappears at $t = t_f$ prior to the occurrence of new large event (i.e., the expected main shock). Note, here, that this is not a complete catalog in that there is only one event that can be forecast, the large ($M \sim 7$) event at the end of the sequence.

Following Mignan et al. (2007), the density of events that composes the background seismicity is δ_b^0 outside the stress shadow and δ_b^- inside the stress shadow, with $\delta_b^0 > \delta_b^-$. If the spatiotemporal limits of the stress shadow are defined by a power law time-to-failure relation as in equation (7), the cumulative number of events $\psi(t)$ that comprise the background seismicity increases as a power-law function through time before the main shock, $\psi(t) \propto -(t_f - t)^{D/3}$, where D is the fractal dimension of the associated fault network (Mignan *et al.*, 2007). Note, again, that Non-Critical PAST also quantifies precursory quiescence similar to the seismicity pattern commonly known as a Mogi Doughnut (Mogi, 1969, 1977, 1979).

Theoretical synthetic catalogues for this study were created based on the above principles of the Non-Critical PAST. In order to form a precursory accelerating seismicity pattern based on an increase of the a value, only the location and time of background events must be included (Mignan *et al.*, 2007). Efforts to include magnitude information and aftershock clustering were implemented in addition to the standard formulation.

The background seismicity density δ_b^0 and the noise ratio δ_b^-/δ_b^0 are adjustable parameters, depending respectively on the regional seismic activity and on the effect of the stress shadow on seismicity. In this study, δ_b^0 is fixed to 1000 background events in the catalog per year. This corresponds to 0.008 events/degree^2/day in a 2000 × 2000 km square grid. The noise level, δ_b^-/δ_b^0 is set at 0.1. The location of background events and the time of occurrence are controlled by the stress contour limit $\sigma^*(x,y,t)$. The stress field $\sigma(x,y,t)$ is calculated for a thrust fault ($L = 200$ km, $W = 10$ km, dip = 45°) located in the center of a grid of 2000 × 2000 km. The time interval is [$t_0 = 0$; $t_f = 20$ year]. For each time step of one year, background events have a uniform random distribution in space and time with a density δ_b^0 outside the stress contour $\sigma^*(x,y,t_i)$ and δ_b^- inside the stress contour $\sigma^*(x,y,t_i)$. Coulomb stress calculations are performed with the source code AlmondX (courtesy of G.C.P. King) (Okada, 1992). The stress field $\sigma(x,y,t_i)$ results from a displacement on the fault at t_0, producing a stress shadow which then decreases through time from the linear superposition of a loading stress field using the back-slip model concept (Savage, 1983). The fault is ready to fail at the end of the catalogue, with a coupled precursory quiescence and activation seismicity pattern defined over the entire 20 year time frame (Mignan *et al.*, 2007).

Background events are implemented randomly in space and time from a Gutenberg-Richter relation with a b value of 1.0 (Gutenberg and Richter, 1944). The only anticipated main shock after the end of the catalog is the only event preceded by Non-Critical PAST seismicity. Here, the length and width are fixed, and the magnitude of the upcoming event is calculated from the relationship $M \sim 4 + \log(L^*W)$.

The minimum magnitude of background events is fixed at $M_{\min} = 1.0$, to ensure that, for the density of events in the catalogue, the maximum magnitude $M_{\max}$ for the region is less than 6.0. This corresponds to an a value of approximately 4.4, lower than in many active seismic regions (e.g., $a \sim 6$ in California), while a minimum magnitude $M_{\min} = 1.0$ also is artificial when compared to actual catalogues. This artificial value is due to the fact that these synthetic catalogues simulate the preparation process of only one large earthquake. However, all background events have the same weight independent of their magnitudes, consequently it is the density of background events ($\sim \delta_b^0$) that is similar to the actual catalogues, not the magnitude window.

Spatiotemporal clustering is included in the simulated background seismicity by following the Epidemic Type Aftershock Sequence Model (ETAS) model (e.g., Ogata, 1992; Ogata and Zhuang, 2006). ETAS describes ordinary seismic activity $\lambda(t)$ by the superposition of a constant rate for background seismicity, μ, and the aftershock sequence of any given event j at time t_j, such that

$$\lambda(t) = \mu + \sum_{\{j:tj<t\}} e^{\alpha(Mj-Mc)} \nu(t - t_j). \tag{8}$$

In this case, the ETAS background rate, μ, is not constant but evolves following the Non-Critical Past (see below). $\nu(t) = K(t + c)^{-p}$ is the Modified Omori function (e.g., Utsu, 1961) and $Ke^{\alpha(Mj-Mc)}$ is the aftershock productivity. In this study, productivity is proportional to $10^{\alpha(Mj-Mc)}$ (Helmstetter *et al.*, 2005; Felzer *et al.*, 2004) for a cutoff magnitude $M_c = 1.0$ ($M_c = M_{\text{cutoff}}$).

The components of the parameter vector $\theta = (\mu, p, \alpha, c, K)$ of the ETAS model (Table 2) are obtained from the literature, except for the regional background seismicity rate μ which is fixed to δ_b^0 (and to δ_b^- inside the stress shadow) (Guo and Ogata, 1997; Felzer *et al.*, 2004; Helmstetter *et al.*, 2003, 2005). K is obtained from $K = n/(b{\cdot}\ln(10){\cdot}(M_{\max} - M_{\min})/(p-1)/c^{(p-1)})$ (Helmstetter *et al.*, 2005). The spatial distribution of aftershocks is assumed to be Gaussian,

Table 2

Parameters used in the synthetic catalogues

Parameter		Value
Magnitude-frequency	a value	4.4
	b value	1
	$M_{\min}$	1
	$M_{\max}$	6
Background	δ_b^0 (events/day/degree^2)	0.008
	δ_b^- (events/day/degree^2)	0.1 δ_b^0
ETAS clustering	p value	1.1
	α (magnitude^{-1})	1
	c (days)	0.001
	K	0.004

such that ~99% of aftershocks occur inside a circle of radius $L = 0.01 \times 10^{0.5M}$ from the main shock.

Aftershocks also obey Båth's law, in which the difference in magnitude between the main shock and its largest aftershock Δm is fixed to 1.0 to ensure that all aftershocks have a magnitude $M < M_{\max}$ (SHCHERBAKOV *et al.*, 2005). Note that the synthetic catalogs are more naturally defined using $\delta_{tot} = \delta_b^0 + \delta_{aft}$, where δ_{aft} represents ETAS clustering. In natural catalogs, this more accurately reflects the total seismicity, including natural aftershock sequences. Different δ_{tot} were tested by changing the cutoff magnitude M_{cutoff} of synthetic catalogues. Variation of $M_{min} = [1, 2, 3]$ (i.e., $\delta_{tot} \sim [4.95, 0.42, 0.038]$ events/degree^2/year), provides the capability to study theoretical precursory seismicity in different active seismic regions.

Figure 10 shows the synthetic seismicity, calculated as detailed above. On the left is shown a synthetic catalog based on the Non-Critical PAST, with a reverse fault ($L = 200$ km, $W = 10$ km, dip = 45°, expected magnitude $M \sim 7.3$) located in the center of an 18-degree square region. On the right is shown a similar synthetic seismicity catalog with no Non-Critical PAST signal, i.e., background seismicity with random ETAS events, as formulated above, for use as a null hypothesis.

Figure 11 shows the normalized inverse TM metric for the synthetic seismicity catalogs shown in Fig. 10. Figure 11a is the inverse TM metric for the Non-Critical PAST catalog of Fig. 10a, while Fig. 11b shows the inverse TM metric for the random catalog of Fig. 10b. Note that both have significant breaks at year 15, related to a large event and its associated ETAS aftershock sequence.

The PI index calculations for an ergodic and a nonergodic time period for the catalog of Fig. 10a are shown in Fig. 12a and Fig. 12c, respectively. Figure 12e shows the PI index calculation for the same ergodic time period as Fig. 11a, but for the random null hypothesis catalog of Fig. 10b. Here the calculation is for events of $M \geq 1$ and a discretization boxsize of 1 degree. Figure 12a is calculated for a forecast time period of year 6 through year 14. Figure 12c is calculated for a forecast time period of year 6 through year 16. Figure 12e is calculated for year 6 through year 14 again, but for the catalog shown in Fig. 10b, as already noted. Red designates all values of the $\mathbf{\Delta P} > 0$. Here we see that the forecast using the ergodic time period (Fig. 12a)

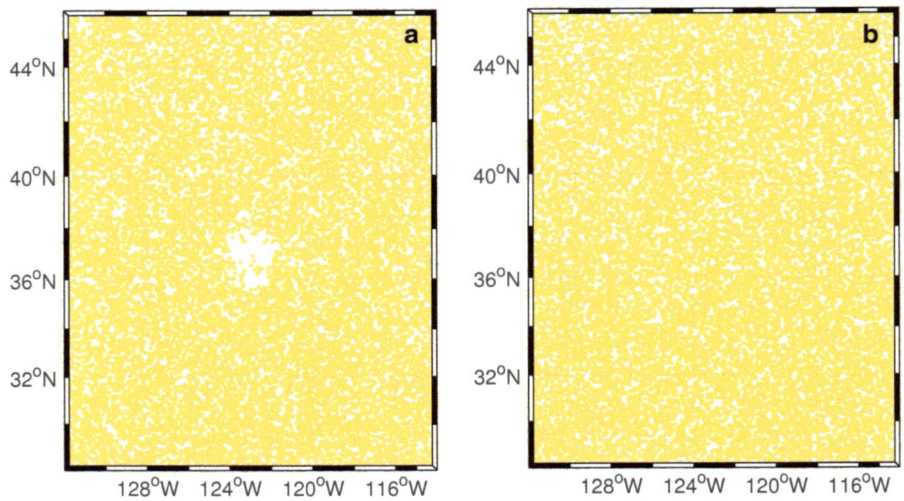

Figure 10
Synthetic seismicity plots. **a** Synthetic catalog based on the Non-Critical PAST. A reverse fault ($L = 200$ km, $W = 10$ km, dip = 45°, expected magnitude $M \sim 7.3$) is located in the center of an 18-degree square region. A stress *shadow* is created at $t = 0$ and decreases in size through time, until $t_f = 20$ years. The density of background events is $\delta_b^0 = 0.008$ event/day/degree (and $\delta_b^- = 0.1\delta_b^0$ inside the stress shadow) such that only events of magnitude $1 < M < 6$ occur in the 20 years time frame. Background events produce aftershocks which can also produce their own aftershocks following the ETAS model and Bath's law ($\Delta m = 1$). **b** A similar synthetic seismicity catalog with no Non-Critical PAST signal, i.e., background seismicity with random ETAS events

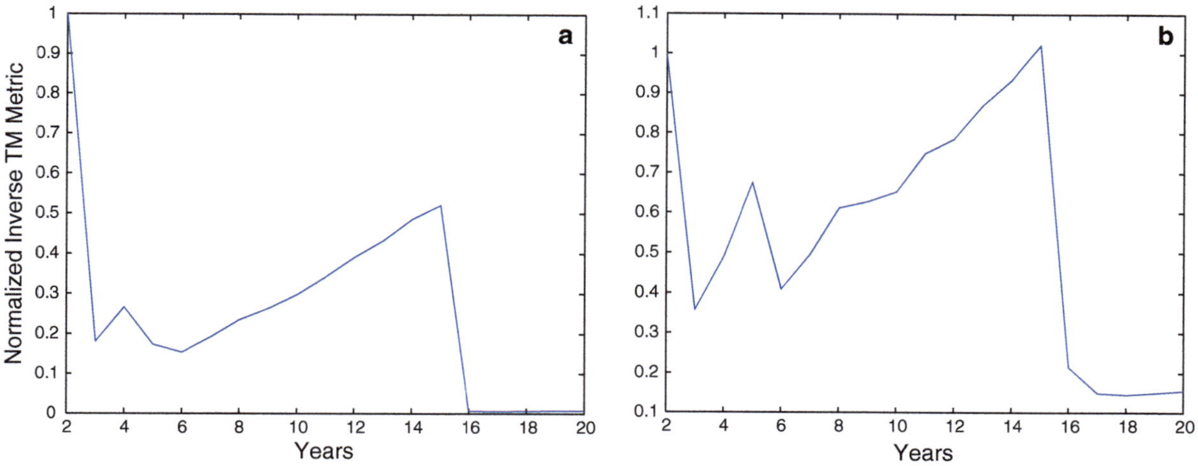

Figure 11
Plot of the inverse TM metric for the synthetic seismicity catalogs shown in Fig. 10. **a** The inverse TM metric, normalized to the first value, for the Non-Critical PAST catalog of Fig. 10a, while **b** shows the inverse TM metric for the random catalog of Fig. 10b

successfully forecasts the location of the upcoming event, while the random catalog for the same time period fails to forecast the event (Fig. 12e). More importantly, the forecast over the nonergodic time period (Fig. 12c) also fails to forecast the upcoming event. Note that we have not computed ROC diagrams for this test, as the catalog ends just prior to the occurrence of the large event at year 20 and such a test would not be relevant for this data set.

Figures 12b, 12d, and 12f show the seismicity rate change, $\Delta \mathbf{S}'$, for the same catalogs and time periods as in Figs. 12a, 12c, and 12e. The color scale is normalized to the maximum value, where red is activation and blue is quiescence. Figure 12b confirms that the PI index is locating the anomalous quiescence associated with the stress shadow from the recurring large earthquake. For the nonergodic time period, Fig. 12d, the absolute value of that quiescence is significantly smaller, relative to the activation in the aftershock sequence from the large event in year 15. This results in the loss of significant power in the PI signal and a failure to forecast the upcoming event in Fig. 12b. This helps to explain the results for Taiwan in Section 3.1.2. Again, the identification of ergodic regions of phase space can significantly improve the forecast capability of the PI method in particular, and seismicity-based forecasting techniques in general.

4. Conclusions

In this work we employ the Thirumalai–Mountain fluctuation metric and data from existing seismic monitoring networks to identify the presence of punctuated ergodicity, an equilibrium property, in the dynamics of the natural earthquake fault system in three different tectonic regions; California, Taiwan, and eastern Canada. Because stationarity is a necessary but not sufficient condition for ergodicity, the identification of the appropriate spatial, temporal, and magnitude parameter ranges for ergodic behavior also ensures that the spatial and temporal averages are stationary and equivalent for those same regions of space and time.

After the identification of the ergodic regions for each catalog, we investigated the behavior of a forecasting technique that uses seismicity data and is linked intrinsically to the statistics of the seismic catalog, the PI index. The application of the PI index to both ergodic and nonergodic minimum magnitude cutoffs in eastern Canada provides evidence that ergodic periods provide significantly better forecasts, and that this is a result of the identification of stable, long-term seismicity rates which result in a reduction in the number of false positives.

We also tested the forecast capability of the PI index for ergodic and nonergodic time periods in Taiwan and synthetic catalogs. This analysis provides

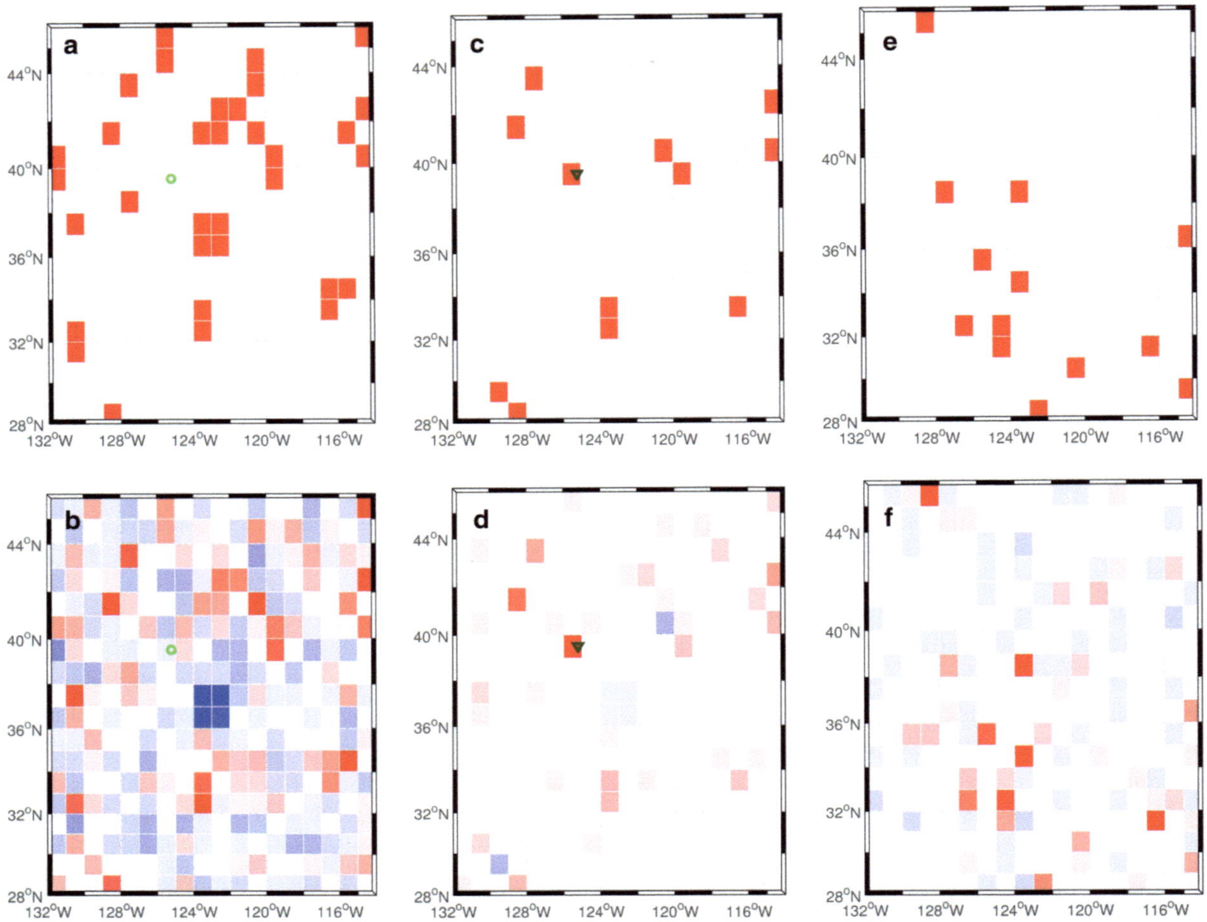

Figure 12
Plot of the PI index, $\Delta \mathbf{P}$, and seismicity rate change, $\Delta \mathbf{S}$, for $M \geq 1$ and a discretization boxsize of 1 degree, for the catalogs shown in Fig. 10. **a** PI index for the Non-Critical PAST catalog of Fig. 10a and a forecast time period of Year 14–Year 6. *Red* designates all values of the $\Delta \mathbf{P} > 0$; **b** $\Delta \mathbf{S}$ for the catalog and time period shown in **a**. The *color scale* is linear, *blue* to *white* to *red*, where *blue* represents relative quiescence and *red* represents relative activation, normalized to the maximum value. **c** $\Delta \mathbf{P}$ for the Non-Critical PAST catalog of Fig. 10a and a forecast time period of Year 16–Year 6; **d** $\Delta \mathbf{S}$ for the catalog and time period shown in **c**; **e** PI index for the random catalog of Fig. 10b and a forecast time period of Year 14–Year 6; and **f** $\Delta \mathbf{S}$ for the catalog and time period shown in **e**

evidence that the affect of larger events in the catalog and their associated aftershocks, which break ergodicity, leads to an overestimation of the background rate and an inability to locate some portion of future events. These misses appear to be linked to anomalous quiescence, which is more difficult to identify as a result of the high activation rate from aftershock occurrence. This affect, associated with aftershock sequence, may also be important for smaller amplitude foreshock sequences, and suggests that using the TM metric to identify ergodic periods will facilitate improvement of the efficacy of forecasting methodologies that focus on activation as well as quiescence.

Finally, comparison with methods that require declustering, also shown in Section 3.1.2, shows that the TM metric is both efficient and accurate in isolating regions of both ergodic and nonergodic behavior. Overall, our results lead to the conclusion that the TM metric provides a clear and useful method for identifying stationary regions of phase space associated with natural earthquake catalogs, and can be used to improve the forecasting capability of the PI index and other seismicity-based forecasting techniques that rely on stable, long-term averages. Future work will include the extension of testing to more realistic and varied synthetic catalogs, attempts to identify methods for further improvement of both the PI index and other seismicity-based forecasting methods, and the development of a declustering algorithm based on the TM metric.

Acknowledgments

The work of KFT and YT was supported by the NSERC and Aon Benfield/ICLR Industrial Research Chair in Earthquake Hazard Assessment. This research also was supported by the Southern California Earthquake Center. SCEC is funded by NSF Cooperative Agreement EAR-0529922 and USGS Cooperative Agreement 07HQAG0008. The SCEC contribution number for this paper is 1213. The research of SZLK-K was funded by the Institute of Catastrophic Loss Reduction. We would like to thank Dr. J. Adams and Dr. S. Halchuck for supplying the SHEEF catalog. HCL is grateful for the effort of CWB to maintain the CWBSN and the support from the National Science Council (ROC), the Institute of Geophysics (NCU, ROC) and the Department of Earth Science (UWO, Canada). Research by HCL is funded by NSC grant 096-2917-1-008-005. The work of CCC was supported by the National Science Council (ROC) and the Department of Earth Sciences (NCU, ROC). Several figures in this presentation are generated by a tool package "MAPI (Mapping and Analysis via Pattern Informatics)" developed by YT (ytoya@uwo.ca). Other images were plotted with the help of GMT software developed and supported by Paul Wessel and Walter H.F. Smith. Other geographical maps in this manuscript were generated using M_Map Toolbox developed by R. Pawlowicz (http://www.eos.ubc.ca:/~rich/). This paper was significantly improved by the comments of two anonymous reviewers.

REFERENCES

Adams, J. E. and Basham, P. W. (1989), *The seismicity and seismotectonics of Canada east of the Cordillera,* Geoscience Canada *16*, 3–16.

Bowman, D. D., Ouillon, G., Sammis, C. G., Sornette, A., and Sornette, D. (1998), *An observational test of the critical earthquake concept,* J. Geophys. Res. *103*, 24,359–24,372.

Brehm, D. J. and Braile, L. W. (1998), *Intermediate-term earthquake prediction using precursory events in the New Madrid Seismic Zone,* Bull. Seismol. Soc. Am. *88*, 564–580.

Bufe, C. G. and Varnes, D. J. (1993), *Predictive modeling of the seismic cycle of the greater San Francisco Bat region,* J. Geophys. Res. *98*, 9871–9883.

Dieterich, J. (1994), *A constitutive law for rate of earthquake production and its application to earthquake clustering,* J. Geophys. Res. *99*, 2601–2618.

Dieterich, J. H., Cayol, V., and Okubo, P. (2000), *The use of earthquake rate changes as a stress meter at Kilauea volcano,* Nature *408*, 457–460.

Dodge, D. A., Beroza, G. C., and Ellsworth, W. L. (1996), *Detailed observations of California foreshock sequences: Implications for the earthquake initiation process,* J. Geophys. Res. *101*, 22371–22392.

Evison, F. F. and Rhoades, D. A. (2004), *Demarkation and scaling of long-term seismogenesis,* Pure Appl. Geophys. *161*, 21–45.

Felzer, K. R., Abercrombie, R. E., and Ekström, G. (2004), *A common origin for aftershocks, foreshocks, and multiplets,* Bull. Seismol. Soc. Am. *94*, 88–98.

Ferguson, C. D., Klein, W., and Rundle, J. B. (1999), *Spinodals, scaling, and ergodicity in a threshold model with long-range stress transfer,* Phys. Rev. E *60*,1359–1374.

Guo, Z. and Ogata, Y. (1997), *Statistical relations between the parameters of aftershocks in time, space, and magnitude,* J. Geophys. Res. *102*, 2857–2873.

Gutenberg, B. and Richter, C. F. (1944), *Frequency of earthquakes in California,* Bull. Seismol. Soc. Am. *34*, 185–188.

Habermann, R. E. (1983), *Teleseismic detection in the Aleutian Island Arc,* J. Geophys. Res. *88*, 5056–5064.

Habermann, R. E. (1987), *Man-made changes of seismicity rates,* Bull. Seism. Soc. Am. *77*, 141–159.

Helmstetter, A., Kagan, Y. Y., and Jackson, D. D. (2005), *Importance of small earthquakes for stress transfers and earthquake triggering,* J. Geophys. Res. *110*, doi:10.1029/2004JB003286.

Helmstetter, A., Ouillon, G., and Sornette, D. (2003), *Are aftershocks of large Californian earthquakes diffusing?* J. Geophys. Res. *108*, doi: 10.1029/2003JB002503.

Holliday, J., Chen, C., Tiampo, K. F., Rundle, J. B., and Turcotte, D. L. (2007), *A RELM earthquake forecast based on pattern informatics,* Seis. Res. Lt. *78*, 87–93.

Holliday, J. R., Nanjo, K. Z., Tiampo, K. F., Rundle, J. B., Tiampo, K. F., and Turcotte, D. L. (2005), *Earthquake forecasting and its verification,* Nonlinear Processes in Geophys. *12*, 965–977.

Holliday, J. R., Rundle, J. B., Tiampo, K. F., Klein, W., and Donnellan, A. (2006a), *Systematic procedural and sensitivity analysis of the Pattern Informatics method for forecasting large ($M > 5$) earthquake events in southern California,* Pure Appl. Geophys., doi:10.1007/s00024-006-0131-1.

Holliday, J. R., Rundle, J. B., Tiampo, K. F., and Turcotte, D. L. (2006b), *Using earthquake intensities to forecast earthquake occurrence times,* Nonlinear Processes in Geophysics *13*, 585–593.

Holmes, P., Lumley, J. L., and Berkooz, G., *Turbulence, Coherent Structures, Dynamical Systems and Symmetry* (Cambridge, University Press, U.K. 1996).

Jiménez, A., Tiampo, K. F., Levin, S., and Posadas, A. (2005), *Testing the persistence in earthquake catalogs: The Iberian Peninsula,* Europhys. Lett., doi:10.1209/epl/i2005-10383-8.

Jolliffe, I. T. and Stephenson, D. B., *Forecast Verification: A Practitioner's Guide in Atmospheric Science* (John Wiley, Hoboken, N. J. 2003).

Keilis-Borok, V. I. (1999), *What comes next in the dynamics of lithosphere and earthquake prediction?* Phys. Earth Planet. Int. *111*, 179–185.

Klein, W., Rundle, J. B., and Ferguson, C. D. (1997), *Scaling and nucleation in models of earthquake faults,* Phys. Rev. Lett. *78,* 3793–3796.

Marsan, D. (2003), *Triggering of seismicity at short timescales following Californian earthquakes*, J. Geophys. Res. *108*, 2266, doi:10.1029/2002JB001946.

Matthews, M.V. and Reasenberg, P. (1987), *Comment on Habermann's method for detecting seismicity rate changes,* J. Geophys. Res. *92*, 9443–9445.

Mignan, A. (2008a), *The Non-Critical Precursory Accelerating Seismicity Theory (NC PAST) and limits of the power-law fit methodology,* Tectonophysics *452,* doi: 10.1016/j.tecto.2008.02.010.

Mignan, A. (2008b), *The stress accumulation model: Accelerating moment release and seismic hazard,* Adv. Geophys. *49*, doi: 10.1016/S0065-2687(07)49002-1.

Mignan, A., King, G. C. P., and Bowman, D. D. (2007), *A mathematical formulation of accelerating moment release based on the Stress Accumulation model,* J. Geophys. Res. *112*, B07308, doi: 10.1029/2006JB004671.

Mignan, A. and Tiampo, K. F. (2010), *Testing the Pattern Informatics index on synthetic seismicity catalogues based on the Non-Critical PAST,* Tectonophysics *483*, 255–268, doi:10.1016/j.tecto.2009.10.023.

Mogi, K. (1969), *Some features of recent seismic activity in and near Japan (2), Activity before and after large earthquakes,* Bull. Earthquake Res. Inst., Univ. Tokyo *47*, 395–417.

Mogi, K. (1977), *Seismic activity and earthquake predictions,* Proc. Symp. on Earthquake Prediction, Seis. Soc. Japan, 203–214.

Mogi, K. (1979), *Two kind of seismic gaps,* Pure Appl. Geophys. *117,* 1172–1186.

Moore, E. F. (1962), *Machine models of self reproduction.* In Proc. Fourteenth Symp. Appl. Math., 17–33.

Mori, H. and Kuramoto, Y. *Dissipative Structures and Chaos* (Springer-Verlag, Berlin 1998).

Mountain, R. M. and Thirumalai, D. (1992), *Ergodicity and loss of dynamics in supercooled liquids,* Phys. Rev. A *45*, 3380–3383.

Ogata, Y. and Zhuang, J. (2006), *Space-time ETAS models and an improved extension,* Tectonophysics *413*, 13–23.

Ogata, Y. (1992), *Detection of precursory seismic quiescence before major earthquakes through a statistical model,* J. Geophys. Res. *97*, 19845–19871.

Okada, Y. (1992), *Internal deformation due to shear and tensile faults in a half-space,* Bull. Seismol. Soc. Am. *82*, 1018–1040.

Orihara, Y., Noda, Y., Nagao, T., and Uyeda, S. (2002), *A possible case of SES selectivity at Kozushima island, Japan,* J. Geodyn. *33*, 425–432.

Papazachos, B. C., Karakaisis, G. F., Papazachos, C. B., and Scordilis, E. M. (2007), *Evaluation of the results for an intermediate-term prediction of the 8 January, 2006 $M_w = 6.9$ Cythera earthquake in the Southwestern Aegean,* Bull. Seismol. Soc. Am. *97*, 347–352, doi: 10.1785/0120060075.

Reasenberg, P. A. (1985). *Second-order moment of central California seismicity, J. Geophys. Res.* 90, 5479–5495.

Rundle, J. B., Klein, W., Tiampo, K. F., and Gross, S. (2000), *Linear pattern dynamics of nonlinear threshold systems,* Phys. Rev. E *61*, 2418–2432.

Rundle, J. B., Tiampo, K. F., Klein, W., and Sá Martins, J. S. (2002), *Self-organization in leaky threshold systems: The influence of near mean field dynamics and its implications for earthquakes, neurobiology and forecasting,* Proc. National Acad. Sci., U.S.A., Suppl. 1, *99*, 2463.

Savage, J. C. (1983), *A dislocation model of strain accumulation and release at a subduction zone,* J. Geophys. Res. *88*, 4984–4996.

Shcherbakov, R., Turcotte, D. L., and Rundle, J. B. (2005), *Aftershock statistics,* Pure Appl. Geophys. *162*, 1051–1076.

Thirumalai, D., Mountain, R. D., and Kirkpatrick, T. R. (1989), *Ergodic behavior in supercooled liquids and in glasses,* Phys. Rev. A *39*, 3563–3574.

Thirumalai, D. and Mountain, R. D. (1993), *Activated dynamics, loss of ergodicity, and transport in supercooled liquids,* Phys. Rev. E *47*, 479–489.

Tiampo, K. F., Rundle, J. B., Klein, W., Holliday, J., Sá Martins, J. S., and Ferguson, C. D. (2007), *Ergodicity in natural earthquake fault networks,* Phys. Rev. E *75*, 066107.

Tiampo, K. F., Rundle, J. B., and Klein, W. (2006a), *Premonitory seismicity changes prior to the Parkfield and Coalinga earthquakes in southern California,* Tectonophysics *413*, 77–86.

Tiampo, K. F., Rundle, J. B., and Klein, W. (2006b), *Stress shadows determined from a phase dynamical measure of historic seismicity,* Pure Appl. Geophys. doi:10.1007/200024-006-0134-y.

Tiampo, K. F., Rundle, J. B., Klein, W., Sá Martins, J. S., and Ferguson, C. D. (2003), *Ergodic dynamics in a natural threshold system*, Phys. Rev. Lett. *91,* 238501.

Tiampo, K. F., Rundle, J. B., McGinnis, S., Gross, S., and Klein, W. (2002), *Mean-field threshold systems and phase dynamics: An application to earthquake fault systems,* Europhys. Lett. *60*, 481–487.

Toda, S., Stein, R. S., and Sagiya, T. (2002), *Evidence from the AD 2000 Izu Islands earthquake swarm that stressing rate governs seismicity*, Nature *419*, 58–61.

Toya, Y., Tiampo, K.F., Rundle, J.B., Chen, C., Li, H., and Klein, W. (2009), *Pattern Informatics approach to earthquake forecasting in 3-D,* Concurrency and Computation: Practice and Experience, doi: 10.1002/cpe.1531.

Utsu, T. (1961), *A statistical study on the occurrence of aftershocks,* Geophys. Mag. *30*, 521–605.

Wiemer, S. (2005), A software package to analyze seismicity: ZMAP. (http://www.earthquake.ethz.ch/software/zmap).

Wiemer, S. and Wyss, M. (2000), *Minimum magnitude of completeness in earthquake catalogs: Examples from Alaska, the Western United States, and Japan,* Bull. Seismol. Soc. Am. *90*, 859–869.

Wyss, M., and Wiemer, S. (1997), *Two current seismic quiescences within 40 km of Tokyo,* Geophys. J. Int. *128*, 459–473.

Wu, Y.-M., Chang, C.-H., Zhao, L., Teng, T.-L., and Nakamura, M., (2008), *A comprehensive relocation of earthquakes in Taiwan from 1991 to 2005,* Bull Seismol. Soc. Am. *98*, 1471–1481.

Zöller, G., Hainzl, S., and Kurths, J. (2000), *Observation of growing correlation length as an indicator for critical point behavior prior to large earthquakes,* J. Geophys. Res. *106*, 2167–2175.

(Received August 1, 2008, revised April 1, 2009, accepted September 16, 2009, Published online March 10, 2010)

Pure Appl. Geophys. 167 (2010), 783–800
© 2010 Birkhäuser / Springer Basel AG
DOI 10.1007/s00024-010-0077-1

Pure and Applied Geophysics

Reverse Tracing of Precursors Applied to the Annual Earthquake Forecast: Retrospective Test of the Annual Consultation in the Sichuan-Yunnan Region of Southwest China

YIZHE ZHAO,[1] ZHONGLIANG WU,[1,2] CHANGSHENG JIANG,[2] and CHUANZHEN ZHU[2]

Abstract—An Annual Consultation on the Likelihood of Earthquakes is regularly undertaken in continental China. It is a special kind of intermediate-term medium-range earthquake forecast and a practical service provided by seismological communities within the capability of modern seismology. One of the problems the Annual Consultation encounters is its false alarm rate, which has been higher than expected and has caused numerous controversies on the methodology and significance of the Annual Consultation. To tackle this problem we use the concept of reverse tracing of precursors (RTP) to analyze the long-term seismic activity in the alarm regions, or regions thought to have increased probability of earthquakes, as identified by the Annual Consultation. We apply the Pattern Informatics (PI) and relative intensity (RI) algorithms to the long-term estimation of seismic hazard in the RTP analysis. A retrospective study of the Sichuan-Yunnan region of southwest China, using the data from 1990 to 2003, shows that, by choosing the threshold proportion of overlapped areas between the long-term estimates and the Annual Consultation, the RTP analysis can correctly remove some of the false-alarm regions from the "black list' and thus reduce the false-alarm rate of the Annual Consultation, without reducing the hit rate. Remarkably, the RTP approach, although originally proposed for short-term earthquake predictions, seems able to contribute to the Annual Consultation as well.

Key words: Reverse tracing of precursors, Intermediate-term medium-range earthquake forecast, Pattern informatics algorithm, Relative intensity algorithm, Sichuan-Yunnan region, Annual consultation.

1. Introduction

Among intermediate-term medium-range earthquake forecasts with variable spatio-temporal ranges, annual earthquake forecasts play a special role, by balancing the capability of time-dependent seismic hazard estimation and the annual plan for earthquake hazard preparedness. In China, such an approach is implemented by the Annual Consultation Meeting on the Likelihood of Earthquakes organized by the China Earthquake Administration (CEA). By combining tectonic, seismic, and other geophysical information, the group of experts draws conclusions about the seismic outlook for the next year and identifies the areas with higher seismic risk (SHI *et al.*, 2000, 2001; ZHENG *et al.*, 2000; ZHANG *et al.*, 2002; MA *et al.*, 2004; WU *et al.*, 2007). Figure 1 shows an example of the output of the Annual Consultation, from which successful forecasts (hits), false-alarms, and failures-to-predict (missed earthquakes) can be seen clearly. Appendix 1 provides some detailed information about the Annual Consultation. Recent studies of statistical seismology reveal that retrospective study of earthquake case histories, although being useful in providing heuristic clues for developing the forecasting/prediction schemes, is intrinsically problematic due to subjective selection of data (e.g., MULARGIA, 1997). The only remedy is forward testing of forecasting performance. The Annual Consultation meeting, although having many problems regarding its science and technology, has the unique scientific merit of being a forward test and, most remarkably, it has been persistently conducted for over three decades.

One of the problems of the Annual Consultation is its false-alarm rate, which has been higher than expected (WANG, 2005). From Fig. 1 it can be seen that there are six false-alarm regions for the year 2002. An extreme case is that annual alarms for a few regions could last for more than 3 years without any earthquake occurring, but have no reason to be excluded

[1] College of Earth Science, Graduate University of Chinese Academy of Sciences, Beijing 100049, China. E-mail: wuzl@cea-igp.ac.cn
[2] Institute of Geophysics, China Earthquake Administration, Beijing 100081, China.

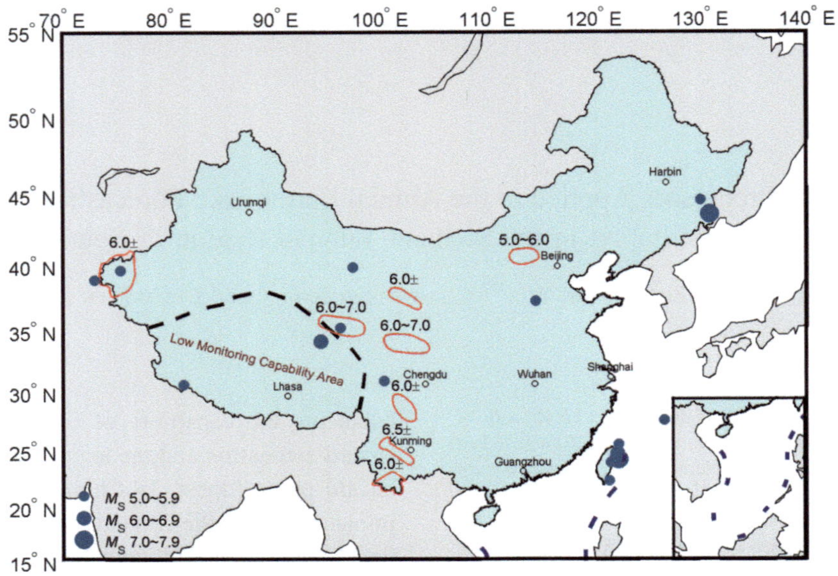

Figure 1

Example of the output of the Annual Consultation: The Annual Consultation for the year 2002. *Red closed full lines* delimit the regions with increased probability of earthquakes, with numbers denoting the expected magnitude range. The *closed black dashed line* in Tibet represents the area with low monitoring capability and is not considered in the Annual Consultation. *Blue dots* illustrate the epicenters of earthquakes which occurred in 2002. Successful hits, false alarms and failures-to-predict are clear in the figure

from the "black list' if a self-consistent criterion is used for the identification of the alarm regions.

To solve this problem we use the reverse tracing of precursors (RTP) approach proposed by Keilis-Borok *et al.* (2004). The idea is to trace long-term anomalies within the regions identified by the Annual Consultation as the alarm regions in the coming year. There are several options for choosing the long-term anomalies (e.g., Field, 2007) in the RTP analysis. Our choice is the Pattern Informatics (PI) and Relative Intensity (RI) algorithms proposed by Rundle *et al.* (2000a, 2003) which are used to analyze the pattern of seismicity. These algorithms have been applied to several regions such as central Japan (Nanjo *et al.*, 2006), Taiwan Island (Chen *et al.*, 2005), and the Sichuan-Yunnan region in southwest China (Jiang and Wu, 2008). Our study is mostly based on the results of Jiang and Wu (2008). This approach is, to some extent, a development of the idea of Keilis-Borok *et al.*, (2004), because originally RTP is based on the analysis of a set of intermediate-term precursors in an area of short-term anomalies, such as the region of long-range activation of seismicity detected by earthquake chains (Shebalin *et al.*, 2004, 2006). In our approach, the RTP is based on the analysis of PI/RI anomalies within the preceding 5 years in an alarm region with increased probability of earthquakes on a 1 year time scale.

Considering the background seismicity, we choose the Sichuan-Yunnan region of southwest China as the testing region. There are two reasons for this choice. First, in this region there were generally two or three alarm areas each year identified by the Annual Consultation, with many false-alarms. Secondly, a retrospective study shows that the PI/RI algorithm performs well in this region, significantly outperforming a random forecast, and being robust against the uncertainties of earthquake parameters and the choice of grids (Jiang and Wu, 2008). We choose the testing period as 1990–2003, a time period which precedes the use of the PI/RI algorithms in the Annual Consultation.

2. *Regions for Study and Data Used for Analysis*

The region considered in this investigation is taken as 20.8°–33.0°N and 97.2°–105.0°E, including Sichuan and Yunnan Provinces and their surrounding regions. The Annual Consultation for this region has taken place since the early 1970s. In the Annual Consultation the factors determining the "alarm regions" include tectonic, geophysical and seismological data, as well as some geochemical and hydrological data, as seen in Appendix 1. Partly due to the distinct tectonic characteristics, with the Xianshuihe, Annninghe, Zemuhe, Xiaojiang, Honghe, Lancangjiang, and Longmenshan fault systems bordering different tectonic blocks (Xu and Deng, 1996), strong earthquakes in this region exhibit a significant component of stochastic clustering, with the recurrence process lacking both ideal periodicity and predictability by the time-predictable or magnitude-predictable models (Xu *et al.*, 2005; Yi *et al.*, 2002). False alarms are also a serious problem in this region. Table 1 shows the statistics of the Annual Consultation from 1990 to 2003. It may be seen that the false alarm rate is quite high, and reducing this rate is one of the important targets for the Annual Consultation.

The earthquake catalogue used in this study is the Monthly Earthquake Catalogue from 1970/01/01 to 2004/01/01 provided by the China Earthquake Networks Center (CENC). The magnitude (M_L) is based on the observations from local seismological networks. The threshold magnitude for completeness has been less than M_L 3.0 since 1970 (Su *et al.*, 2003; Yi *et al.*, 2006). Figure 2a shows the distribution of earthquakes with surface-wave magnitude $M_S \geq 5.0$ in this region for the period from 1970/01/01 to 2004/01/01. The temporal distribution of the cumulative number of earthquakes with $M_L \geq 3.0$ from 1970 is shown in Fig. 2b. The consistent slope shows the homogeneity of monitoring capability over time. The step-like change in 1976 is due not to the change of monitoring capability, but to the aftershock sequence of the 1976/08/16 and 1976/08/23 Songpan-Pingwu M_S 7.2 earthquakes. Figure 2c shows the frequency-magnitude distribution for the region from 1970/01/01 to 2004/01/01. The depth of earthquakes cannot be determined accurately in the routine analysis of the seismological centers. Therefore depth is not considered in the analysis—we just use all the shallow earthquakes.

Table 1

Statistics of the Annual Consultation in the Sichuan-Yunnan region from 1990 to 2003

Year	No. of alarm regions	No. of successful alarms	No. of false alarms	Ratio of false alarms (%)
1990	3	1	2	67
1991	2	0	2	100
1992	2	1	1	50
1993	2	1	1	50
1994	2	0	2	100
1995	3	1	2	67
1996	3	1	2	67
1997	3	2	1	33
1998	3	0	3	100
1999	3	1	2	67
2000	3	0	3	100
2001	2	1	1	50
2002	3	0	3	100
2003	3	1	2	67

3. *Reverse Tracing of Precursors*

One of our objectives is to remove the false alarm regions using the reverse tracing of the precursors (RTP) approach. In this approach, seismicity patterns are analyzed in the reverse order of their appearance: Short-term patterns are analyzed first, although they emerge later. This approach has a natural Earth-specific conceptual explanation (Keilis-Borok, 2002; Keilis-Borok *et al.*, 2004; Shebalin *et al.*, 2004, 2006) that a strong earthquake is the result of a lasting large-scale process whose different stages involve different parts of the fault network. Physically, the RTP approach tries to identify a rare small-scale phenomenon that carries a memory of the larger scale history of the system. It is regarded as promising for forecasting the approach to criticality in a wide class of complex systems, not limited to earthquakes (Shebalin *et al.*, 2004, 2006).

The original version of RTP (Shebalin *et al.*, 2004, 2006) uses earthquake chains to identify the short-term anomalous region. The earthquake chains, which depict an area of a long-range short-term activation of seismicity, have false alarms because not all chains precede large earthquakes. To eliminate false alarms, intermediate-term precursors identified using a pattern recognition approach were used. This idea can be directly applied to the Annual Consultation. In our approach, we consider the alarm regions

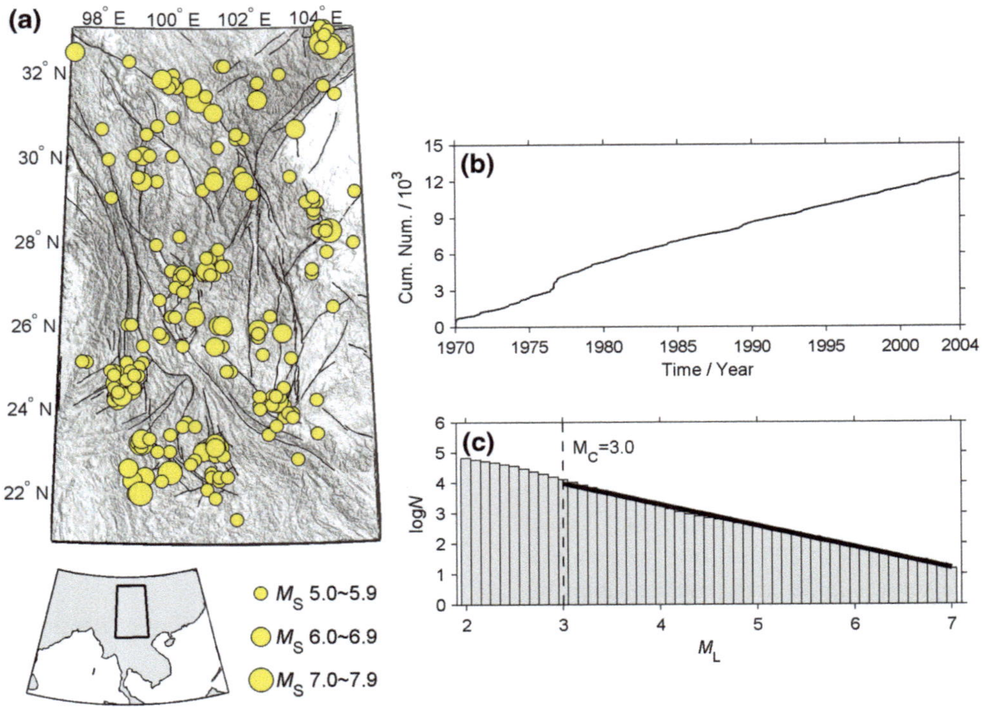

Figure 2

a Epicenter map of earthquakes with $M_S \geq 5.0$ in the Sichuan-Yunnan region for the period from 1970/01/01 to 2004/01/01 with active faults indicated by dark gray lines. The region under study is shown in the lower map. **b** Temporal distribution of the cumulative number of earthquakes with $M_L \geq 3.0$ from 1970 in the studied region. The step-like change in 1976 is due not to the change of monitoring capability but to the aftershock sequence of the 1976/08/16 and 1976/08/23 Songpan-Pingwu earthquakes, both registering magnitude M_S 7.2. **c** Cumulative frequency-magnitude distribution for the studied region from 1970/01/01 to 2004/01/01. The cutoff magnitude selected in the figure is $M_C = 3.0$, actually higher than the real completeness magnitude. The *bold black line* shows the fitting of the Gutenberg-Richter law above M_L 3.0

identified by the Annual Consultation as the anomalous regions with shorter time scale, i.e., the "candidates" (for the annual alarms) according to the terminology of Shebalin *et al.* (2004, 2006). Then we consider the intermediate-term anomalies in these alarm regions to explore whether the false-alarms could be eliminated.

To eliminate false alarms, we investigate whether or not intermediate-term anomalies or "precursors" have occurred within each candidate alarm region within a few, say five, years preceding its appearance. Figure 3 depicts the time scales considered in the analysis. Note that intermediate-term forecast methods generally have the "forecast window" equal in length to the "learning window". Operationally, reverse tracing is to consider the anomalies emerging from the "learning window", with the "candidate" alarm time window beginning at the end of the "learning window". The "forecast window" for long-term anomalies makes no sense if a retrospective test of the long-term forecast is not to be conducted.

4. PI/RI "Hotspots' as Intermediate-Term Anomalies in the RTP Approach

There are several options for the selection of the intermediate-term anomalies in the RTP approach. In this study we use the Pattern Informatics (PI) and Relative Intensity (RI) algorithm, which is also a description of the approach of catastrophic events in a complex system (Chen *et al.*, 2005; Holliday *et al.*, 2006a, 2007). As proposed by Rundle *et al.* (2002, 2003), PI "hotspots" depict the fluctuation of seismicity, including both quiescence and activation, while RI "hotspots" depict activation only. Both mappings specify the locations of the highest variation of seismic activity at the smallest magnitudes, because a black-or-white pixel grid is used in the calculation. The intermediate-term forecasts are based on the hypothesis that future large earthquakes will occur where the activity of smaller earthquakes has a large fluctuation. The underlying physics of the algorithm is that, based on the statistical physics of complex systems, seismogenetic dynamics is regarded as a "threshold system" driven by persistent forces or currents. The energy dissipation in such a complex system is approximately stable in the long-term, with stochastic Boltzmann fluctuations. Cooperative behavior can be identified by considering the correlation properties of the system (Rundle *et al.*, 2000a, 2002; Tiampo *et al.*, 2002). "Hotspots", of either PI or RI, describe the increase of the probability of earthquakes at an intermediate-term time scale, say, 5 years. Generally "hotspots" describe the top 30% of the normalized probability increase; a somewhat arbitrary but effective representation describing the alarm regions. Appendix 2 gives details of the PI and RI algorithm.

We use the PI/RI code provided by Rundle's group to analyze the seismicity in the Sichuan-Yunnan region. Figure 4 shows an example of such an analysis on the 5-year forecast time interval 1989/01/01 to 1994/01/01. Details of the analysis are described in Jiang and Wu (2008), which show by a retrospective test that the PI/RI forecast significantly outperforms random guessing and is robust against the uncertainties of earthquake parameters and the selection of spatio-temporal parameters in the calculation. Repeating the description of Jiang and Wu (2008): the spatial grid is taken as $D = 0.2°$ and the depth of earthquakes is not considered in the parameterization; the "anomaly training window" and the "forecast window" both have a length of 5 years; the starting point of the catalogue subject to analysis is taken as 15 years before the starting time of the "forecast window"; the cutoff magnitude of the catalogue is taken as M_L 3.0, satisfying the requirement that it has to be at least two magnitude units less than that of the "target earthquake"—in the retrospective test, the "target magnitude" is M_S 5.0+, as in the Annual Consultation; the logarithm amplitude $\log(\Delta P/\Delta P_{max})$ is used to represent the relative probability increase for M_S 5.0+ earthquakes; only the top 30% of values are considered as the "hotspots" and shown in the PI/RI mapping; a receiver operating characteristic (ROC) test (Swets, 1973; Molchan, 1997) is conducted against real earthquake activity. Figure 4c shows the selection of the time instants t_0, t_1, t_2, and t_3 in the analysis. The aim is to exploit the results of Jiang and Wu (2008) in order to reduce the false-alarm rate of the Annual Consultation.

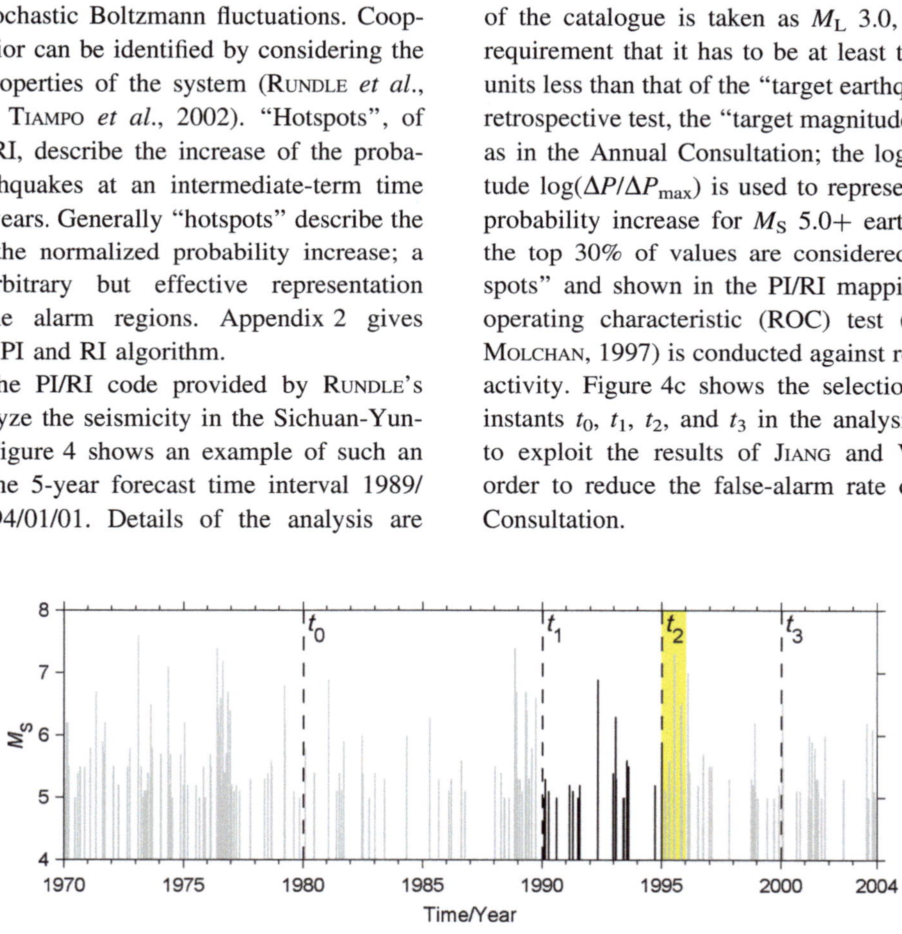

Figure 3
PI/RI algorithm used in the RTP of Annual Consultation. For the selection of parameters see Appendix 2. The *three vertical dashed lines to the right* represent t_1 the starting time of the ''anomaly training window'', t_2 the ending time of the ''anomaly training window'' and the starting time of the ''forecast window'', and t_3 the ending time of the ''forecast window'' of the PI/RI algorithm. The catalogue selected starts from t_0. The yellow zone delimits the forecast window in the Annual Consultation. Physically, for eliminating the false alarms in the Annual Consultation for the year 1995, the RTP approach is to use the catalogue from 1980 to 1994 to identify the long-term anomaly during the period 1990/01/01 to 1994/12/31, as indicated by *dark gray lines* showing the earthquakes

5. PI/RI Used in Application of RTP to the Annual Consultation

Figures 5 and 6 depict the application of PI and RI results, respectively, to the RTP of the Annual Consultation for the Sichuan-Yunnan region. These results are obtained by applying the following three steps. Step 1: plot on the same map the alarm regions produced by the Annual Consultation and the PI/RI ''hotspots''. Step 2: calculate the proportion of overlapped area between each alarm region and the PI/RI ''hotspots''; empirically determine a threshold proportion for the detection of the false alarm regions. Step 3: eliminate the regions which have a proportion of ''overlapped area'' lower than the threshold, that is, eliminate the false-alarm regions.

Figure 5a shows the result of the Annual Consultation for the year 1995/01/01 to 1996/01/01, in which a black bold line delimits the alarm region with increased probability of earthquakes as identified by the group of experts considering different factors such as geophysical anomalies. The same figure shows also the "hotspot" distribution corresponding to the forecast window 1995/01/01 to 2000/01/01, obtained from the learning period 1990/01/01 to 1995/01/01. As a reference, the earthquakes that occurred in the period 1995/01/01 to 1996/01/01 are also plotted (blue circles), together with the earthquakes that occurred in the "anomaly training window" 1990/01/01 to 1995/01/01 (gray reverse triangles). Figure 5b shows the overlapping between the PI "hotspots" and the alarm regions from the Annual Consultation, with numbers denoting the proportion of the overlapped area. In this figure, since a threshold proportion of 7% is chosen, the northernmost alarm region is to be eliminated. Analogously, Fig. 6 shows the use of the RI algorithm

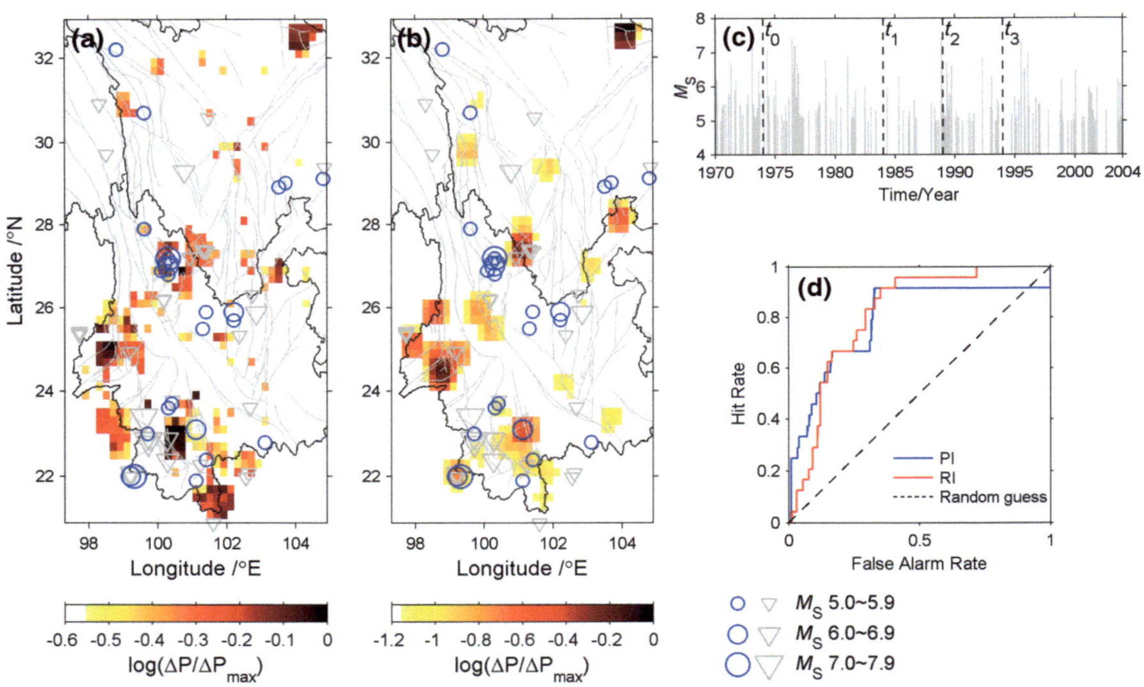

Figure 4
Retrospective test of the PI/RI algorithm for the Sichuan-Yunnan region, with 5-year forecast time window from 1989/01/01 to 1994/01/01. *Color-coded ''hotspots''* highlight the relative probability increase for earthquakes above M_S 5.0, with spatial grid size 0.2°. *Blue circles* are earthquakes above M_S 5.0 occurring within the ''forecast window'', and gray reverse triangles are earthquakes above M_S 5.0 within the ''anomaly training window''. **a** PI method. **b** RI method. **c** Temporal distribution of earthquakes with $M_S \geq 5.0$ in the studied region from 1970. In the figure, t_1 the starting time of the ''anomaly training window'', t_2 the ending time of the ''anomaly training window'' and the starting time of the ''forecast window'', and t_3 the ending time of the ''forecast window'' are displayed by vertical dashed lines. The catalogue is selected as starting from t_0. **d** ROC test. The *blue solid line* is the ROC curve for the PI forecast, the *red solid line* is the ROC curve for the RI forecast, and black dashed line is that for random guessing

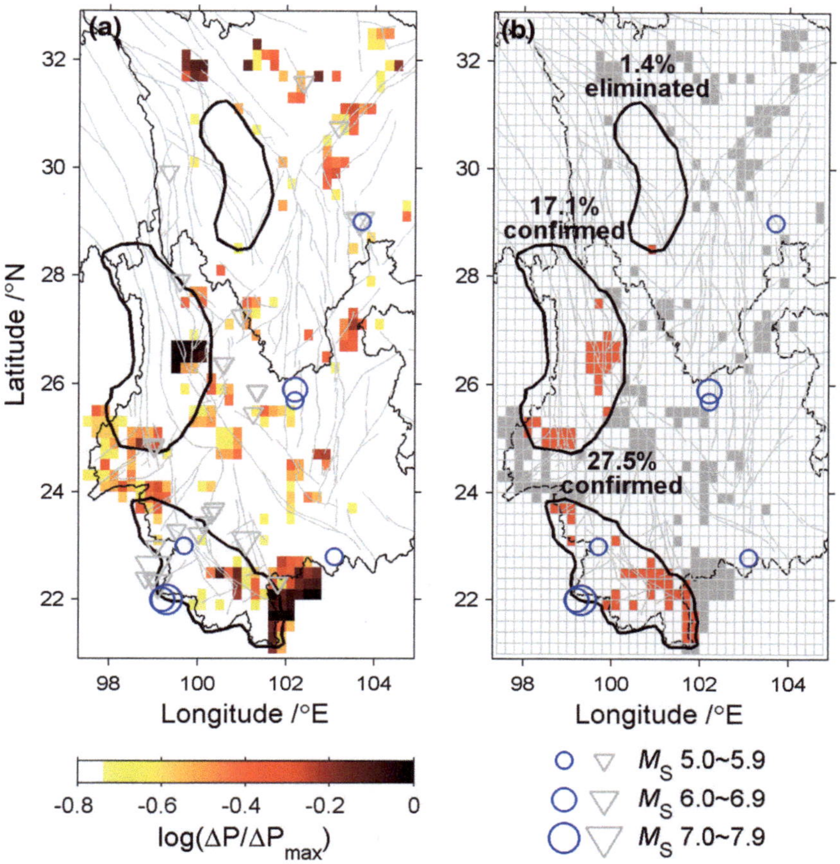

Figure 5
a The ''hotspot'' distribution using the PI method for the forecast period 1995/01/01 to 2000/01/01 and the result of the Annual Consultation for the year 1995/01/01 to 1996/01/01 (delimited by *black bold lines*). *Blue circles* show the ''target earthquakes'' in the period 1995/01/01 to 1996/01/01, and *gray reverse triangles* show the earthquakes in the ''anomaly training window''. **b** Proportion of overlapping area between the PI ''hotspots'' and the alarm region from the Annual Consultation. *Text* shows the proportion of the overlapped area and the decision for the removal or the confirmation of each alarm region when the threshold proportion is 7%

for the RTP. The threshold proportion for RI is taken as 13%. The reasons for these choices are explained in the next paragraph.

The selection of the threshold proportion is done empirically, because in practice one has to balance the correct and incorrect removal of the false alarm regions (the latter resulting in failures-to-predict). Figure 7 displays the number of false-alarm regions removed versus the threshold proportion of overlapped area, for the period 1990–2001 in the Sichuan-Yunnan region. In the figure, gray positive bars show the number of false-alarm regions correctly removed, and white negative bars show the number incorrectly removed. Comparison of these two numbers leads to the empirical selection of the threshold. In the figure, a vertical dashed line shows the 'optimal' choice of the threshold proportion of overlapped area. In this approach we adopt a conservative decision-making criterion to correctly remove false-alarm regions without causing additional more failures-to-predict. The figure shows that for the PI algorithm, the threshold is 7%, and for the RI algorithm the threshold is 13%, the thresholds used in Figs. 5 and 6. Table 2 shows the removal of false alarm regions for each year by using PI and RI algorithms, for the year 1990–2001. Since we take the most conservative criterion as shown in Fig. 7, there are no extra failures-to-predict caused by the removal. It can be seen that the RI algorithm seems more powerful for removing false alarms. The reasons for this may be

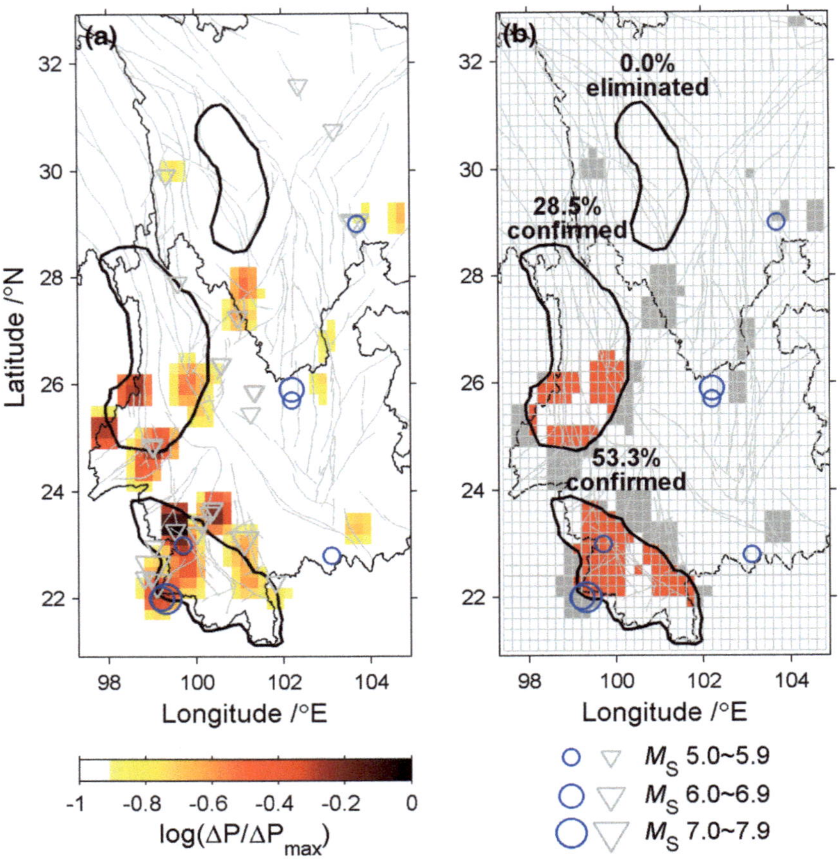

Figure 6

a The ''hotspot'' distribution using the RI method for the forecast period 1995/01/01 to 2000/01/01 and the result of the Annual Consultation for the year 1995/01/01 to 1996/01/01 (delimited by *black bold lines*). **b** Proportion of overlapping area between the RI 'hotspots' and the alarm region from the Annual Consultation. For further details see Fig. 5. In this figure the threshold proportion of overlapped area is taken as 13%

related to the characteristics of seismicity in this region—for example, clustering properties significantly affect in the annual seismic activity, as shown by the result of Jiang and Wu (2008). In any case, both PI and RI can remove some of the false alarm regions, and thus can be used in the practical Annual Consultation.

We divided the analysis into two stages: the "learning" stage (1990–2001), as shown in Fig. 7 and Table 2, which determines the threshold proportion; and the "testing" stage (2002 and 2003), in which we use the threshold proportion obtained in the "learning" stage to remove the false alarms. Figure 8 and Table 3 shows the result of false-alarm removal for these 2 years. Details of the Annual Consultation for these 2 years are provided in Appendix 1.

6. Conclusions and Discussion

One of the problems of the Annual Consultation, a special form of intermediate-term earthquake forecast, is the false-alarms in its outputs. To tackle this problem we try to use the concept of reverse tracing of precursors (RTP) by considering the long-term seismic activities in the alarm regions, or the regions thought to have increased probability of earthquakes, as identified by the Annual Consultation. We apply the Pattern Informatics (PI) and Relative Intensity (RI) algorithms to the long-term estimation of seismic hazard in the RTP analysis. We select the Sichuan-Yunnan region of southwest China, where high seismicity makes the PI/RI algorithms perform well, and select the period 1990–2003, when the

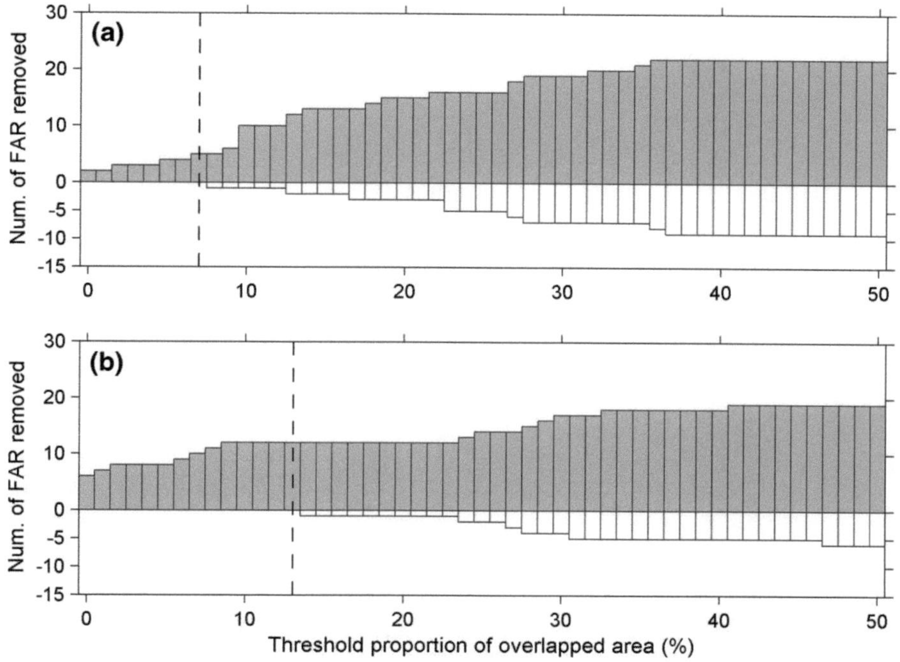

Figure 7
Number of false-alarm areas removed versus the threshold proportion of overlapped area. *Gray positive bars* show the number of false-alarm regions (FAR) correctly removed, and white negative bars show the number of ''false-alarm'' regions (FAR) incorrectly removed—that is, there were earthquake/s which occurred in the forecast window within the region. *Vertical dashed line* shows the ''optimal'' choice of the threshold proportion of overlapped area. In this approach a conservative decision is made to correctly remove the false-alarm regions without causing more failures-to-predict due to the removal

Table 2

Removal of false alarm regions in the Annual Consultation for Sichuan-Yunnan from 1990 to 2001 by RTP

Year	No. of false alarm regions	No. of RTP-removed false alarm regions by PI	No. of RTP-removed false alarm regions by RI
1990	2	1	2
1991	2	0	1
1992	1	0	1
1993	1	0	0
1994	2	1	1
1995	2	1	1
1996	2	0	1
1997	1	1	1
1998	3	0	1
1999	2	0	1
2000	3	1	2
2001	1	0	0

Annual Consultation had not started to consider PI/RI results, for the retrospective study. A retrospective test shows that the RTP analysis using the PI/RI results can correctly remove some of the false-alarm regions from the 'black list' and thus reduce the false-alarm rate of the Annual Consultation, without reducing the hit rate. This is implemented by choosing the threshold proportion of overlapped areas between the long-term estimates and the Annual Consultation. In our investigation, 7 and 13% are the thresholds for the PI and RI algorithms, respectively.

Our understanding of the physical basis of the successful removal is to date only conceptual. In the RTP analysis (KEILIS-BOROK *et al.*, 2004), an earthquake is generated by two interacting processes in a fault network: An accumulation of energy that the earthquake will release, and a rise of instability triggering this release. Energy is carried by the stress field, and instability is carried by the difference between the stress and strength fields. Both processes can be detected and characterized by 'premonitory' patterns of seismicity or other relevant fields (SHEBALIN *et al.*, 2006). The RTP method is based on the hypothesis that a 'real' short-term precursor

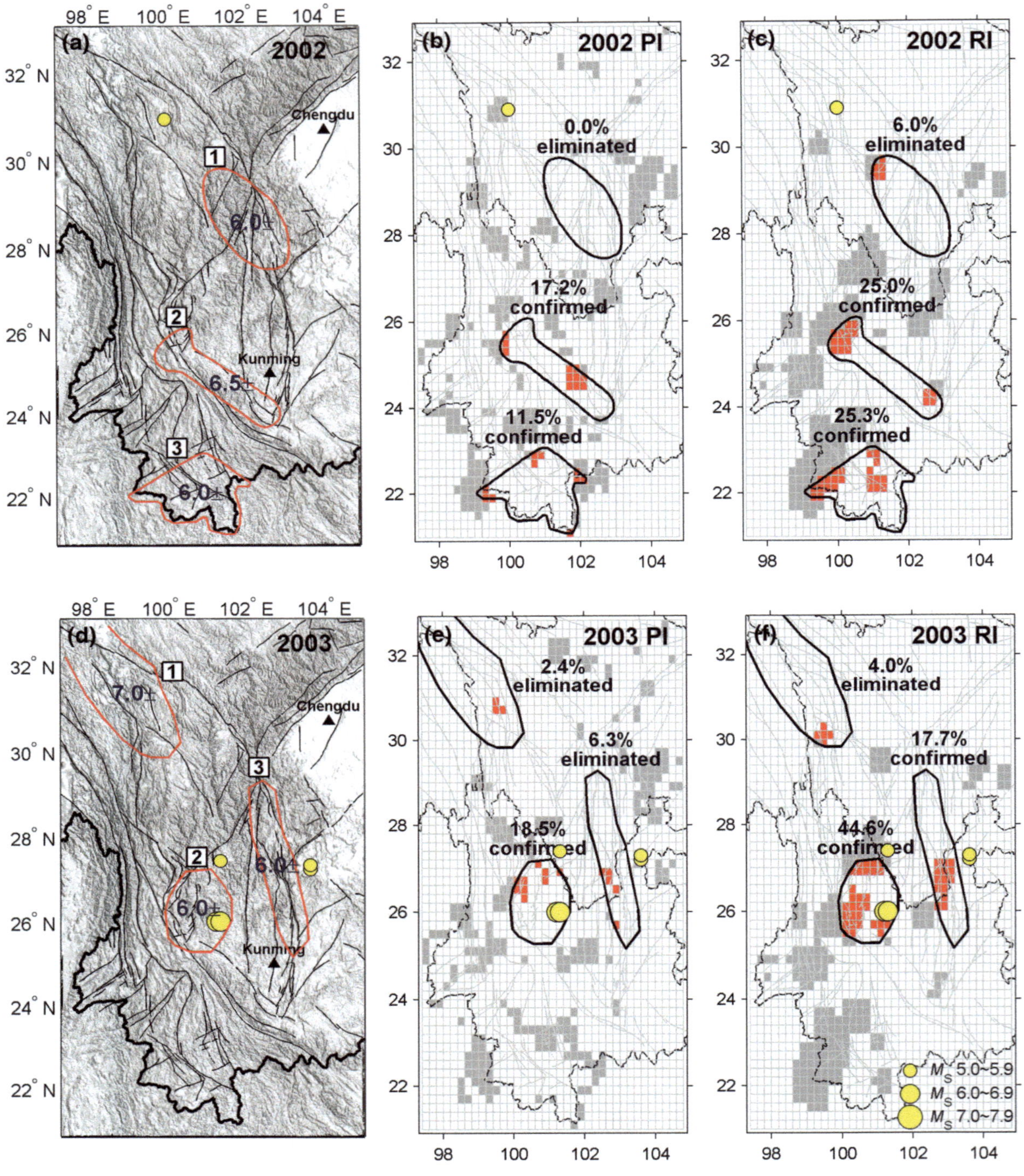

Figure 8

Annual Consultation results in Sichuan-Yunnan region for the years 2002 and 2003, respectively, and the RTP processing to remove false alarms. *Red closed full lines* delimit the regions with increased probability of earthquakes, with *numbers* denoting the expected magnitude range. *Yellow dots* show the epicenters of earthquakes occurring in 2002 and 2003, respectively. *Framed black numbers* represent the areas in Table 3. Other details are similar to those in Figs. 5 and 6

Table 3

Removal of false alarm regions in the Annual Consultation for Sichuan-Yunnan for 2002 and 2003 by RTP

Year	Label of alarm regions as shown in Fig. 8	Forecasting result	Proportion of the overlapped area by PI (%)	Decision for alarm region	No. of RTP-removed false alarm regions by PI	Ratio of false alarm before and after RTP by PI	Proportion of the overlapped area by RI (%)	Decision for alarm region	No. of RTP-removed false alarm regions by RI	Ratio of false alarm before and after RTP by RI
2002	1	False alarm	0.0	Eliminated		100%	6.0	Eliminated		100%
	2	False alarm	17.2	Confirmed	1	vs.	25.0	Confirmed	1	vs.
	3	False alarm	11.5	Confirmed		67%	25.3	Confirmed		67%
2003	1	False alarm	2.4	Eliminated		67%	4.0	Eliminated		67%
	2	Hit	18.5	Confirmed	2	vs.	44.6	Confirmed	1	vs.
	3	False alarm	6.3	Eliminated		0%	17.7	Confirmed		33%

(as opposed to a false alarm) should be preceded by intermediate-term precursors in its vicinity. In this case, the RTP analysis has dual advantages: It specifies the areas in which to look for intermediate-term precursors; and it identifies a small-scale phenomenon that carries a memory of the larger scale history of the system. In contrast, traditional analysis generally starts with tracing of the intermediate-term patterns hidden in the background seismicity. Most of them are not local, and involve arbitrary selection of spatio-temporal borders. This conceptual hypothesis explains why RTP seems useful for removing the false alarm regions, and, more importantly, why different methods for identifying short-term and intermediate-term anomalies can be accommodated by the RTP approach. Table 4 compares the short-term and intermediate-term anomalies used originally by Keilis-Borok *et al.* (2004) and by this study.

Because of regional dependence on the characteristics of seismic activity, further investigation is needed to determine whether the results of this study hold for other regions. In this approach, however, there is a little extension of the RTP concept: the RTP approach originally proposed for short-term earthquake predictions seems also able to contribute to the Annual Consultation.

Acknowledgments

We thank Prof. J. Rundle and Dr. C. C. Chen for helpful discussion on the PI/RI algorithm and providing the computer code. Catalogue of earthquakes is from China Earthquake Networks Center (CENC). Annual Consultation (1990–2003) results are provided by the Working Group for Promoting Earthquake Forecast Study of the Institute of Geophysics, China Earthquake Administration. We thank Prof. D. Vere-Jones, Prof. D. Rhoades, and the anonymous reviewers for their assistance in improving the manuscript.

Appendix 1: Annual Consultation

The Annual Consultation Meeting on the Likelihood of Earthquakes in the next year has been

Table 4

Comparison of anomalies used in RTP

	Short-term ("candidate")		Intermediate-term	
	Time scale	Anomaly	Time scale	Anomaly
KEILIS-BOROK *et al.* (2004)	6 months	Precursory chain	Several years	TIP
This study	1 year	Annual Consultation	5 years	PI or RI

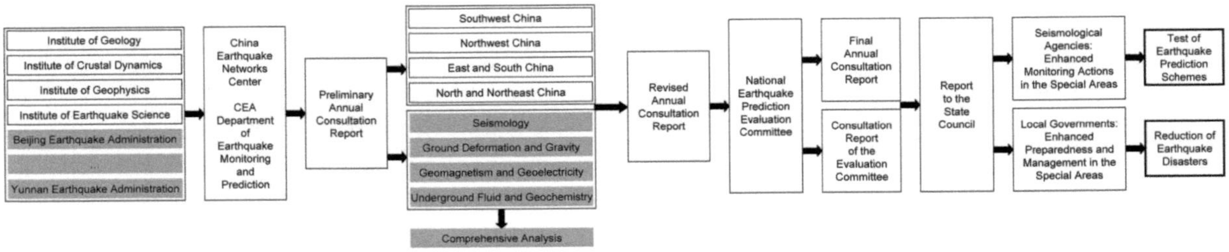

Figure 9
Sketch showing the organization of the Annual Consultation

organized by the Chinese seismological agency since the 1970s. At the turn of the year, seismologists use their methods of precursor/anomaly identification and earthquake forecast/prediction to identify the areas with increased probability of earthquake occurrence in the coming year. An ad hoc group of experts conducts an evaluation of the proposed precursors/anomalies and the forecast/prediction. By combining tectonic, seismic, and other geophysical information, the group of experts draws conclusions about the seismic tendency in the next year and identifies the areas with higher seismic risk. This working group writes a report to the China Earthquake Administration (CEA), and CEA in turn reports these results to the central government. In the regions which are deemed to have increased probability of earthquakes, seismological monitoring will be strengthened, and some engineering/social countermeasures will be taken. The organization of the Annual Consultation is as shown in Fig. 9.

Only a few results of the annual consultation meetings have been published in widely accessible academic journals. Accordingly the results and methodology as well as the significance of this approach are still not well known to the international seismological community. In the early 1990s, some foreign seismologists were invited to participate in the meeting. American seismologist FRANCIS WU (1997) gave an introduction to the history and state-of-the-art of the annual consultation. The Seismological Press in Beijing has published some research results (in Chinese) related to the annual consultation, with public release after 3 years because of the sensitivity and uncertainty of the present study of earthquake predictability.

The study of annual earthquake tendency is basically organized in a matrix. Academically, the studies on annual earthquake prediction are divided into five subjects: seismology; geomagnetism and geo-electricity; ground deformation and gravity; underground fluid and geochemistry, geothermal; and comprehensive analysis. The China Earthquake Administration deployed and is operating many observatories monitoring changes of geomagnetic field, geo-electric field, underground water and water content, and is organizing mobile measurements of deformation, gravity, geomagnetic and geo-electric field, with organized quality-control systems and formulated standards for routine data processing. The precursors considered are almost the same as those studied elsewhere, such as seismicity patterns, geomagnetic and geo-electric precursors, ground deformation and gravity changes, and changes of underground water level and its chemical content,

Table 5

Areas in Sichuan-Yunnan region with increased probability of earthquakes identified by the Annual Consultation for the years 2002 and 2003, respectively, together with the basis for the identification

Area	Expected magnitude range	Anomalies observed	Comprehensive analysis tools used	Confidence level evaluated by the expert group
2002-1	6±	*Seismic*: 'signal earthquake'[a]; 'seismic belt'[b]; seismic gaps; enhanced seismicity; concentration of moderate earthquakes; seismic quiescence; seismic strain release; 'seismic modulating ratio'[c]; '*C* value (seismic concentrative degree)'[d]; '*Y* value (load/unload response ratio)'[e] *Deformation*: 'ground tilt'[f] and 'observational data of extensometer'[g] in Panzhihua; 'short leveling observation'[h] and 'short baseline observation'[g] in Tangjiaping; 'short leveling observation'[h] in Ningnan; 'leveling observation'[h] in Xundian *Geomagnetism*: 'E_{NS} (natural geoelectric potential)'[i] in Xichang and 'E_{NW} (natural geoelectric potential)'[i] in Mianning; 'ρ_{NW} and ρ_{NS} (apparent resistivity)'[j] in Mianning; 'geomagnetic anomalies'[k] in Qiaojia *Underground fluid*: 'water level anomalies'[l] in Huili, Panzhihua, Xichang, Shimian and Luzhou; 'change characters of gas CO_2 released from a fault'[m] in Shimian; 'water temperature anomaly'[n] in Xuanwei.	'Algorithmic Complexity'[o]; Time and Space of Increased Probability (TSIP); Genetic-Neural Network; 'Linear Synthetic Probability Method'[p]; 'New Generation Expert System for Earthquake Prediction (NGESEP)'[q]	0.7
2002-2	6.5±	*Seismic*: seismic belts[b]; enhanced seismicity; seismic frequency; seismic strain release; 'seismic modulating ratio'[c]; '*D* value (seismic risk degree)'[r]; 'GL value (seismic inhomogeneous degree)'[s]; '*Y* value (load/unload response ratio)'[e] *Deformation*: 'ground tilt'[f] in Chuxiong and Yongsheng; 'baseline observation'[g] and 'leveling observation'[h] in Tonghai and Shiping *Geomagnetism*: 'geomagnetic anomalies'[k] in Lijiang and Yongsheng *Underground fluid*: 'water level anomalies'[l] in Dayao, Zhongdian and Eeryuan; 'anomaly of Hydro-Radon'[t] in Eryuan; 'anomaly of water mercury'[u] in Mile, Chuxiong, Eryuan and Jianshui; 'water temperature anomaly'[n] in Eshan and Dayao; 'outflow'[v] in Xiaguan.	Grey modeling; 'Algorithmic Complexity'[o]; Time and Space of Increased Probability (TSIP); Genetic-Neural Network	0.8
2002-3	6±	*Seismic*: seismic belts[b]; concentration of moderate earthquakes; seismic frequency; *b* value; seismic strain release; 'seismic modulating ratio'[c]; η value; '*D* value (seismic risk degree)'[r]; 'GL value (seismic inhomogeneous degree)'[s] *Underground fluid*: calcium and magnesium concentration vibration in Simao	Grey modeling; Genetic-Neural Network	0.6
2003-1	7±	*Seismic*: 'grouping earthquakes'[w]; enhanced seismicity based on seismic quiescence; strong earthquake gap *Deformation*: 'baseline observation (A1–C1)'[g] in Xianshuihe; 'baseline observation (3-X)'[g] in Laoqianning; 'ground tilt'[f] in Guza *Underground fluid*: shallow and low temperature anomaly in Qumalai.	Threshold Regression; Genetic-Neural Network; 'Expert System for Earthquake Prediction (NGESEP)'[q]	0.6
2003-2	6±	*Seismic*: seismic quiescence; seismic frequency; precursory earthquake swarm *Deformation*: 'locked sections of active fault'[x]; 'short leveling observation'[h] and 'short baseline observation'[g] in Shiping *Underground fluid*: 'water level anomalies'[l] in Dayao, Nanhua and Wuding; 'co-seismic responses of water level of some wells'[y]; 'change characters of gas CO_2 released from a fault'[m] in Xiaguan, Midu, Heqing and Eryuan	'Expert System for Earthquake Prediction (ESEP)'[z]; 'New Generation Expert System for Earthquake Prediction (NGESEP)'[q]	0.7

Table 5 continued

Area	Expected magnitude range	Anomalies observed	Comprehensive analysis tools used	Confidence level evaluated by the expert group
2003-3	6±	*Seismic*: seismic quiescence; seismic belts[b]; seismic gaps; 'sequential migration pattern of moderate earthquakes'[β]; enhanced seismicity; concentration of moderate earthquakes; seismic quiescence; seismic strain release; 'seismic modulating ratio'[c]; '*C* value (seismic concentrative degree)'[d]; '*Y* value (load/unload response ratio)'[e] *Deformation*: 'leveling observation (A–B)'[h] in Anshunchang; 'leveling observation (7–8)'[h] in Mianxie; 'leveling observation (1–2)'[h] in Mianning; 'leveling observation (4–5)'[h] in Ningnan; 'leveling observation (A–B)'[h] and 'baseline observation (A–B)'[g] in Tangjiaping *Geomagnetism*: 'ρ_{NS} (apparent resistivity)'[j] in Mianning *Underground fluid*: 'water level anomalies'[l] in Xichangchuan-03 well, Panzhihua-05 well and Huilichuan-18 well; increasing 'water level coseismic response'[y] in Xichangchuan-03 well, Shimianchuan-02 well, Panzhihua-05 well and Qionglaichuan-22 well; decreasing 'water level coseismic response'[y] in Nanxichuan-12 well and Shifang-k2 well; 'change characters of gas released from a fault'[m] in Mianning	'Expert System for Earthquake Prediction (ESEP)'[z]	0.6

Well-known anomalies are simply listed in the table. Lack of original references in the documents of the Annual Consultation prevents from the correct tracing of the related approaches. The cited publications (in Chinese with English abstract) with * and without title in the reference list, are at best for indirect reference

[a] Yao *et al.* (2006)

[b] Long *et al.* (2008)

[c] Wang *et al.* (2001)

[d] Wang *et al.* (1999)

[e] Yin *et al.* (2004)

[f] Wang *et al.* (2008)

[g] Ma *et al.* (2005)

[h] Feng *et al.* (2006)

[i] Zheng (1995)

[j] Tang *et al.* (1998)

[k] Zhang *et al.* (2007)

[l] Hu and Wang (2008)

[m] Guan *et al.* (2002)

[n] Qiu (2006)

[o] Lü *et al.* (1997)

[p] Yu *et al.* (2006)

[q] Wang *et al.* (1998)

[r] Dai *et al.* (1997)

[s] Li *et al.* (2006)

[t] Fang (2007)

[u] Li *et al.* (2008)

[v] Gao *et al.* (2002)

[w] Fu *et al.* (2005)

[x] Che *et al.* (1996)

[y] Chen *et al.* (2007)

[z] Wang *et al.* (2004)

[β] Cai *et al.* (2002)

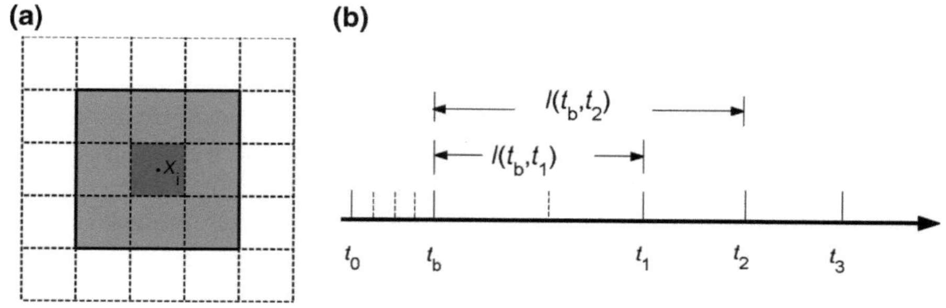

Figure 10
Sketch map showing the spatial and temporal range selection of earthquake time series. **a** Sketch map of spatial range selection: The spatial range of the time series for site x_i includes the *box* centered on x_i and the eight nearest neighbor boxes (the Moore neighborhood), as shown by *gray squares*. **b** The times and time windows related to earthquake time series of point x_i

among others. Geographically, continental China is divided into four regions for study: The north- and northeast-China region; the northwest-China region; the southwest-China region; and the east- and south-China region. When considering these regions, the surrounding areas of continental China which are tectonically related are also considered.

In the comprehensive analysis which combines the information from different disciplines, Chinese seismologists prefer the combination of quantitative calculation and qualitative experience. Some of the calculation tools, such as fuzzy logic, pattern recognition, artificial neural networks and expert systems, are used for learning and predicting. On the other hand, experienced experts play an important role in forming the final conclusions. From Table 5, one is offered a bird's-eye view of the basis for seismologists to identify the areas with increased probability of earthquakes.

One of the widely cited criteria for the evaluation of the annual forecast is the R score, somewhat similar to the ROC test (Swets, 1973; Molchan, 1997), proposed by Xu (1989), being represented by $R =$ (successful predictions/total number of earthquakes) − (false alarm area/total aseismic area). When all the predictions are successful, $R = 1$; When all the predictions are wrong, $R = -1$; For random guessing, $R = 0$ (Shi *et al.*, 2000, 2001). Evaluation of the performance of the annual consultation meeting was conducted by several groups (Shi *et al.*, 2000, 2001; Zheng *et al.*, 2000; Zhang *et al.*, 2002; Ma *et al.*, 2004). On average the R value is no more than 0.3, nonetheless this is still significantly better than random guessing. This apparent success is largely dependent on the probabilistic estimation of background seismicity (Shi *et al.*, 2004).

Appendix 2: PI Algorithm and 'Hotspot/s' Representation

In the PI algorithm (Rundle *et al.*, 2000b, 2002), according to the magnitude of the target earthquakes to be forecasted, the seismogenetic region is binned into boxes or 'pixels' with size $D \times D$ centered at a point x_i. Each point x_i is associated with a time series $N_i(t)$, where $N_i(t)$ is the time-dependent average rate of earthquakes with magnitude greater than the cutoff magnitude M_c in box i and its eight nearest neighbor boxes (the Moore neighborhood), as seen in Fig. 10a. $N_i(t)$ is calculated for box i within a period starting from time t_b to time t ($t > t_b$), as seen in Fig. 10b.

As shown in Fig. 10b, there are five key times during the construction of the time series in every box: t_0 the starting time of the catalogue, t_b the starting time of the sliding window, t_1 the starting time of the ''anomaly training window'', t_2 the ending time of the ''anomaly training window'' and the starting time of the ''forecast window'', and t_3 the ending time of the ''forecast window''. There are three time windows which are important in the calculation: (1) time window for comparison t_b–t_1; (2) second time window t_b–t_2, or ''anomaly training window'' t_1–t_2; and (3) forecast window t_2–t_3.

The ''seismic active intensity'' function of box i is defined as the average rate of occurrence of earthquakes:

$$I_i(t_b, t) = \frac{1}{t - t_b} \sum_{t'=t_b}^{t} N_i(t') \cdot \quad (1)$$

This seismic active intensity is normalized in time by subtracting the temporal mean and dividing by the temporal standard deviation. The normalized statistical property of box i in time window t_b–t can then be obtained by:

$$\hat{I}_i(t_b, t) = \frac{I_i(t_b, t) - \langle I_i(t_b, t) \rangle}{\sigma(t_b, t)} \cdot \quad (2)$$

Next, the difference of the normalized intensity functions for different time windows is computed as the measurement of seismic activity anomaly for the ith box:

$$\Delta I_i(t_b, t_1, t_2) = \hat{I}_i(t_b, t_2) - \hat{I}_i(t_b, t_1) \cdot \quad (3)$$

The average of intensity change over time $t_0 < t_b < t_1 - (t_2 - t_1)$ is then computed:

$$\overline{\Delta I_i(t_0, t_1, t_2)} = \frac{1}{t_0 - t_1} \sum_{t_b=t_0}^{t_1} \Delta I_i(t_b, t_1, t_2) \cdot \quad (4)$$

The probability of a future strong earthquake in box i, $P_i(t_0, t_1, t_2)$, is defined as the square of the average intensity change:

$$P_i(t_0, t_1, t_2) = \overline{\Delta I_i(t_0, t_1, t_2)}^2. \quad (5)$$

Finally, subtract the mean probability over all boxes and denote this change as the probability increase of future earthquakes via

$$\Delta P_i(t_0, t_1, t_2) = P_i(t_0, t_1, t_2) - \langle P_i(t_0, t_1, t_2) \rangle. \quad (6)$$

Increased probability of strong earthquakes is shown by 'hotspots'. The hotspots are defined to be the boxes where $\Delta P_i(t_0, t_1, t_2)$ is positive, or, the probability function $P_i(t_0, t_1, t_2)$ is larger than the background level. Since the intensity function has a quadratic form, either seismic activation or quiescence can be reflected by the PI map. Generally in the PI map, only the top 30% of the hotspots are considered to be associated with significantly increased earthquake probability.

PI results are often compared with Relative Intensity (RI) results (Chen *et al.*, 2005; Holliday *et al.*, 2006b). In the comparison, the RI forecast is used as a null hypothesis to test the necessity of using the PI algorithm, but RI itself is also a tool for earthquake-rate forecasting. The RI algorithm assumes that strong earthquakes tend to occur in regions where more earthquakes occurred before, i.e., earthquake occurrence has the property of 'clustering'. Similar to the PI algorithm, the RI statistics are normalized, and the 'alarm threshold' is determined by the relative value of the RI statistics ranging within [0,1]. For direct comparison with the PI forecast, only the top 30% of the RI increases are regarded as 'hot spots' and are shown on the map.

To evaluate the performance of forecasts, especially the performance compared to random guessing, a receiver operating characteristic (ROC) test (Swets, 1973; Molchan, 1997) can be conducted. The ROC test systematically varies the alarm threshold of the forecast region and counts the 'hit rate' and 'false alarm rate' based on real earthquake activity. A 'hit' is defined as the case in which the target event occurs within the alarmed cell or its nearest neighbors. In the ROC diagram, the horizontal axis is the false-alarm-rate, and the vertical axis is the hit-rate. Random guessing gives a diagonal from (0,0) to (1,1), and a reasonably effective forecast must give a curve well above this diagonal.

References

Cai, J. G., Li, Y. L., Liu, L. F., Xu, Z. Y., Qin, J. Z., Su Y. J., and Fu, H. (2002), *Sequential migration pattern of sesimicity in Yunnan Region*, Seismol. Geol. *24*, 335–345. (*)

Che, Z. H., Liu, S. H., Liu, T. H., and Liu, X. J. (1996), Crustal Deform. Earthq. *16*, 46–51. (*)

Chen, C. C., Rundle, J. B., Holliday, J. R., Nanjo, K. Z., Turcotte, D. L., Li, S. C., and Tiampo, K. F. (2005), *The 1999 Chi-Chi, Taiwan, earthquake as a typical example of seismic activation and quiescence*, Geophys. Res. Lett. *32*, L22315, doi: 10.1029/2005GL023991.

Chen, D. Q., Liu, Y. W., Yang, X. H., and Liu, Y. M. (2007), Seismol. Geol. *29*, 122–132. (*)

Dai, W. L., Wang, W., Song, J. G., and Zhou, Y. H. (1997), Northwestern Seismol. J. *19*, 1–6. (*)

Fang, H. F. (2007), Northwestern Seismol. J. *29*, 187–191. (*)

Feng, W., Yao, W. Z., and Xu, C. C. (2006), J. Coll. Disaster Prev. Tech. *8*, 56–61. (*)

FIELD, E. D. (ed.) (2007), *Special Issue: Regional earthquake likelihood models*, Seism. Res. Lett. *78*, 1–140.

FU, Z. X., JIN, X. S., SHAO, H. C., and DING, X. (2005), Earthquake *25*, 65–74. (*)

GAO, X. Q., XU, Q. L., LAN, L., CHEN, L., FAN, B. Q., and WANG, D. (2002), Inland Earthq. *16*, 346–354. (*)

GUAN, Z. J., WEN, X. Z., and CHENG, W. Z. (2002), Earthq. Res. Sichuan *105*, 22–27. (*)

HOLLIDAY, J. R., RUNDLE, J. B., TIAMPO, K. F., and TURCOTTE, D. L. (2006a), *Using earthquake intensities to forecast earthquake occurrence times*, Nonlin. Process. Geophys. *13*, 585–593.

HOLLIDAY, J. R., RUNDLE, J. B., TIAMPO, K. F., KLEIN, W., and DONNELLAN, A. (2006b), *Modification of the pattern informatics method for forecasting large earthquake events using complex eigenfactors*, Tectonophysics *413*, 87–91.

HOLLIDAY, J. R., CHEN, C. C., TIAMPO, K. F., RUNDLE, J. B., TURCOTTE, D. L., and DONNELLAN, A. (2007), *A RELM earthquake forecast based on pattern informatics*, Seism. Res. Lett. *78*, 87–93.

HU, J. L., and WANG, X. (2008), Seismol. Res. Northeast China *24*, 35–39. (*)

JIANG, C. S. and WU, Z. L. (2008), *Retrospective forecasting test of a statistical physics model for earthquakes in Sichuan-Yunnan region*, Sci. China. Ser. D *51*, 1401–1410.

KEILIS-BOROK, V. (2002), *Earthquake prediction: State-of-the-art and emerging possibilities*, Annu. Rev. Earth Planet. Sci. *30*, 1–33.

KEILIS-BOROK, V., SHEBALIN, P., GABRIELOV, A., and TURCOTTE, D. (2004), *Reverse tracing of short-term earthquake precursors*, Phys. Earth Planet. Inter. *145*, 75–85.

LI, X., GUO, A. X., and ZHANG, H. F. (2006), J. Seismol. Res. *29*, 230–233. (*)

LI, D. X., LI, W., LUO, Y. X., and LU, Z. (2008), North China Earthq. Sci. *26*, 52–55. (*)

LONG, H. Y., NIE, X. H., and GAO, G. Y. (2008), Northwestern Seismol. J. *30*, 56–61. (*)

LÜ, Y. J., LU, Y. Z., and ZHENG, Y. J. (1997), Earthquake *17*, 25–33. (*)

MA, H. S., LIU, J., WU, H., and LI, J. F. (2004), *Scientific evaluation of annual earthquake prediction efficiency based on R-value*, Earthquake *24*, 31–37 (in Chinese with English abstract).

MA, G. Q., LI, X. P., ZHAO, C. H., AN, H. S., ZHANG, H. L., JIANG, Y. J., and WANG, W. (2005), North China Earthq. Sci. *23*, 36–39. (*)

MOLCHAN, G. M. (1997), *Earthquake predictions as a decision-making problem*, Pure Appl. Geophys. *149*, 233–247.

MULARGIA, F. (1997), *Retrospective validation of the time association of precursors*, Geophys. J. Int. *131*, 500–504.

NANJO, K. Z., HOLLIDAY, J. R., CHEN, C. C., RUNDLE, J. B., and TURCOTTE, D. L. (2006), *Application of a modified pattern informatics method to forecasting the locations of future large earthquakes in central Japan*, Tectonophysics *424*, 351–366.

QIU, Y. P. (2006), South China J. Seismol. *26*, 86–91. (*)

RUNDLE, J. B., KLEIN, W., TIAMPO, K., and GROSS, S. (2000a), *Linear pattern dynamics in nonlinear threshold systems*, Phys. Rev. E *61*, 2418–2431.

RUNDLE, J. B., KLEIN, W., TURCOTTE, D. L., and MALAMUD, B. D. (2000b), *Precursory seismic activation and critical-point phenomena*, Pure Appl. Geophys. *157*, 2165–2182.

RUNDLE, J. B., TIAMPO, K. F., KLEIN, W., and MARTINS, J. S. S. (2002), *Self-organization in leaky threshold systems: The influence of near-mean field dynamics and its implications for earthquakes, neurobiology, and forecasting*, Proc. Natl. Acad. Sci. USA *99*, 2514–2521.

RUNDLE, J. B., TURCOTTE, D. L., SHCHERBAKOV, R., KLEIN, W., and SAMMIS, C. (2003), *Statistical physics approach to understanding the multiscale dynamics of earthquake fault systems*, Rev. Geophys. *41*, 1019, doi:10.1029/2003RG000135.

SHEBALIN, P., KEILIS-BOROK, V., ZALIAPIN, I., UYEDA, S., NAGAO, T., and TSYBIN, N. (2004), *Advance short-term prediction of the large Tokachi-oki earthquake, September 25, 2003, M = 8.1 A case history*, Earth Planets Space *56*, 715–724.

SHEBALIN, P., KEILIS-BOROK, V., GABRIELOV, A., ZALIAPIN, I., and TURCOTTE, D. (2006), *Short-term earthquake prediction by reverse analysis of lithosphere dynamics*, Tectonophysics *413*, 63–75.

SHI, Y. L., LIU, J., and ZHANG, G. M. (2000), *The evaluation of Chinese annual earthquake prediction in the 90s*, J. Grad. Sch. Acad. Sinica *17*, 63–69 (in Chinese with English abstract).

SHI, Y. L., LIU, J., and ZHANG, G. M. (2001), *An evaluation of Chinese annual earthquake predictions, 1990–1998*, J. Appl. Probab. *38A*, 222–231.

SHI, Y. J., WU, Z. L., and BAI, L. (2004), *Evaluation of annual prediction of seismicity tendency using Pearson test*, J. Grad. Sch. Chin. Acad. Sci. *21*, 248–253 (in Chinese with English abstract).

SU, Y. J., LI, Y. L., LI, Z. H., YI, G. X., and LIU, L. F. (2003), *Analysis of minimum complete magnitude of earthquake catalog in Sichuan-Yunnan region*, J. Seismol. Res. *26*, 10–16 (in Chinese with English abstract).

SWETS, J. A. (1973), *The relative operating characteristic in psychology*, Science *182*, 990–1000.

TANG, J., ZHAO, G. Z., WANG, J. J., LI, W. J., and ZHAN, Y. (1998), Seismol. Geol. *20*, 164–171. (*)

TIAMPO, K. F., RUNDLE, J. B., MCGINNIS, S., GROSS, S. J., and KLEIN, W. (2002), *Mean-field threshold systems and phase dynamics: an application to earthquake fault systems*, Europhys. Lett. *60*, 481–488.

WANG, H. T. (2005), *Some preliminary think on prediction of annual seismic risk region*, Rec. Dev. World Seismol. *317*, 103–105 (in Chinese with English abstract).

WANG, W., WU, G. F., HUANG, B. S., ZHUANG, K. Y., JIANG, C. X., LI, D. S., and ZHOU, P. L. (1998), J. Seismol. Res. *21*, 33–41. (*)

WANG, W., DAI, W. L., SONG, J. G., and ZHOU, Y. H. (1999), J. Seismol. Res. *22*, 51–56. (*)

WANG, W., SONG, X. Y., XIE, D., and WANG, Z. Z. (2001), Earthquake *21*, 7–12. (*)

WANG, W., WU, G. F., ZHANG, B. F., LI, S. L., ZHENG, Z. B., and LU, Y. Z. (2004), Earthq. Res. China *20*, 126–133. (*)

WANG, Y. Y., WANG, A. D., ZHAO, Y. Z., and JIA, X. D. (2008), Seismol. Res. Northeast China *24*, 1–7. (*)

WU, F. T. (1997), *The Annual Earthquake Prediction Conference in China (National Consultative Meeting on Seismic Tendency)*, Pure Appl. Geophys. *149*, 249–264.

WU, Z. L., LIU, J., ZHU, C. Z., JIANG, C. S., and HUANG, F. Q. (2007), *Annual consultation on the likelihood of earthquakes in continental China: Its scientific and practical merits*, Earthq. Res. China *21*, 365–371.

XU, S. X. (1989), *The evaluation of earthquake prediction ability.* In Department of Science, Technology and Monitoring, State Seismological Bureau (ed.), The Practical Research Papers on Earthquake Prediction Methods (Seismicity Section), Beijing, Seismo. Press, 586–589 (in Chinese).

Xu, X. W. and Deng, Q. D. (1996), *Nonlinear characteristics of paleoseismicity in China*, J. Geophys. Res. *101*, 6209–6231.

Xu, X. W., Zhang, P. Z., Wen, X. Z., Qin, Z. L., Chen, G. H., and Zhu, A. L. (2005), *Features of active tectonics and recurrence behaviors of strong earthquake in the western Sichuan province and its adjacent regions*, Seismol. Geol. *27*, 446–461 (in Chinese with English abstract).

Yao, B. S., Sha, H. J., and Fu, Z. Z. (2006), North China Earthq. Sci. *24*, 47–51. (*)

Yi, G. X., Wen, X. Z., and Xu, X. W. (2002), *Study on recurrence behaviors of strong earthquakes for several entireties of active fault zones in Sichuan-Yunnan region*, Earthq. Res. China *18*, 267–276 (in Chinese with English abstract).

Yi, G. X., Wen, X. Z., Wang, S. W., Long, F., and Fan, J. (2006), *Study on fault sliding behaviors and strong-earthquake risk of the Longmenshan-Minshan fault zones from current seismicity parameters*, Earthq. Res. China *22*, 117–125 (in Chinese with English abstract).

Yin, X. C., Mora, P., Peng, K. Y., Wang, Y. C., and Weatherley, D. (2004), Rec. Dev. World Seismol. *305*, 24–30. (*)

Yu, W., Zhang, S. F., Sun, Q., Qu, Y. S., and Wu, X. P. (2006), North China Earthq. Sci. *24*, 20–23. (*)

Zhang, G. M., Liu, J., and Shi, Y. L. (2002), *An scientific evaluation of annual earthquake predication ability*, Acta Seismol. Sinica *15*, 550–558.

Zhang, J. G., Cao, Y., Zhang, F., and Sun, Q. (2007), North China Earthq. Sci. *25*, 20–27. (*)

Zheng, Z. B. (1995), Northwestern Seismol. J. *17*, 57–62. (*)

Zheng, Z. B., Liu, J., Li, G. F., Qian, J. D., and Wang, X. Q. (2000), *Statistical simulation analysis of the correlation between the annual estimated key regions with a certain seismic risk and the earthquakes in China*, Acta Seismol. Sinica *13*, 575–584.

(Received January 8, 2008, revised March 31, 2009, accepted April 23, 2009, Published online March 17, 2010)

Pure Appl. Geophys. 167 (2010), 801–818

DOI 10.1007/s00024-010-0064-6

Pure and Applied Geophysics

Repeated Intermittent Earthquake Cycles in the San Francisco Bay Region

Mark S. Bebbington,[1] David S. Harte,[2] and Steven C. Jaumé[3]

Abstract—Forecasts of future earthquake hazard in the San Francisco Bay region (SFBR) are dependent on the distribution used for the possible magnitude of future events. Based on the limited observed data, it is not possible to statistically distinguish between many distributions with very different tail behavior. These include the modified and truncated Gutenberg–Richter distributions, and a composite distribution assembled by the Working Group on California Earthquake Probabilities. There is consequent ambiguity in the estimated probability of very large, and hence damaging, events. A related question is whether the energy released in earthquakes is a small or large proportion of the stored energy in the crust, corresponding loosely to the ideas of self-organized criticality, and intermittent criticality, respectively. However, the SFBR has experienced three observed accelerating moment release (AMR) cycles, terminating in the 1868 Hayward, 1906 San Andreas and 1989 Loma Prieta events. A simple stochastic model based on elastic rebound has been shown to be capable of producing repeated AMR cycles in large synthetic catalogs. We propose that such catalogs can provide the basis of a test of a given magnitude distribution, via comparisons between the AMR properties of the real and synthetic data. Our results show that the truncated Gutenberg–Richter distribution produces AMR behavior closest to the observed AMR behavior. The proviso is that the magnitude parameters b and $m_{\max}$ are such that a sequence of large events that suppresses activity for several centuries is unlikely to occur. Repeated simulation from the stochastic model using such distributions produces 30-year hazard estimates at various magnitudes, which are compared with the estimates from the 2003 Working Group on California Earthquake Probabilities.

Key words: Earthquake magnitude, accelerating moment release, truncated Gutenberg–Richter, tail behavior, hazard estimates.

[1] IFS-Statistics, Massey University, Private Bag 11222, Palmerston North, New Zealand. E-mail: m.bebbington@massey.ac.nz

[2] Statistics Research Associates, P.O. Box 12649, Thorndon, Wellington, New Zealand. E-mail: david@statsresearch.co.nz

[3] Department of Geology and Environmental Geosciences, College of Charleston, Charleston, SC 29424, USA. E-mail: jaumes@cofc.edu

1. Introduction

There is considerable interest in estimating the medium-long term (30 years) earthquake hazard in the San Francisco Bay Region (SFBR); see Working Group on California Earthquake Probabilities, 2003 [henceforth 'Working Group (2003)']. A major determinant in this hazard is the distribution of possible magnitudes for great ($M > 7.5$) earthquakes. As there is only one such earthquake in the historical catalog, there is little to go on. A number of approaches based on empirical laws and fault segment models (see, for example, Ward, 2000; Andrews and Schwerer, 2000; Working Group, 2003) have been proposed, but a means of testing them is lacking—the data set is too small to determine the best model based on the usual measures of fit such as the log-likelihood and AIC. In this paper we describe a possible approach to constructing such a test, using the fact that the SFBR has been shown (Sykes and Jaumé, 1990) to produce repeated cycles culminating in a large earthquake, which fit the Accelerated Moment Release (AMR) hypothesis. Using simulations of long sequences, we will see that the different magnitude distributions impose very different sample path characteristics on the simulated sequences. These differences are difficult to detect when fitting the models to relatively short sequences like that of the SFBR.

AMR is not a model. It is more of a decision or prediction rule, based on empirical observation, for the time when the current sequence will terminate with a large event. A sequence of events is said to exhibit AMR behavior if a curve fitted to the cumulative seismic release has sufficient curvature. In the SFBR, there have been three such 'AMR cycles' since 1850 (SYKES and JAUMÉ, 1990). Such cycles are always analyzed individually because it is not specified how to fit one 'AMR model' to the complete data set. For forecasting purposes, the AMR characteristics need to be embedded into a stochastic model. A 'model' in this sense is a procedure or algorithm with sufficient specification to enable data to be simulated *ad infinitum*. Such a simple stochastic model known to be consistent with AMR (JAUMÉ and BEBBINGTON, 2004) is the Stress Release Model (SRM) of VERE-JONES (1978). The SRM is not dissimilar to the AMR concept; 'Stress' (not the tensor stress, rather a scaler such as Benioff strain or seismic moment) increases over time, and then is released as some function of event magnitude size. Hence the magnitude distribution is also a critical determinant of the model's behavior.

Thus we can elucidate our testing procedure: We will simulate the SRM using various candidate magnitude distributions, and examine the resulting synthetic catalogs statistically. Apart from validating the model, a simple analysis of the magnitudes produced tells us nothing, as the distribution will be exactly the same as that of the input. What is needed is a different metric. The AMR formulation provides such a metric, as we can examine such characteristics as the proportion of AMR sequences, the estimated parameters in the formula used to describe AMR, the average length (number of events) in each cycle, and so on, for consistency with the observed data. The class of possible models is made up of variations on the magnitude distribution and whether the total seismic moment in the SFBR is conserved or not conserved.

We shall next describe the SFBR, and models for the distribution of the earthquake magnitudes. The following section defines AMR, and describes the SRM. We test whether the event rate in the latter for the SFBR is best parameterized as a function of the Benioff strain or the seismic moment. Details of simulation procedures, and summary statistics of the simulated catalogs, follow. While the local fits of various magnitude distributions to a short data set like that of the SFBR differ little, the sample path properties when viewed over 10,000 years are quite different. A number of the possible models can be eliminated at this point because of extremely unlikely sample path properties. An additional available metric to discriminate between the remaining contenders is that the model should produce AMR behavior, as observed in the SFBR data set. In the penultimate section we conduct a detailed statistical analysis of the AMR characteristics of the simulated catalogs, discuss the implications, and identify those magnitude distributions incompatible with the observed AMR sequences in the SFBR. The final section discusses the results, and provides forecasts for the likelihood of large earthquakes in the SFBR.

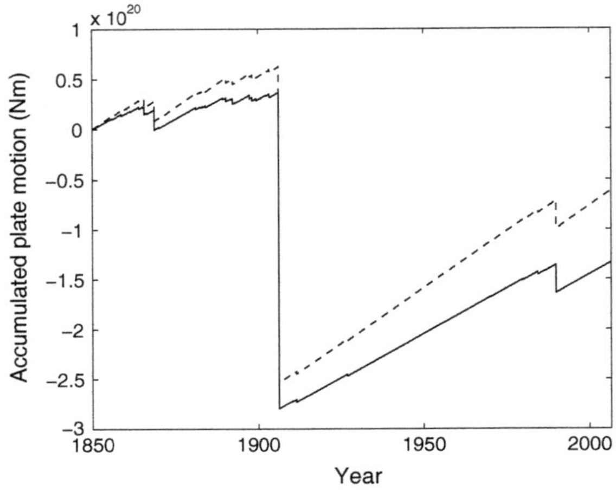

Figure 1
Accumulated seismic moment for the San Francisco Bay Region (*dashed line*), and that available for release in earthquakes according to BAKUN (1999) (*solid line*)

2. *Magnitude Distributions and the San Francisco Bay Region*

We follow BAKUN (1999) in defining the SFBR to be a 245-km-long section of coastal California from south of San Juan Bautista to north of Santa Rosa that spans the San Andreas fault and the other strands of the San Andreas fault system (see BAKUN, 1999, Fig. 1).

BAKUN (1999) compiled a catalog (modified from the Ellsworth catalog) containing 53 events of magnitude at least 5.5. Five events identified as aftershocks (BAKUN, 1999) were removed to decluster the catalog for the SRM, leaving 48. The magnitude of the 1906 San Francisco earthquake was reduced to

7.6 as representing the moment within the San Francisco Bay Region (WALD *et al.*, 1993; THATCHER *et al.*, 1997; SONG *et al.*, 2008). See Table 7 in the Appendix for the final catalog. According to BUFE and VARNES (1993), and SYKES and JAUMÉ (1999) this contains three AMR sequences terminating respectively in the 1868 Hayward, 1906 San Andreas and 1989 Loma Prieta events. ELLSWORTH *et al.* (1981) expressed this in terms of the seismic cycle. The graph (Fig. 1) shows the accumulated moment (minus seismic moment release) calculated by BAKUN (1999) using a geodetic input rate of 2.30×10^{18} Nm/year relative to an arbitrary median. BAKUN (1999) estimates that only 80% of plate-motion movement accumulation is available for release in earthquakes which is also shown in Fig. 1.

There are a number of possible distributions for the frequency–magnitude distribution. Along with the simple Gutenberg–Richter power law, we will consider the *truncated Gutenberg–Richter* (TGR) distributions (BURROUGHS and TEBBENS, 2002) and *tapered Pareto* (or *Kagan*) (VERE-JONES *et al.*, 2001; DALEY and VERE-JONES, 2003, Section 7.3). More specific to the SFBR, there are also the empirical (observed) distribution, and the SFBR Earthquake Source Model (WORKING GROUP, 2003).

The tapered Pareto form has the distribution function (assuming that the lower curvature, corresponding to catalog incompleteness, is immaterial to our purposes and can be ignored)

$$F(y) = 1 - \left(\frac{y}{y_0}\right)^{-\alpha} \exp[-(y - y_0)/U], \qquad y > y_0, \tag{1}$$

where y is the 'size' (to be defined) of an event. For small and intermediate y, this is close to power-law form, but for y large it is dominated by the exponential taper. Increasing the turning point U increases the mean of the distribution, and thus the average event size. Distributions of this type are commonly used when it is desirable for the body of the distribution to approximate a power law, but where the moments must remain finite, as is the case with earthquake sizes. For more on the distribution (1) and its properties, see KAGAN (1991; 1997; 2002), VERE-JONES *et al.* (2001), and KAGAN and SCHOENBERG (2001). In the geophysical literature the tapered Pareto distribution usually appears as the modified Gutenberg–Richter distribution (SORNETTE and SORNETTE, 1999; BIRD *et al.*, 2000)

$$F(M) = 1 - (M_t/M)^{\beta} \exp[(M_t - M)/M_c] \tag{2}$$

for seismic moments, where M_t is the observational completeness threshold (typically presumed known), and M_c the upper turning point.

The truncated Gutenberg–Richter distribution (BURROUGHS and TEBBENS, 2002) for magnitude has the simpler distribution

$$F(M) = \frac{1 - \exp[b(m_0 - M)\ln 10]}{1 - \exp[b(m_0 - m_{\max})\ln 10]}. \tag{3}$$

In this case the upper cutoff magnitude $m_{\max}$ sets a strict limit on the size of the largest event, as opposed to the 'softer' upper magnitude limit of (1). Following VERE-JONES *et al.* (2001), we will let the random variable following (1) be the Benioff strain release ε_i, which is then converted to magnitude as

$$M = m_0 + \frac{4}{3}\log_{10} \epsilon_i \tag{4}$$

(BENIOFF, 1951; GUTENBERG and RITCHER, 1956). Thus the 'b value' in the power-law range of the distribution is equal to 0.75α, and we will specify U in (1) through the *equivalent magnitude* $\gamma = (\log_{10} U - 2.4)/0.75$, which makes for easier comparison with $m_{\max}$ in (3).

Fitting the distributions in (1) and (3) to the data in Table 7 gives maximum likelihood estimates $\widehat{\alpha} = 1.2324, \widehat{\gamma} = 7.6772$ (b value ≈ 0.9243) and $\widehat{b} = 0.9236, \widehat{m}_{\max} = 7.6$. The resulting distributions are compared with the observed empirical distribution in Fig. 2. The most important observation here is that there is very little difference between the fitted distributions, except in the tail. Which of these is the better distribution to use cannot be answered on the basis of conventional (likelihood-based) statistical techniques, especially as it is these tail probabilities that are of most interest.

With the small sample represented by the catalog, the fitting procedure is dominated by the magnitude of the 1906 San Francisco earthquake (cf. ANDREWS and SCHWERER, 2000). Hence we will examine the possible parameter values by means of a sensitivity analysis on this magnitude. The resulting maximum

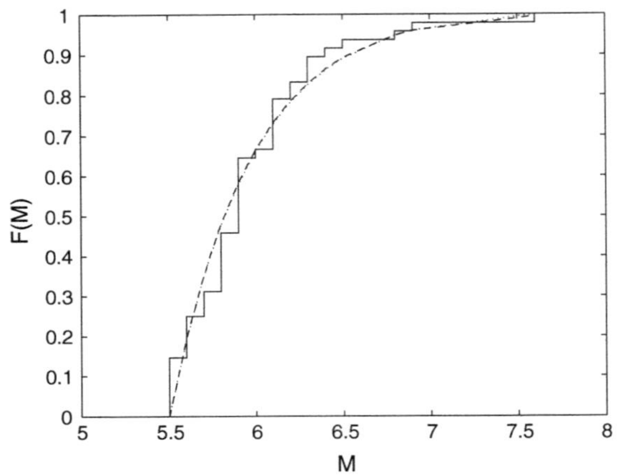

Figure 2
Fitted magnitude distributions for the San Francisco Bay Region. The *solid line* is observed magnitudes, *dashed* the tapered Pareto distribution with $\alpha = 1.2324$, $\gamma = 7.6772$ and *dotted* the truncated Gutenberg–Richter with $b = 0.9236$, $m_{max} = 7.6$

likelihood estimates of the parameters are shown in Fig. 3. Note that the maximum likelihood estimate of m_{max} is of course the largest magnitude in the catalog. The main observation here is that the equivalent magnitude, γ in the tapered Pareto distribution, is by far the most sensitive. Because of the likelihood of producing a very large simulated value, it will also strongly affect the results.

The WORKING GROUP (2003, Table 4.8) provided a magnitude distribution that is a combination of (semi-) characteristic above magnitude 6.7, and Gutenberg–Richter with $b = 0.9$ below this cutoff. The resulting magnitude–frequency plot is shown in Fig. 4. Note the peak at approximately magnitude 7.85. BAKUN (1999) used a Gutenberg–Richter distribution with $b = 0.99$ for $M > 5.5$, but this is not feasible as a simulation distribution due to the finite probability of getting an arbitrarily large event. WARD (2000) used a truncated Gutenberg Richter with $b = 1.1$, $M_{max} = 7.75$, and ANDREWS and SCHWERER (2000) considered a truncated Gutenberg Richter with $b = 0.9$, $M_{max} = 7.78$ and a 'rolloff distribution', for which we shall substitute the tapered Pareto, with $\gamma = 8.0$. The WORKING GROUP (2003) also used a truncated Gutenberg Richter with $b = 0.9$, $M_{max} = 8.25$, which appears to be the best truncated Gutenberg–Richter fit to the distribution in Fig. 4.

A final important consideration pertaining to the conservation of energy, or how much of the

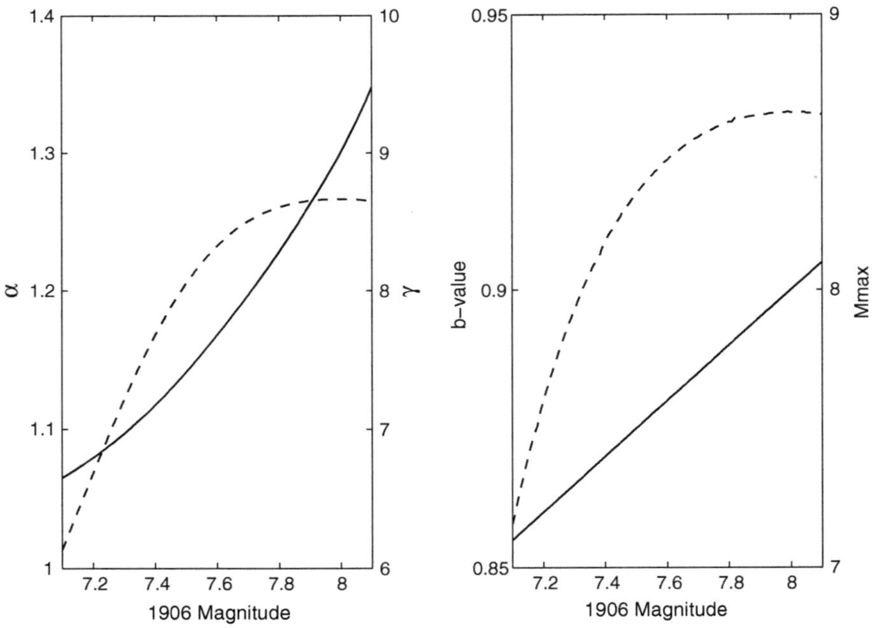

Figure 3
Sensitivity analysis where we assume different magnitudes for the 1906 event to see the effect on the estimated parameters of the magnitude distributions fitted to the SFB region data. Parameter estimates for Tapered Pareto are in the *left plot*, the *solid line* is γ, the *dashed* α. Parameter estimates for the truncated Gutenberg–Richter are in the *right plot*, the *solid line* is m_{max}, the *dashed* the b value

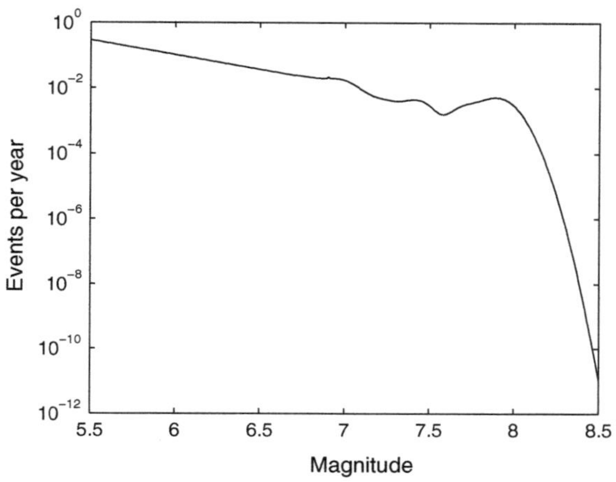

Figure 4
Working Group 2003 frequency–magnitude relation

accumulated strain is available for release in earthquakes. Empirical observations such as the Gutenberg–Richter Magnitude–Frequency relationship and the small magnitude of dynamic stress drops relative to predicted crustal stresses have suggested to some that the seismogenic crust is constantly in a state of "self-organized criticality" (SOC), where the initiation of any small event has the potential to grow into a large earthquake that spans the full dimension of the tectonic fault system (KAGAN, 1994; GELLER *et al.*, 1997). Hence there must always be a large reservoir of energy stored in the crust, and the seismic coupling coefficient (see BEN-ZION and LYAKHOVSKY (2006) for a discussion on the context of damage rheology) must be small. In contrast, the Accelerating Moment Release (AMR) hypothesis, based on the post-event study of seismic catalogs, is that the regional rate of seismic moment or strain release accelerates prior to the occurrence of many moderate to large earthquakes in deforming continental regions (see, for example, JAUMÉ and SYKES, 1999, and references therein).

3. AMR and the Stress Release Model

The acceleration in seismic release is most often described by a power-law time to failure equation (VARNES, 1989; BUFE and VARNES, 1993):

$$\Omega(t) = A + B(t_f - t)^m, \tag{5}$$

where $\Omega(t)$ is the observed quantity of interest (cumulative seismic moment, Benioff strain, or event count), t_f is the time of the main-shock event that ends the sequence, and A, B, and m are parameters that describe the acceleration. Seismic moment is often considered the preferred quantity to use on theoretical grounds (MAIN, 1999; VERE-JONES *et al.*, 2001) but Benioff strain is most often used in empirical work because it appears to provide better 'post-dictions' of t_f (BUFE and VARNES, 1993). In terms of earthquake quantities the *cumulative Benioff strain release* is

$$\Omega(t) = \sum_{i=1}^{N(t)} E_i^{1/2} = \sum_{i=1}^{N(t)} \epsilon_i = \sum_{i=1}^{N(t)} 10^{2.4+0.75M_i} \tag{6}$$

(BENIOFF, 1951; GUTENBERG and RICHTER, 1956), where $N(t)$ is the number of events in $(0, t)$, and E_i, ε_i and M_i are the energy release, Benioff strain release, and magnitude, respectively, of the ith event.

BOWMAN *et al.* (1998) suggested that the presence or absence of AMR in an earthquake sequence can be characterized by the value of the ratio

$$C = \frac{\text{power law fit RMS error}}{\text{linear fit RMS error}}, \tag{7}$$

with a threshold at $C = 0.7$ marking the transition from AMR to non-AMR.

POLLITZ *et al.* (2004), using a regional viscoelastic coupling model (POLLITZ and NYST, 2005), estimated the evolution of regional stress for the SFBR from 1838. This indicted that the majority of $M > 5.5$ earthquakes since 1906 are consistent with stress triggering from previous earthquakes. Thus the SFBR appears to be a reasonable candidate to be examined in isolation for repeated AMR sequences.

Proper quantification of earthquake occurrence forecasts from AMR requires some way of simulating sequences of events from the model. For example, large synthetic catalogs provide the means to examine inference properties of parameter estimates, and thus obtain some idea of the expected precision of forecasts (cf., ROBINSON *et al.*, 2005; ZHOU *et al.*, 2006). VERE-JONES *et al.* (2001) consider simulating directly from the AMR formulation (5), identifying a spectrum of methods for doing so. However, a difficulty with the formulation of the AMR idea in (5) is that it only covers a single sequence, terminating in a major event. There is no indication of the mechanism between sequences, and thus the model is not capable of being simulated indefinitely to produce synthetic catalogs. Thus a model that provides for repeated AMR sequences is desirable. JAUMÉ and BEBBINGTON (2004) showed that AMR could be generated from the simple Stress Release Model (SRM). The same idea of AMR resulting from previous stress shadows was pursued by MIGNAN *et al.* (2007) and MIGNAN (2008) in their formulation of the stress accumulation model.

The SRM (VERE-JONES, 1978) is a simple point process intensity model, with a conditional intensity that increases over time, and decreases as a result of earthquakes. The SRM is thus a stochastic version of the elastic rebound model. We should first note that the quantity termed 'stress' in the model is not stress, but rather a scalar denoting cumulative Benioff strain or seismic moment. The term 'stress' is used as a shorthand to denote the quantity, S, which is accumulated to form the underlying system variable, or *level*, of the stochastic process. Combining the Gutenberg–Richter energy relation $\log_{10} E = 1.5M + const.$, where M is magnitude, with the form $S \propto E^{\eta/1.5}$ engenders the empirical relationship

$$S = 10^{\eta(M - M_{\min})}, \tag{8}$$

where $M_{\min}$ is a minimum, or reference, magnitude. Thus the process tracks the level of, for example, Benioff strain ($\eta = 0.75$) or seismic moment ($\eta = 1.5$) accumulated in a large tectonic region.

The region size for which the stress release model is appropriate (ZHENG and VERE-JONES, 1991, 1994) is comparable to those identified by BOWMAN *et al.* (1998) as exhibiting AMR behavior. For the purposes of this study we will simply assume that the region modeled by the SRM corresponds to one exhibiting AMR behavior. Thus we abstract the spatial dimension, leaving only the temporal dimension, which can thus be modeled by the SRM. HARDEBECK *et al.* (2008) examined the fitting procedure of BOWMAN *et al.* (1998), using simulated catalogs from a spatial ETAS model (OGATA, 1998). The investigation focussed on the degrees of freedom available to the AMR fitting procedure in optimizing the spatial region, and in selecting the time origin of the AMR sequence. We note that these degrees of freedom do not exist in this study, as the spatial region is fixed, and the origin for the subsequent AMR sequence is the termination of the previous one. The SRM is without aftershocks (although it can generate them, see BOROVKOV and BEBBINGTON, 2003; BEBBINGTON, 2008). Hence the simulated data are perfectly declustered, which is appropriate for the AMR hypothesis.

The level of the process, $X(t)$, evolves as

$$X(t) = X(0) + \rho t - S(t), \tag{9}$$

where $X(0)$ is the initial value, ρ is a constant loading rate from external tectonic forces, and $S(t)$ is the accumulated seismic release from earthquakes within the region over the period $(0, t)$, that is, $S(t) = \sum_{t_i < t} S_i$, where t_i and S_i are the origin time and the seismic release (8) associated with the ith earthquake.

Earthquakes above the threshold size, $M_{\min}$, are assumed to occur stochastically, where the probability of an event occurring in the time interval $(t, t + \Delta)$ is $\Psi(X(t))\,\Delta + o(\Delta)$ for small Δ. Thus the history controls the rate of events through the current level. Obviously the function $\Psi(X(t))$ must be nondecreasing. For example, $\Psi(X(t))$ = constant results in a random (Poisson process) model of occurrences, while

$$\Psi(X(t)) = \begin{cases} 0, & X(t) \leq x_c, \\ \infty, & X(t) > x_c, \end{cases}$$

corresponds to a time-predictable model, supposing a failure strength x_c. An effective compromise (ZHENG and VERE-JONES, 1991, 1994) between these extremes of behavior is $\Psi(X(t)) = \exp[\mu + \nu X(t)]$. This can also represent the behavior that might be expected from a region with a locally heterogeneous strength. We interpret μ as representing the background seismicity rate, although this cannot be distinguished from the initial level $X(0)$ using observed data. Note that although $X(t)$ can become negative in (9), which is intuitively unattractive for a physical quantity, the inclusion of μ in the formulation means that $X(t)$ is in fact merely a perturbation relative to some unknown level. Meanwhile, the exponential form of $\Psi(X(t))$ ensures that the probability of an event remains positive. The 'sensitivity' parameter ν is interpreted as an amalgam of the distribution of fault strengths in the region.

Statistical analysis is made possible by treating the data in historical earthquake catalogs as a point process with history-dependent conditional intensity function

Table 1

SRM parameters and fit (standard errors from Hessian in parentheses), log *L denotes the* log *likelihood*

	$\eta = 0.75$ (Benioff strain)	$\eta = 1.5$ (Seismic moment)
a	−0.083 (0.222)	−0.569 (0.270)
b	0.005 (0.012)	0.015 (0.010)
c	3.402 (6.372)	0.123 (0.049)
log L	−90.85	−86.05

$$\lambda(t) = \Psi(X(t)) = \exp[\mu + \nu(X(0) + \rho t - S(t))] \\ = \exp[a + b(t - cS(t))]. \quad (10)$$

Estimates of the parameters can then be found by maximizing the log-likelihood function (see, for example, DALEY and VERE-JONES, 2003, Section 7.2),

$$\log L = \sum_i \log \lambda(t_i) - \int_0^T \lambda(t)dt \quad (11)$$

over the observation interval (0, T). The two terms in (11) reward high intensities at times of earthquake occurrence, and penalize high intensities elsewhere, respectively. For a more extensive review of the SRM, see BEBBINGTON and HARTE (2003). Software for fitting the model is provided in HARTE (2009).

The results of fitting the Bakun catalog by the SRM are given in Table 1. The standard deviation of the estimates is obtained from the Hessian of the log likelihood (11) (see BEBBINGTON and HARTE, 2001). The fitted levels for the model with $\eta = 0.75$ and $\eta = 1.5$ are shown in Fig. 5. Obviously the fit using the seismic moment is far superior, confirming the observation of JAUMÉ and BEBBINGTON (2004) that AMR sequences can be best modeled using the SRM with seismic moment as the basic variable. Furthermore, the fitted seismic moment in Fig. 5 shows a striking similarity to the plot of the 80% of accumulated moment available for earthquakes (BAKUN, 1999) shown in Fig. 1. We shall thus assume henceforth that our underlying state variable is seismic moment rather than Benioff strain.

If the seismic coupling coefficient is high, then a period of high activity may temporarily exhaust the available reservoir of strain, reducing the possible size of subsequent events. Alternatively, if the seismic coupling coefficient is small, there need be no reduction in the size of possible events. The former

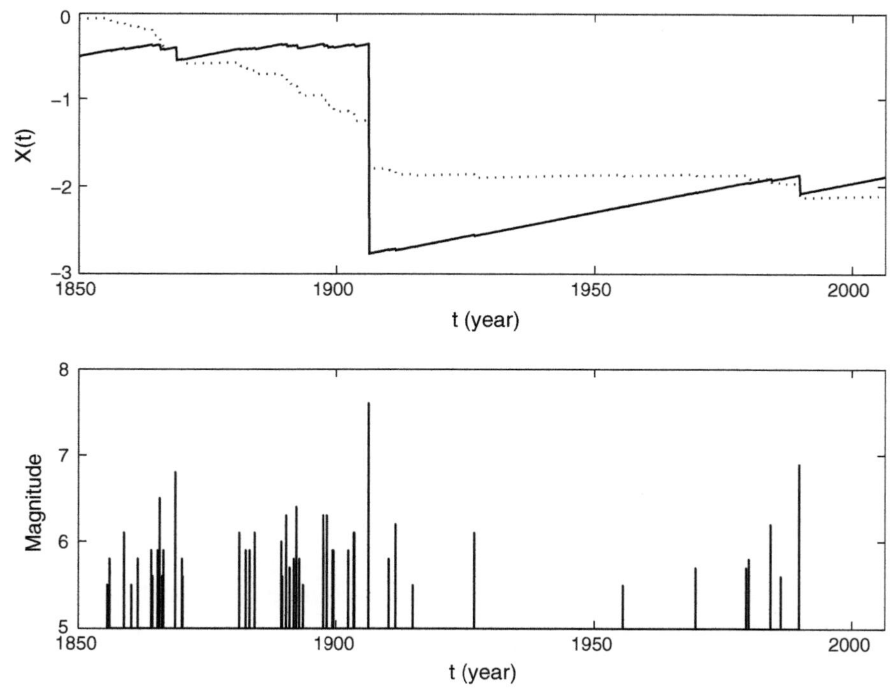

Figure 5
Fitted Benioff strain (*dotted line*) and seismic moment (*solid line*) for the San Francisco Bay Region by the stress release model

Table 2

Magnitude distribution scenarios for the San Francisco Bay Region

No.	Distribution	α/b-value	γ/m max	Note
1	Empirical	N/A	N/A	Table 7
2	Truncated Gutenberg–Richter	1.1	7.75	WARD (2000)
3	Truncated Gutenberg–Richter	0.9	7.78	ANDREWS and SCHWERER (2000)
4	Truncated Gutenberg–Richter	0.9	8.25	WORKING GROUP (2003)
5	Truncated Gutenberg–Richter	0.9236	7.6	MLE
6	Tapered Pareto	1.2	8.0	analogous to ANDREWS and SCHWERER (2000)
7	Tapered Pareto	1.2324	7.6772	MLE
8	WORKING GROUP (2003)	(0.9)	(Table 4.8)	Semi-Characteristic

implies that there is a minimum value (which can be set at zero by rescaling a) below which the level $X(t)$ cannot fall. The latter implies that $X(t)$ is on an arbitrary scale without limits. In order to represent the first case, we need to estimate the constant $X(0)$ in (9). A method of doing this is to fit the model to the observed events, and find the minimum seismic moment by replacing the 1906 event by the hypothesized maximum magnitude event. Using the value of $M_{\max} = 8.25$ estimated by the WORKING GROUP (2003), we arrive at a seismic moment remaining, immediately after the 1906 event, equivalent to approximately an 8.06 magnitude earthquake.

The relations between low or high seismic coupling coefficient, and between $\eta = 0.75$ and $\eta = 1.5$ are critical assumptions in that they determine the way we reset the model after a large event. These assumptions represent the physics that are absent in the AMR specifications, but are critical (regardless if we use the SRM or not) if a model capable of simulating forecast hazard is to be constructed. Given such a model framework, we can simulate synthetic catalogs for the SFBR using a number of different magnitude distributions.

4. Simulations

From the discussion in the previous section, we can assemble the magnitude scenarios in Table 2. We will append an 'A' to the scenario number to indicate the 'seismic moment conservation' criterion of energy equivalent to a magnitude 8.06 remaining immediately after the 1906 event. The empirical magnitudes are included as a nonparametric benchmark.

The simulation procedure is relatively straightforward (see JAUMÉ and BEBBINGTON, 2004 for more details) and proceeds as follows for the tapered Pareto distribution. The truncated Gutenberg–Richter distribution works similarly.

1. Using a variant of the Shedler-Lewis thinning method (LEWIS and SHEDLER, 1976; OGATA, 1981) we generate the time of the next event using the time-varying intensity $\lambda(t)$ in (10). Note that $X(t)$ increases between events.
2. We then generate a strain release from the distribution (1), using the fact (VERE-JONES *et al.*, 2001) that the distribution is the minimum of an exponential and a Pareto random variable. The threshold y_0 is set such that the resulting magnitude distribution has $M_{\min} = 5.0$.
3. This strain release is then converted into a magnitude M via (4). If we are assuming conservation of seismic moment, and if the generated magnitude is greater than the calculated seismic moment release remaining $X(t)$, reduce it to this amount.
4. The level $X(t)$ is then reduced by an amount $10^{1.5(M-M_{\min})}$.
5. After reducing $X(t)$ by the generated amount, we find the next event time.

To validate the model(s), we can repeatedly simulate with the candidate distribution for 158 years (1850–2008), and examine various statistics, such as the maximum magnitude, the number of earthquakes above specified magnitudes, and the coefficient of variation (standard deviation / mean) of the interevent times. The WORKING GROUP (2003) estimate the number of events per year as $N_{M\geq 6.7} = 0.031$ per year and $N_{M>7} = 0.010$. The corresponding figure from WARD (2000) is

Table 3

Simulated results for 100 runs of 158 years each. Tabulated are the minimum and maximum parentheses (over the 100 simulations) of various statistics (columns) for the different magnitude models (rows). A '+' indicates a censored value

No.	$\bar{M}$	$M_{\max}$	$N_{M \geq 5.5}$	$N_{M \geq 6.7}$	$N_{M > 7}$	CV (T)	Max (T)
Target	5.95	7.6	48	3–5.9	1–2.5	1.76	28.87
1	(5.83,6.40)	(6.8,7.6)	(11,125)	(1,7)	(0,2)	(0.33,1.28)	(5.20, 89.91)
2	(5.81,6.12)	(6.73,7.62)	(21,127)	(1,7)	(0,4)	(0.45,1.23)	(5.47,48.44)
2A	(5.81,6.08)	(6.79,7.62)	(25,135)	(1,8)	(0,4)	(0.48,1.26)	(4.59,43.32)
3	(5.86,6.35)	(6.96,7.77)	(10,109)	(1,8)	(0,4)	(0.35,1.30)	(5.58,85.13)
3A	(5.80,6.17)	(6.85,7.62)	(18,109)	(1,8)	(0,4)	(0.38,1.28)	(5.63,76.51)
4	(5.84,6.97)	(6.83,8.24)	(2,118)	(1,8)	(0,5)	(0.37,1.47)	(1.56,143.78)
4A	(5.86,6.34)	(6.94,7.62)	(11,96)	(1,8)	(0,4)	(0.46,1.26)	(5.87,52.33)
5	(5.83,6.29)	(6.87,7.59)	(14,107)	(1,8)	(0,4)	(0.51,1.38)	(5.86,44.15)
5A	(5.85,6.15)	(6.84,7.60)	(17,111)	(1,9)	(0,5)	(0.42,1.25)	(5.61,55.32)
6	(5.85,8.20)	(6.83,8.84)	(1,106)	(1,8)	(0,4)	(0.32,1.43)	(1.03,113.65+)
6A	(5.83,6.19)	(6.91,7.61)	(19,96)	(1,9)	(0,4)	(0.45,1.35)	(6.56,55.31)
7	(5.83,8.03)	(6.85,8.23)	(1,120)	(1,8)	(0,5)	(0.28,1.93)	(4.66,101.72+)
7A	(5.81,6.17)	(6.74,7.62)	(12,126)	(1,8)	(0,4)	(0.30, 1.28)	(5.71,55.22)
8	(5.87,6.55)	(6.95,8.06)	(3,122)	(1,9)	(0,4)	(0.24,1.56)	(2.77,133.80)
8A	(5.86,6.21)	(6.94,7.61)	(12,121)	(1,10)	(0,3)	(0.40,1.16)	(5.91,60.80)

$N_{M>7} = 0.016$, and from ANDREWS and SCHWERER (2000), $N_{M \geq 6.7} = 0.038$. The observed numbers in 150 years are $N_{M \geq 5.5} = 48 (= 0.31$ per year), $N_{M \geq 6.7} = 3 (= 0.02$ per year) and $N_{M>7} = 1$ (= 0.006 per year). The observed coefficient of variation for the interevent times CV(T) between events of magnitude at least 5.5 was 1.76, and the longest interval [Max(T)] was 28.87 years. These and other 'targets', along with the simulated results (minimum and maximum) for 100 runs of 158 years are given in Table 3. We see that the simulated catalogs are all reasonable except for consistent understatement of the tendency for events to cluster. In the observed catalog, this is the result of short intervals early on, followed by very long intervals due to the stress shadow cast by the 1906 event. This combination, which requires a very large event around the middle of the period, is rare in the simulations. There may also be some effect from triggering between faults (cf., POLLITZ *et al.*, 2004), which is not a factor included in the simple SRM. This should not affect conclusions regarding the larger earthquakes.

5. Results and Analysis

We will now test the candidate magnitude distributions by means of their sample path properties. In particular we will simulate long SRM sequences using these distributions and compare their AMR properties with those observed in the SFBR data.

The various models were simulated forward for 10,000 years from present. The intensities fitted to the simulated events are shown in Figs. 6 and 7. These show that the tapered Pareto (scenarios 6 and 7), and the WORKING GROUP (2003) (scenarios 4 and 8), distributions require a conservation of seismic moment if stretches of thousands of years without events (where the fitted intensity is approximately zero) are to be avoided.

Using a cutoff of $M_{\text{main}} = 6.8$, the seismic strain release between earthquakes with $M \geq M_{\text{main}}$ was fitted to (5), constrained by the t_f and $A = \Omega(t_f)$ of the main shock. The values of B and m were found by minimizing the least-squares error between (5) and the observed seismic strain release. The minimum number of events in an AMR sequence was required to be at least 10, and the parameter m was restricted to the interval (0, 0.8) to ensure some degree of (positive) curvature. The degree of AMR behavior is then measured by the value C from (7), and in particular the proportion of events with $M > M_{\text{main}}$ preceded by AMR sequences with $C < 0.7$ (BOWMAN *et al.*, 1998). In order to separate the effects of magnitude distribution and stress release model, the catalogs were randomized by repeatedly assigning, to the generated magnitudes, occurrence times uniformly distributed in

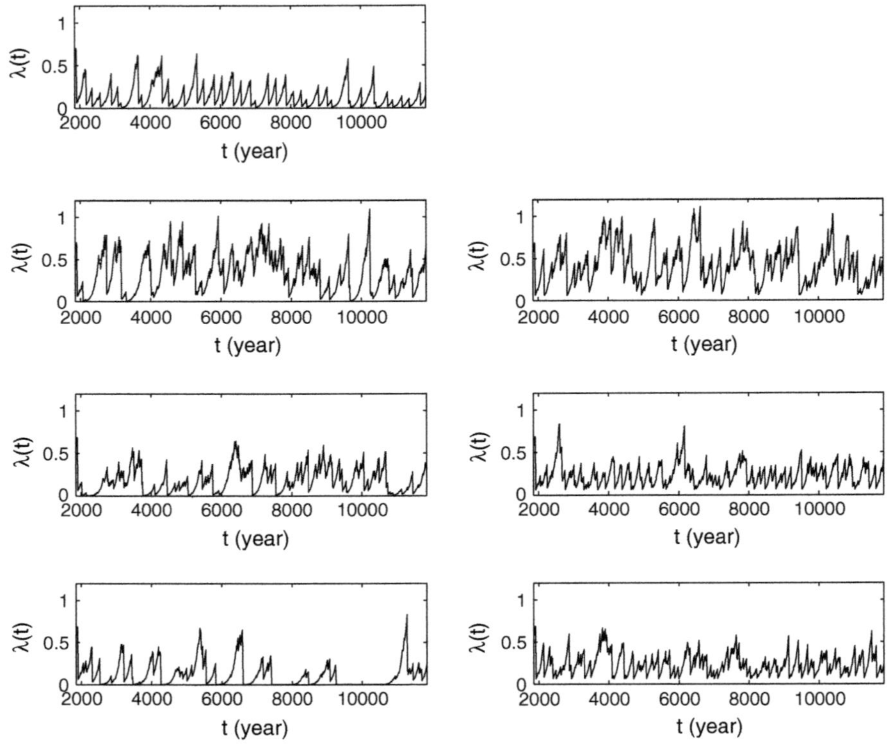

Figure 6
Fitted intensity to simulated catalogs from magnitude Scenario 1 (*top*) to Scenario 4 (*bottom*). Seismic moment nonconserved (*left*) and conserved (*right*)

the 10,000 year window, and recalculating the AMR sequences.

Figure 8 shows how the relative proportions of sequences in the unrandomized and randomized catalogs vary with the curvature C. We see that, in general, the conserved moment scenarios exhibit stronger AMR character than the nonconserved scenarios. The exception is Scenario 7A, where moment conservation appears to create 'anti-AMR'. The observed values of C from the 1868 Hayward, 1906 San Andreas and 1989 Loma Prieta events are 0.56, 0.33 and 0.46, respectively. We see that the empirical magnitude distribution, Scenario 1, has a peak in the ratio above these values, indicating incompatibility with the observations. The two strongest scenarios (in an AMR sense) appear to be 4A and 5A.

The distributions of the fitted parameters $\log(-B)$ and m, and the numbers of events in the sequences n_s are shown in Figs. 9, 10 and 11. We see from Fig. 9 that the fitted values of B from the AMR sequences in the conserved moment scenarios are all consistent with the observed values. The nonconserved moment scenarios are not quite as good, but still consistent. The non-AMR sequences clearly differ from the AMR-sequences, except in the case of Scenario 1, the empirical distribution. The impression from Fig. 10 regarding the fitted parameter m is very similar. Finally, Fig. 11 shows that, apart from Scenario 1, all the lengths of the AMR sequences are consistent with the observations, but that non-AMR sequences tend to be longer than AMR-sequences. Overall, Scenario 1 is inconsistent, Scenarios 4, 6 and 7A appear a little worse than average, and 2A, 3, 3A, 4A, 5A and 7 slightly better than average.

We are also interested in the relationships between the fitted parameters $\log(-B)$ and m, and the magnitudes M_f of the main events, and the numbers of events in the sequences, n_s. The correlations are given in Table 4. Noting that none of the observed correlations (from only 3 sequences) are significant, we can still see that the correlations with n_s appear to be of different character from the observed in some of

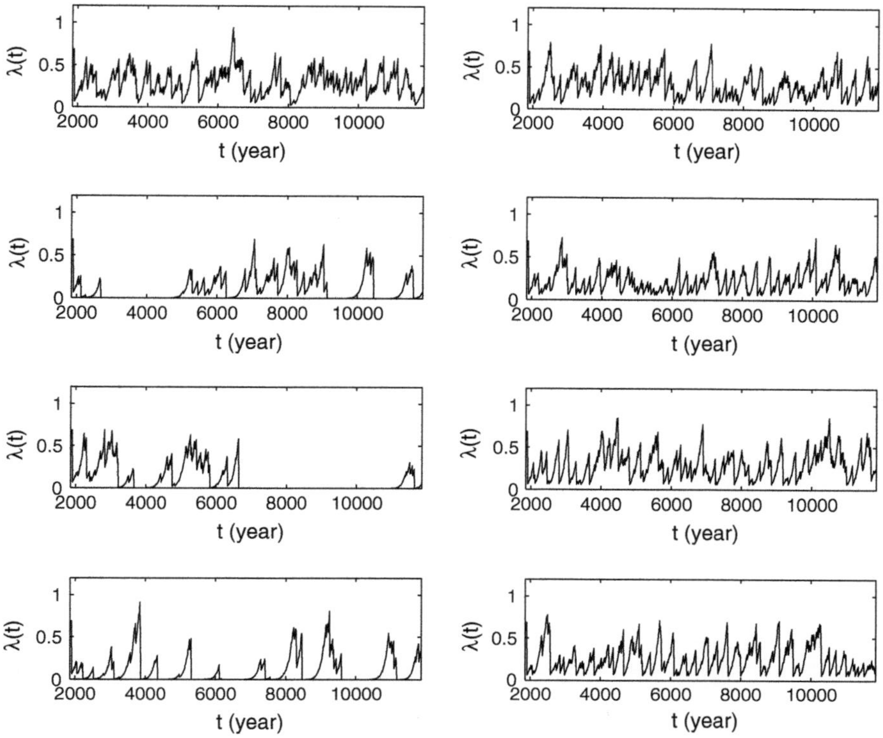

Figure 7
Fitted intensity to simulated catalogs from magnitude Scenario 5 (*top*) to Scenario 8 (*bottom*). Seismic moment nonconserved (*left*) and conserved (*right*)

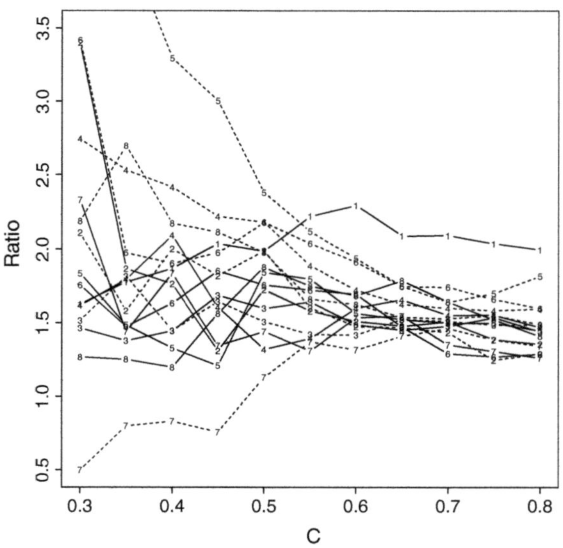

Figure 8
Ratio (simulated/randomized) of proportion of AMR sequences with curvature less than C. Nonconserved scenarios are shown by *solid lines*, conserved by *dashed lines*

the simulated sequences. This is presumably related to the absence, remarked on above, in the model of the local triggering effect between faults. Nevertheless, strong positive correlations between $\log(-B)$ and n_s, and between M_f and n_s, such as in scenarios 2,6 and 7 might be considered desirable, while the strong negative correlations in Scenario 4 are undesirable.

6. Discussion

We have shown that repeated AMR cycles can be well modeled by the simple SRM using seismic moment as the system state. Any frequency–magnitude distribution can be 'plugged in', and exerts considerable influence on the AMR properties of the model. Repeated simulation can thus provide estimates of variability in the AMR parameters, which

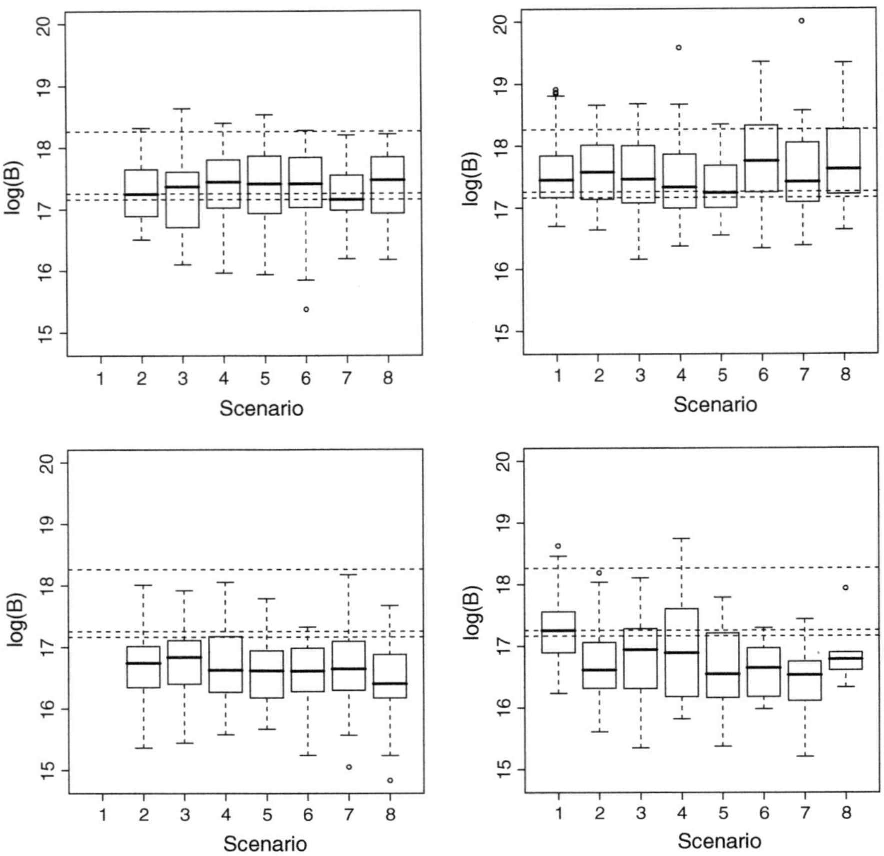

Figure 9
Boxplots of the fitted parameter B: (*Top*) $C <$ 0.7, (*bottom*) $C >$ 0.7, (*left*) magnitude conserved, (*right*) magnitude nonconserved. Values from the three observed sequences shown as *horizontal dashed lines*

are not available from observation, and these can be used as a test statistic for the magnitude distribution.

For the SFBR, Scenario 1, the empirical magnitude distribution is not realistic, as there is no reason that the only possible large magnitudes should be 6.8, 6.9 and 7.6. We are able to reject Scenarios 4, 6, 7 and 8 by visual analysis of the simulated catalogs. Scenario 7A is rejected from the antithetical (to AMR) behavior exhibited in Fig. 8. From the remainder, the Scenarios most consistent with the observed values of B, m and n_s in Figs. 9 to 11, and the correlations in Table 4 are Scenarios 2 and 4A, with 3 and 5/5A in the next rank. Scenarios 4A and 5A appear to have the strongest AMR character. These are all truncated Gutenberg–Richter distributions, and the conclusion appears to be that, apart from the possibility of too large an event suppressing all activity for unrealistically long periods, this is a robust distribution from an AMR perspective. A similar conclusion was reached by ANDREWS and SCHWERER (2000).

We will now simulate Scenarios 2, 3 and 4A, and for comparison, Scenario 8A, to obtain estimates of 30 year hazard. Each scenario is simulated 1000 times for 30 years from present, the largest simulated earthquake in each run recorded, and a probability of an earthquake with magnitude exceeding a given level calculated. This is repeated 100 times in order to derive a mean, and a 95% confidence interval for the probability. The results are shown in Fig. 12. For ease of comparison with the results of the WORKING GROUP (2003), selected probabilities (and 95%

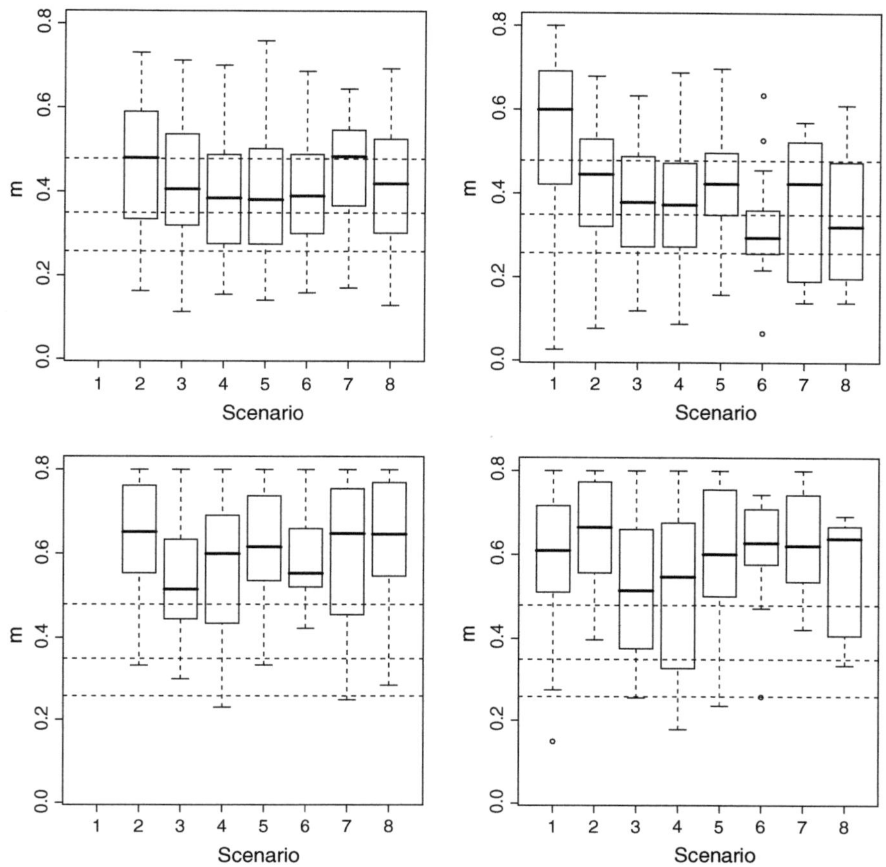

Figure 10
Boxplots of the fitted parameter m: (*Top*) $C < 0.7$, (*bottom*) $C > 0.7$, (*left*) magnitude conserved, (*right*) magnitude nonconserved. Values from the three observed sequences shown as *horizontal dashed lines*

confidence intervals) are listed in Table 5. We see that the estimated 30 year probabilities are consistently around or below the lower end of the 95% confidence interval calculated by the WORKING GROUP (2003). The reason seems to be that the state trajectory (Fig. 5) fitted to the catalog is still relatively low due to the 1906 event. Also, apart from the conservation of seismic moment condition, which makes large events slightly less likely, there is no link between the level of the system and the magnitude of an event. A possible link is discussed at the end of this section. The range of probabilities from the various magnitude scenarios could of course form the basis of a weighted estimate.

The simulation procedure was repeated for a 100 year interval, with results (probability of an earthquake of at least a given magnitude, plus 95% confidence intervals) given in Table 6. Overall, scenarios 3,4A and 8A are reasonably consistent, apart from the likelihood of a great ($M > 7.5$) earthquake. Scenario 2 has a significantly different character, and this comes through in the forecasts.

We see that, with the exception of Scenario 2 with a b value of 1.1, the magnitude scenario makes little difference to the short and medium term probability forecasts from the model. The differences are in the long-term behavior, in particular, the details of the AMR behavior. The latter generally prefers models with some constraint on the likelihood of a series of large events close together in time, be that through a high b value, or a conservation condition on the seismic moment. An alternative to fitting an invariant

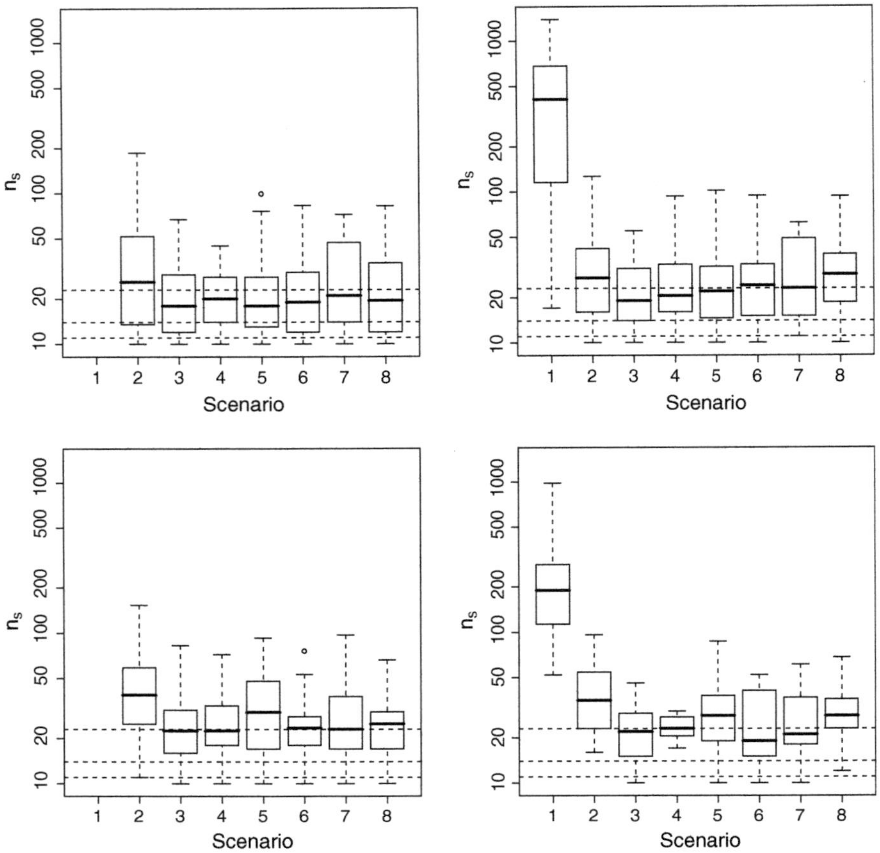

Figure 11
Boxplots of the sequence lengths n_s: *(Top)* $C < 0.7$, *(bottom)* $C > 0.7$, *(left)* magnitude conserved, *(right)* magnitude nonconserved. Values from the three observed sequences shown as *horizontal dashed lines*

Table 4

Correlation coefficients of fitted AMR sequences

Scenario	$C > 0.7$								$C < 0.7$			
	B/m	M_f/m	M_f/B	n_s/m	n_s/B	n_s/M_f	B/m	M_f/m	M_f/B	n_s/m	n_s/B	n_s/M
1	−0.69	−0.39	0.75	0.68	−0.15	0.02	−0.71	−0.58	0.83	0.56	−0.06	−0.11
2	−0.88	−0.62	0.81	0.60	−0.33	−0.26	−0.79	−0.44	0.73	0.03	0.38	0.36
2A	−0.79	−0.52	0.62	0.65	−0.21	−0.31	−0.82	−0.42	0.65	0.41	−0.07	0.14
3	−0.87	−0.64	0.84	0.55	−0.26	−0.11	−0.84	−0.48	0.72	0.27	0.03	0.30
3A	−0.85	−0.50	0.76	0.56	−0.18	0.00	−0.79	−0.41	0.66	0.31	0.04	−0.02
4	−0.81	−0.67	0.90	0.47	−0.05	−0.07	−0.98	−0.98	0.97	0.63	−0.58	−0.54
4A	−0.88	−0.66	0.81	0.50	−0.21	−0.17	−0.90	−0.61	0.73	0.16	0.10	0.29
5	−0.74	−0.44	0.72	0.62	−0.16	−0.12	−0.87	−0.31	0.60	0.48	−0.15	0.35
5A	−0.88	−0.60	0.72	0.63	−0.32	−0.29	−0.77	−0.15	0.50	0.57	−0.17	0.34
6	−0.76	−0.61	0.87	0.40	0.10	−0.08	−0.72	0.10	0.47	0.00	0.56	0.85
6A	−0.81	−0.59	0.79	0.45	−0.03	−0.08	−0.80	−0.46	0.61	0.44	−0.14	−0.04
7	−0.81	−0.67	0.86	0.50	0.02	−0.04	−0.80	−0.33	0.69	0.19	0.22	0.39
7A	−0.75	−0.31	0.64	0.66	−0.16	0.09	−0.86	−0.42	0.62	0.39	−0.17	0.03
8	−0.83	−0.72	0.93	0.54	−0.23	−0.28	−0.61	−0.66	0.98	0.61	0.05	0.08
8A	−0.84	−0.59	0.82	0.57	−0.23	−0.16	−0.85	−0.50	0.60	0.32	0.01	−0.06
Observed	(None are significantly different from zero)						−0.02	−0.21	0.98	−0.14	0.99	0.94

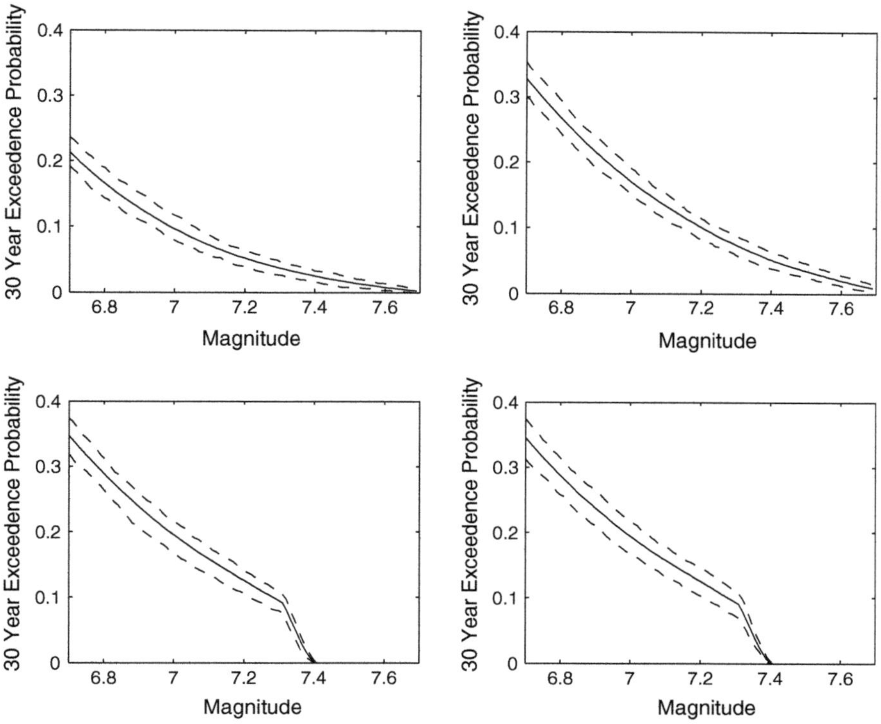

Figure 12
Estimated mean probability (*solid line*) and 95% confidence intervals (*dashed lines*) of an earthquake exceeding a given magnitude within 30 years. *Top left* Scenario 2, *top right* Scenario 3, *bottom left* Scenario 4A, *bottom right* Scenario 8A

Table 5

Probability of an earthquake exceeding a given magnitude within 30 years. (95% confidence interval in parentheses).

Scenario	Magnitude		
	6.7	7.0	7.5
Working Group (2003)	0.62 (0.38,0.85)	0.35 (0.17,0.56)	0.10 (0.02,0.19)
2	0.21 (0.19,0.24)	0.10 (0.08,0.12)	0.02 (0.01,0.03)
3	0.33 (0.30,0.36)	0.17 (0.15,0.20)	0.04 (0.02,0.05)
4A	0.35 (0.31,0.37)	0.19 (0.17,0.22)	0
8A	0.34 (0.30,0.37)	0.16 (0.13,0.19)	0

Table 6

Probability of an earthquake exceeding a given magnitude within 100 years. (95% confidence interval in parentheses).

Scenario	Magnitude		
	6.7	7.0	7.5
Working Group (2003)	0.96 (0.92,0.99)		
2	0.68 (0.65,0.72)	0.37 (0.34,0.40)	0.07 (0.05,0.08)
3	0.85 (0.83,0.87)	0.57 (0.54,0.60)	0.13 (0.11,0.15)
4A	0.87 (0.84,0.89)	0.62 (0.59,0.65)	0.01 (0.00,0.02)
8A	0.86 (0.84,0.88)	0.55 (0.52,0.58)	0.01 (0.01,0.02)

magnitude distribution is to use the ASR model of Jaumé and Bebbington (2004). This generates a change in the rate of seismic release, by increasing their average size (Jaumé, 2000). As observational evidence (Jaumé, 2000) is that the b value of the lower magnitude part of the distribution does not change during ASR sequences, this is done by controlling the large magnitude tail of the distribution, in order to increase the average event size. Thus the equivalent magnitude in the tapered Pareto distribution might be parameterized as

$$\gamma = u_0 + e^{u_1 X}, \tag{12}$$

where $X = X(t)$ is the level of the stochastic process (the SRM in this case), and u_0 and u_1 are parameters. A similar procedure can be followed for the truncated

Gutenberg Richter distribution, varying m_{max} in like manner. JAUMÉ and BEBBINGTON (2004) found that AMR can be accomplished in the SRM by an increase in the rate of events, an increase in their average size, or a combination of the two. BEN-ZION and LYAKHOVSKY (2002) made a similar suggestion, based on model simulations with damage rheology, that AMR can consist of an early period of elevated rate of moderate events that is followed a few years later by a phase of increasing average event size.

Acknowledgments

This work was supported by the Marsden fund, administered by the Royal Society of New Zealand.

Appendix

Table 7

Table 7

Modified Ellsworth Catalog (BAKUN, 1999)

No.	Latitude	Longitude	Magnitude	Time (years since 1850)
1	38.50	−122.00	5.5	5.653
2	37.00	−122.00	5.5	6.003
3	37.50	−122.30	5.8	6.123
4	37.50	−121.90	6.1	8.902
5	37.00	−122.00	5.5	10.292
6	37.80	−122.00	5.8	11.505
7	37.10	−121.70	5.9	14.154
8	37.70	−122.00	5.9	14.176
9	37.50	−122.00	5.6	14.385
10	37.00	−122.00	5.9	15.394
11	37.00	−122.00	6.5	15.770
12	37.00	−122.00	5.6	16.233
13	37.50	−121.30	5.9	16.535
14	37.70	−122.10	6.8	18.807
15	37.20	−122.10	5.8	20.131
16	37.77	−122.16	5.6	20.252
17	37.40	−121.40	6.1	31.274
18	37.10	−121.76	5.9	32.488
19	36.91	−121.61	5.9	33.243
20	37.10	−122.20	6.1	34.232
21	38.00	−121.90	6.0	39.381
22	37.71	−122.16	5.6	39.581
23	36.90	−121.60	6.3	40.312
24	37.30	−121.76	5.7	41.005
25	38.30	−122.44	5.8	41.778

Table 7 continued

No.	Latitude	Longitude	Magnitude	Time (years since 1850)
26	38.40	−122.00	6.4	42.299
29	36.80	−121.50	5.8	42.869
30	38.47	−122.72	5.5	43.605
31	37.00	−121.50	6.3	47.469
32	38.20	−122.40	6.3	48.246
33	37.00	−122.00	5.9	49.329
34	37.50	−122.00	5.5	49.417
35	37.20	−121.50	5.9	49.512
36	38.35	−121.95	5.9	52.378
37	37.40	−121.90	6.1	53.440
38	37.30	−121.80	6.1	53.584
39	37.70	−122.50	7.6	56.293
40	36.90	−121.80	5.8	60.188
41	37.25	−121.75	6.2	61.496
42	37.00	−122.00	5.5	64.853
43	36.62	−122.35	6.1	76.805
46	37.37	−121.78	5.5	105.675
48	38.47	−122.72	5.7	119.751
49	37.12	−121.52	5.7	129.595
50	37.83	−121.78	5.8	130.064
51	37.33	−121.65	6.2	134.313
52	37.48	−121.68	5.6	136.245
53	37.04	−121.88	6.9	139.795

REFERENCES

ANDREWS, D.J. and SCHWERER E. (2000), *Probability of rupture of multiple fault segments*, Bull. Seismol. Soc. Am. *90*, 1498–1506

BAKUN, W.H. (1999), *Seismic activity of the San Francisco Bay region*, Bull. Seismol. Soc. Am. *89*, 764–784

BEBBINGTON, M. (2008), *Estimating rate- and state-friction parameters using a two-node stochastic model for aftershocks.* Tectonophysics *457*, 71–85

BEBBINGTON, M. and HARTE, D. (2001), *On the statistics of the linked stress release process*, J. Appl. Probab. *38A*, 176–187

BEBBINGTON, M. and HARTE, D. (2003), *The linked stress release model for spatio-temporal seismicity: Formulations, procedures and applications*, Geophys. J. Int. *154*, 925–946

BENIOFF, H. (1951), *Earthquakes and rock creep, Part 1: Creep characteristics of rocks and the origin of aftershocks.* Bull. Seismol. Soc. Am. *41*, 31–62

BEN-ZION, Y. and LYAKHOVSKY, V. (2002), *Accelerated seismic release and related aspects of seismicity patterns on earthquake faults*, Pure Appl. Geophys. *159*, 2385–2412

BEN-ZION, Y. and LYAKHOVSKY, V. (2006), *Analysis of aftershocks in a lithospheric model with seismogenic zone governed by damage rheology*, Geophys. J. Int. *165*, 197–210

BIRD, P., KAGAN, Y.Y. and JACKSON, D.D. (2000), *Frequency–magnitude distribution, effective lithosphere thickness, and seismic efficiency of oceanic transforms and spreading ridges*, EOS Trans. AGU 81, (abstract) WP147

BOROVKOV, K. and BEBBINGTON, M. (2003), *A stochastic two-node stress transfer model reproducing Omori's law*, Pure Appl. Geophys. *160*, 1429–1455.

BOWMAN, D.D., OUILLON, G., SAMMIS, C.G., SORNETTE, D. and SORNETTE, A. (1998), *An observational test of the critical earthquake concept*, J. Geophys. Res. *103*, 24,359–24,372

BUFE, C.G. and VARNES, D.J. (1993), *Predictive modeling of the seismic cycle in the greater San Francisco Bay region*, J. Geophys. Res. *98*, 9871–9883

BURROUGHS, S.M. and TEBBENS, S.F. (2002), *The upper-truncated power law applied to earthquake frequency–magnitude distributions: Evidence for a time-independent scaling parameter*, Bull. Seismol. Soc. Am. *92*, 2983–2993.

DALEY, D.J. and VERE-JONES, D., *An Introduction to the Theory of Point Processes*, volume 1, second edition (Springer, New York 2003)

ELLSWORTH, W.L., LINDH, A.G., PRESCOTT, W.H. and HERD, D.G., *The 1906 San Francisco earthquake and the seismic cycle, In Earthquake Prediction: An International Review*, Maurice Ewing volume 4 (eds. Simpson, D.W. and Richards, P.G.)(AGU, Washington DC 1981) pp. 126–140

GELLER, R.J., JACKSON, D.D., KAGAN, Y.Y., and MULARGIA, F. (1997), *Geoscience - Earthquakes cannot be predicted*, Science *275*, 1616–1617

GUTENBERG, B. and RICHTER, C.F. (1956), *Earthquake magnitude, intensity, energy and acceleration*, Bull. Seismol. Soc. Am. *46*, 105–145

HARDEBECK, J.L., FELZER, K.R. and MICHAEL, A.J. (2008), *Improved tests reveal that the accelerating moment release hypothesis is statistically insignificant*, J. Geophys. Res. *113*, B08310, doi: 10.1029/2007JB005410

HARTE, D. (2009), *PtProcess: Time Dependent Point Process Modelling. R package version 3.1-3.* Statistics Research Associates, Wellington. URL: http://cran.at.r-project.org/web/packages/PtProcess

JAUMÉ, S.C., *Changes in earthquake size-frequency distributions underlying accelerating seismic moment/energy release*, In *GeoComplexity and the Physics of Earthquakes*, AGU Geophysical Monograph 120 (eds. Rundle, J.B., Turcotte, D.L. and Klein, W.) (AGU, Washington DC 2000) pp. 199–210

JAUMÉ, S.C. and BEBBINGTON, M.S. (2004), *Accelerating seismic release from a self-correcting stochastic model*, J. Geophys Res. *109*, B12301, doi:10.1029/2003JB002867

JAUMÉ, S.C. and SYKES, L.R. (1999), *Evolving towards a critical point: A review of accelerating seismic moment/energy release prior to large and great earthquakes*, Pure Appl. Geophys. *155*, 279–305

KAGAN, Y.Y. (1991), *Seismic moment distribution*, Geophys. J. Int. *106*, 123–134

KAGAN, Y.Y. (1994), *Observational evidence for earthquakes as a nonlinear dynamic process*, Physica D *77*, 160–192

KAGAN, Y.Y. (1997), *Seismic moment-frequency relationship for shallow earthquakes; Regional comparisons*, J. Geophys. Res. *102*, 2835–2852

KAGAN, Y.Y. (2002), *Seismic moment distribution revisited: I. Statistical results*, Geophys. J. Int. *148*, 520–541

KAGAN, Y.Y. and SCHOENBERG, F. (2001), *Estimation of the upper cutoff parameter for the tapered Pareto distribution*, J. Appl. Probab. *38A*, 158–175

LEWIS, P.A.W. and SHEDLER, G.S. (1976), *Simulation of nonhomogeneous Poisson processes with loglinear rate function*, Biometrika *63*, 501–506

MAIN, I.G. (1999), *Applicability of time-to-failure analysis to accelerated strain before earthquakes and volcanic eruptions*, Geophys. J. Int. *139*, F1–F6

MIGNAN, A. (2008), *The stress accumulation model: Accelerating moment release and seismic hazard*, Adv. Geophys. *49*, 67–201

MIGNAN, A., KING, G.C.P. and BOWMAN, D.D. (2007), *A mathematical formulation of accelerating moment release based on the stress accumulation model*, J. Geophys. Res. *112*, B07308, doi:10.1029/2006JB004671

OGATA, Y. (1981), *On Lewis' simulation method for point processes*, IEEE Trans. Inform. Theory IT-30, 23–31

OGATA, Y. (1998), *Space-time point-process models for earthquake occurrences*, Ann. Inst. Statist. Math. *50*, 379–402

POLLITZ, F., BAKUN, W.H. and NYST, M. (2004), *A physical model for strain accumulation in the San Francisco Bay region: Stress evolution since 1838*, J. Geophys. Res. *109*, B11408, 2004

POLLITZ, F. and NYST, M. (2005), *A physical model for strain accumulation in the San Francisco Bay region*, Geophys. J. Int. *160*, 302–317

ROBINSON, R., ZHOU, S., JOHNSTON, S. and VERE-JONES, D. (2005), *Precursory accelerating moment release (AMR) in a synthetic seismicity catalog: A preliminary study*, Geophys. Res. Lett. *32*, L07309

SAMMIS, C.G. and SMITH, S.W. (1999), *Seismic cycles and the evolution of stress correlation in cellular automaton models of finite fault networks*, Pure Appl. Geophys. *155*, 307–334

SONG, S.G., BEROZA, G.C. and SEGALL, P. (2008), *A unified source model for the 1906 San Francisco earthquake*, Bull. Seismol. Soc. Am. *98*, 823–831

SORNETTE, D. and SORNETTE, A. (1999), *General theory of the modified Gutenberg–Richter law for large seismic moments*, Bull. Seismol. Soc. Am. *89*, 1121–1130

SYKES, L.R. and JAUMÉ, S.C. (1990), *Seismic activity on neighboring faults as a long-term precursor to large earthquakes in the San Francisco Bay area*, Nature *348*, 595–599

THATCHER, W., MARSHALL, G. and LISOWSKI, M. (1997), *Resolution of fault slip along the 470-km-long rupture of the great 1906 San Francisco earthquake and its implications*, J. Geophys. Res. *102*, 5353–5367

VARNES, D.J. (1989), *Predicting earthquakes by analyzing accelerating precursory seismic activity*, Pure Appl. Geophys. *130*, 661–686

VERE-JONES, D. (1978), *Earthquake prediction: a statistician's view*, J. Phys. Earth *26*, 129–146

VERE-JONES, D., ROBINSON, R. and YANG, W. (2001), *Remarks on the accelerated moment release model: Problems of model formulation, simulation and estimation*, Geophys. J. Int. *144*, 517–531

WALD, D.J., KANAMORI, H., HELMBERGER, D.V. and HEATON, T.H. (1993), *Source study of the 1906 San Francisco earthquake*, Bull. Seismol. Soc. Am. *83*, 981–1019

WARD, S.N. (2000), *San Francisco Bay Area earthquake simulations: a step towards a standard physical earthquake model*, Bull. Seismol. Soc. Am. *90*, 370–386

WORKING GROUP ON CALIFORNIA EARTHQUAKE PROBABILITIES (2003), *Earthquake Probabilities in the San Francisco Bay Region: 2002–2031*, USGS Open-File Report 03-214

Zheng, X. and Vere-Jones, D. (1991), *Applications of stress release models to earthquakes from North China*, Pure Appl. Geophys. *135*, 559–576

Zheng, X. and Vere-Jones, D. (1994), *Further applications of the stochastic stress release model to historical earthquake data*, Tectonophysics *229*, 101–121

Zhou, S., Johnston, S., Robinson, R. and Vere-Jones, D. (2006), *Tests of the precursory accelerating moment release using a synthetic seismicity model for Wellington, New Zealand*, J. Geophys. Res. *111*, B05308, doi:10.1029/2005JB003720

(Received September 16, 2008, revised April 7, 2009, accepted June 9, 2009, Published online March 2, 2010)

Pure Appl. Geophys. 167 (2010), 819–830

DOI 10.1007/s00024-010-0065-5

Pure and Applied Geophysics

Using the ETAS Model for Catalog Declustering and Seismic Background Assessment

R. Console,[1] D. D. Jackson,[2] and Y. Y. Kagan[2]

Abstract—The concept of background seismicity is strictly related to the identification of spontaneous and triggered earthquakes. The definition of foreshocks, main shocks and aftershocks is currently based on procedures depending on parameters whose values are notoriously assumed by subjective criteria. We propose a method for recognizing the background and the induced seismicity statistically. Rather than using a binary distinction of the events in these two categories, we prefer to assign to each of them a probability of being independent or triggered. This probability comes from an algorithm based on the ETAS model. A certain degree of subjectivity is still present in this procedure, but it is limited by the possibility of adjusting the free parameters of the algorithm by rigorous statistical criteria such as maximum likelihood. We applied the method to the seismicity of southern California and analyzed the sensitivity of the results to the free parameters in the algorithm. Finally, we show how our statistical declustering algorithm may be used for mapping the background seismicity, or the moment rate in a seismic area.

Key words: Seismic background, declustering, epidemic model, maximum likelihood.

1. Introduction

Temporal clustering, as commonly observed during aftershock sequences, constitutes strong evidence for time-dependent behavior of the seismic process and its departure from a simple spatially-variable, time-independent Poisson process. Nevertheless, seismicity is often modeled by a process with a rate constant in time, and most earthquake damage mitigation measures in seismic areas are based on this assumption. The tectonic processes causing seismic activity apparently change only on a geologic time scale, so that the driving stress and seismic moment rate can be considered constant over a long-term observation.

Regardless of the behavior of the seismicity in time, it is widely recognized that the magnitude distribution of earthquakes follows the well-known Gutenberg–Richter (G–R) law (Gutenberg and Richter, 1944) modeled by Eq. 5 in the Appendix. The value of the b parameter in the G–R law is typically close to 1.0 in different areas of the world and for different magnitude ranges, so expressing a universal property of the seismicity (Bird and Kagan, 2004).

The validity of the G–R law is limited to a range of magnitude where the lower limit is generally connected to the minimum magnitude reported with completeness by the observation system, and the upper limit is related to the maximum size of the rupture produced by the seismogenic process in a given region. Without an upper limit to the magnitude distribution, the total moment rate of any region would be infinite. Therefore, several modifications to the linear trend have been proposed (see Kagan, 2002 and Bird and Kagan, 2004 for a thorough analysis of this issue).

Seismologists have traditionally labeled earthquakes as "foreshocks", "main shocks" or "aftershocks", assuming that only the main shocks can be described by a time-independent stationary process (see e.g., Gardner and Knopoff, 1974 and Reasenberg, 1985). However, a quantitative physical definition of main shock has never been given, and the separation of the above mentioned three classes of earthquakes is based on empirical subjective definitions that are checked against the stationarity of the main shock rate in time.

In order to overcome the above-mentioned problems, a number of people (e.g., Kagan and Knopoff,

[1] Istituto Nazionale di Geofisica e Vulcanologia, Rome, Italy. E-mail: console@ingv.it
[2] University of California, Los Angeles, USA.

1976; KAGAN, 1991; KAGAN and JACKSON, 2000; ZHUANG *et al.*, 2002, 2004, 2008; ZHUANG and OGATA, 2006) have already described some form of stochastic declustering. Their methods are based on the application of models whose parameters are fit by maximum likelihood criteria. While KAGAN and KNOPOFF (1976) used a branching model of earthquake occurrence, ZHUANG *et al.* (2002) made use of a similar model called Epidemic Type Aftershock Sequence (ETAS), in which the background and the clustered structure are estimated in non-parametric and parametric ways, respectively (see Sect. 3 for more details on this model). In this study we used the same kind of method used by ZHUANG *et al.* (2002), with the only difference being that they used variable kernel functions to estimate the backround rate, while our method uses FRANKEL'S (1995) method. More recently, MARSAN and LONGLINÉ (2008) introduced an iterative procedure that finds the best declustering algorithm in a way that the result is somewhat independent of the initial maximum likelihood model parameters. In their algorithm the background and the clustering structure are both estimated in non-parametric ways.

Based on criteria already applied in previous studies, our method makes use of an iterative procedure for obtaining a spatially variable model of the seismicity, suitable for seismological applications and seismic hazard estimates. However, we are not aiming at producing a catalog where some of the events are removed, but rather a catalog where the events receive a weight proportional to the probability of being independent.

The data used for all the numerical applications are drawn from the seismic catalog of the Southern California Earthquake Data Center (SCEDC) (http://www.data.scec.org/catalog_search/date_mag_loc.php). Southern California is one of the most densely sampled seismic regions of the world.

2. *A Smoothing Algorithm for a Spatially Variable Poisson Model*

Figure 1a shows the epicenter distribution of 60,480 earthquakes with magnitude equal to or larger than 2.0 reported by SCEDC from 1984 to 2002. The area taken for the analysis is a rectangle 360 km × 440 km wide, centered on the point of geographical coordinates 34.5°N and 117.0°W. As clearly shown in Fig. 1b, the catalog can be considered complete in the period of time chosen for the analysis. Even if some small magnitude events are missing, especially soon after other larger magnitude earthquakes, the

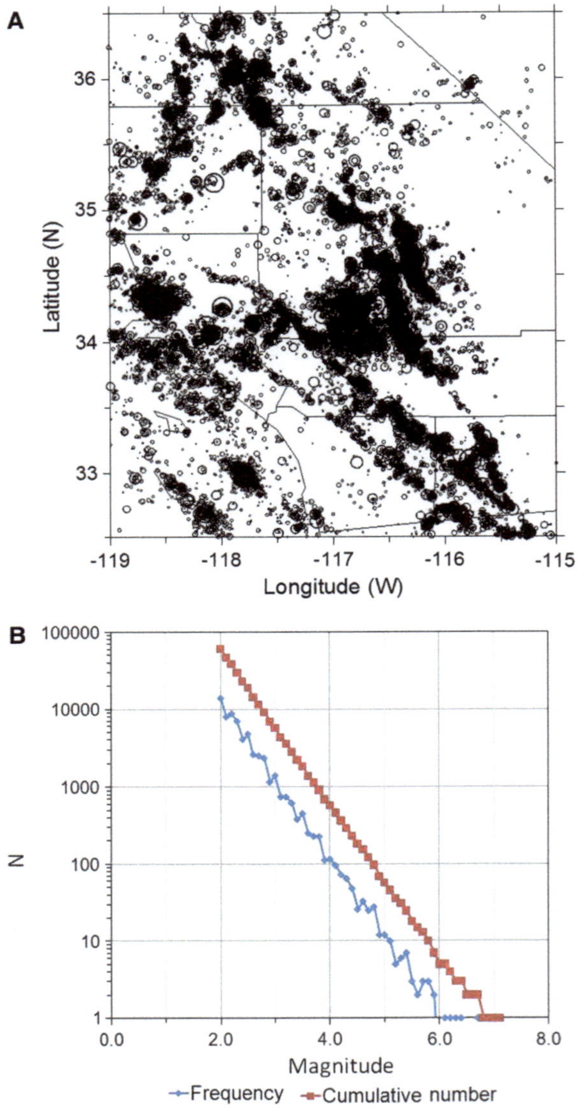

Figure 1
a Epicentral distribution of the earthquakes with magnitude $M \geq 2.0$ reported by the Southern California Earthquake Data Center in the time period 1984–2002. **b** Frequency-magnitude distribution of the earthquake catalog considered in this study

circumstance does not affect the results in terms of background seismicity distribution. This is because, as explained in the introduction, and discussed later on in this paper, these early events, if recorded, would be strongly down-weighted by the algorithm for the computation of the background seismicity.

In order to obtain a continuous rate density $\lambda_0(x,y)$, we applied an algorithm for smoothing the discrete epicenter distribution. Applications of algorithms making use of smoothing kernels for seismic hazard assessment can be found in KAGAN and JACKSON (1994, 2000), JACKSON and KAGAN (1999), and HELMSTETTER *et al.* (2007), whose ideas are largely reflected in this study.

Here we compute a gridded smooth geographical distribution of the seismic rate density at each node k of a regular grid through the method introduced by FRANKEL (1995):

$$\tilde{N}_k = \frac{\sum_l N_l \exp\left(-\Delta_{kl}^2/d^2\right)}{\sum_l \exp\left(-\Delta_{kl}^2/d^2\right)}, \tag{1}$$

where N_l is the number of events in each cell centered on the lth node, Δ_{kl} is the distance between nodes k and l, and d is a free parameter.

The spatially variable rate density $\lambda_0(x,y)$ (used in Eq. 5 of the Appendix) is computed dividing by the observation time interval the value obtained from linear interpolation of $\tilde{N}_k$ among the four nearest grid nodes. The free parameter d is determined by maximizing the likelihood of the seismicity contained in the later half of the catalog under the model obtained from the earlier half. In this study the best fit of the d value has been carried out by maximizing the likelihood of the SCEDC earthquake catalog ($M \geq 2.0$) from 1993 to 2002 under the time-independent Poisson model obtained from the same catalogue in the period from 1984 to 1992. Figure 2 shows the epicenter distributions for the two separate time periods. These two maps show a remarkable similarity, indicating that the spatial distribution of the seismic activity is rather stable in time, though the map of the first period exhibits stronger activity in the area of the Landers (1992) earthquake and the map of the second period shows concentration of epicenters in the areas of the Northridge (1994) and Hector Mine (1999) earthquakes.

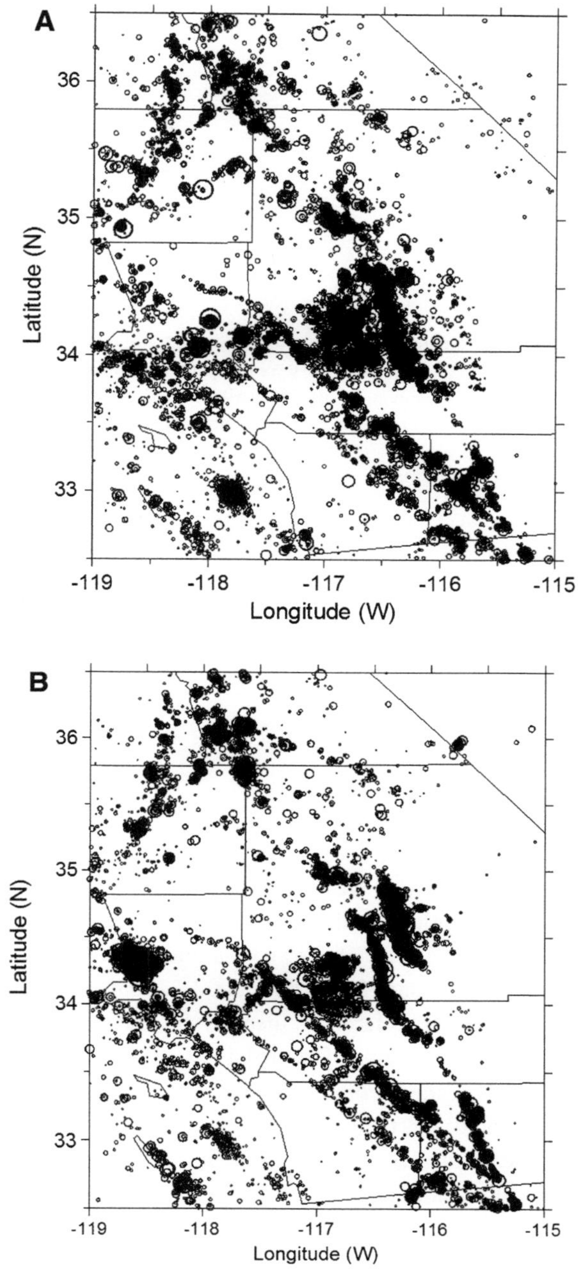

Figure 2

a Epicentral distribution of the earthquakes with magnitude $M \geq 2.0$ observed in Southern California in the time period 1984–1992. **b** As in **a**, for the time period 1993–2002

Figure 3 shows the plot of the likelihood of the second part of the catalog estimated from the first part, as a function of the correlation distance d. A maximum is visible around $d = 5.0$ km, which is the value taken for the following analysis. The smoothed

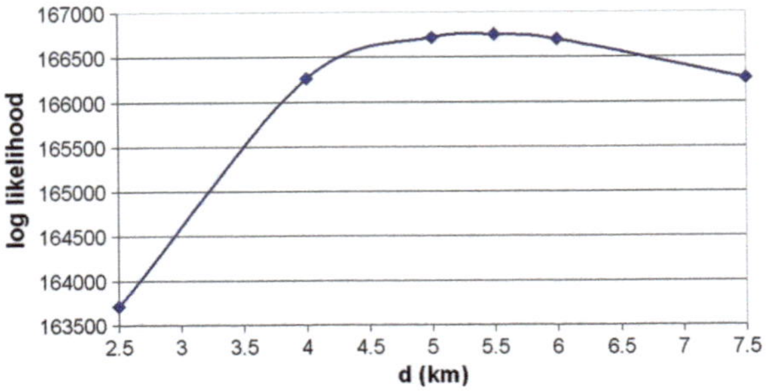

Figure 3
Plot of the likelihood of the Southern California 1993–2002 earthquake catalog under the time-independent Poisson model representing the smoothed spatial distribution of the 1984–1993 catalog, versus different values of the d parameter of the smoothing algorithm

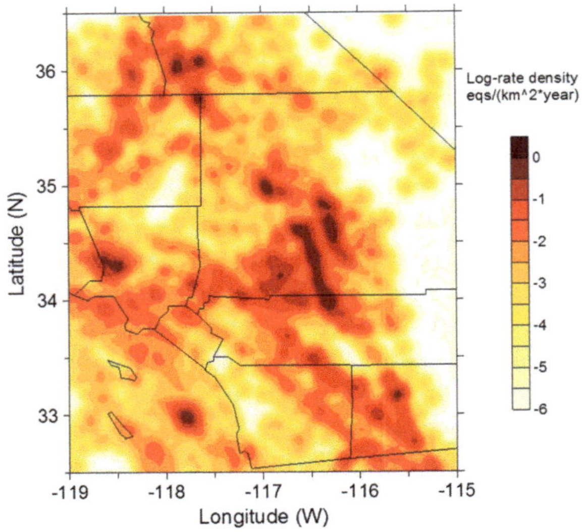

Figure 4
Smoothed distribution of the Southern California (1984–2002) seismicity obtained by the smoothing algorithm of Eq. 1 with a correlation distance $d = 5.0$ km. The color scale represents the number of events occurred in cells 2 km × 2 km wide over the total duration of the catalog (i.e., 19 years)

geographical distribution so obtained for the whole data set is shown in Fig. 4.

3. *Modeling the Earthquake Catalog by the ETAS Model*

As shown in Fig. 5 (line a), the cumulative number of all the earthquakes in the catalog is strongly irregular, because of the presence of aftershock sequences. We may suppose, in agreement with assumptions underlying the ETAS model (see e.g., OGATA, 1998, 1999 CONSOLE and MURRU, 2001; HELMSTETTER and SORNETTE, 2002, 2003; CONSOLE *et al.*, 2007 and references therein) that the seismicity has two components. The first is assumed to have a time-independent Poisson behavior, and the second is represented by the triggered earthquakes. A short outline of the ETAS model is reported in Appendix. Note that stationarity of the background rate and the parameters of the ETAS model is assumed in this, as in many other studies, though it cannot be really validated.

We aim at a spatial distribution that does not include the triggered component of the seismicity, still preserving the total seismic moment released by the seismicity, which is approximately proportional to the total number of earthquakes. With this purpose, we applied an iterative process based on a clustering (epidemic) time dependent model (see also MARSAN and LONGLINÉ, 2008):

1. We started by finding the maximum likelihood set of free parameters using the initial distribution of the smoothed seismicity $\lambda_0(x,y)$ by interpolation of the gridded distribution obtained through Eq. 1;
2. We computed the probability of independence p_i as the ratio between the independent component $f_r \cdot \lambda_0(x_i,y_i,m_i)$ and the composite rate density $\lambda(x_i,y_i,m_i,t_i)$ for every event i in Eq. 4 of the Appendix;
3. Then, without adjusting the d parameter again, we computed a new distribution of $\lambda_0(x,y)$ through Eq. 1 as in step 1, but introducing the weights p_i as a multiplying factor for each event, and dividing by f_r to normalize the result to the total number of events in the catalog;
4. The new smoothed distribution was used in a new maximum likelihood best fit of the free parameters;
5. We proceeded again from step 2, and so on, until a reasonable convergence was reached.

In order to limit the time needed for the maximum likelihood best fit, which is rather computer demanding, step 1 of the above-mentioned process was carried out using a threshold of magnitude 3.0 both for the triggering and the triggered earthquakes

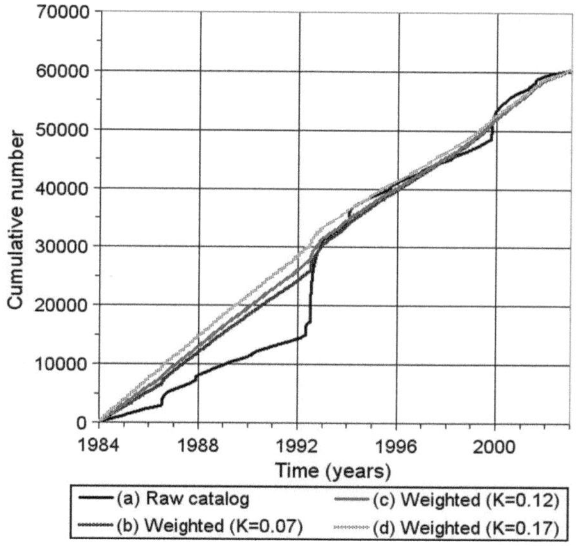

Figure 5

a Cumulative distribution of the number of events in the Southern California (1984–2002) earthquake catalog. **b** Cumulative distribution of the weights p_i over the time spanned by the Southern California (1984–2002) earthquake catalog, normalized to the total number of events; see point 2 in Sect. 3 of the text for the definition of p_i; the model parameters used for obtaining this plot are those shown in the last column of Table 1. **c** As in **b**; the model parameters used for obtaining this plot are those shown in the last column of Table 1, with the substitution of the value $K = 0.12$ for this parameter. **d** As in **b**; the model parameters used for obtaining this plot are those shown in the last column of Table 1, with the substitution of the value $K = 0.17$ for this parameter

(for a total of 5,733 events). The time required for carrying out the same computations, but with a magnitude threshold one unit lower, would have been about 100 times longer. The smoothed distribution of the spatial density, however, was computed on the basis of the entire data set with a lower magnitude threshold equal to 2.0. The best fit values of the parameters obtained after five iterations of the procedure described above are reported in Table 1. The parameters affected by the largest variations are those modeling the spatial distribution (d_0 and q), while the parameters related to the time decay are considerably more stable. Moreover, there is a negative correlation (as expected) between the productivity parameter K and d_0, because both are related to the total number of triggered events. Parameter d_0 is probably influenced also by the location errors of the epicenters reported in the catalog, which is not taken in consideration in our simple algorithm.

Figure 6 shows the map of the new smoothed distribution, using the algorithm described in step 3 and the model parameters obtained after the final iteration. Comparing this map with that shown in Fig. 4, we note a more diffuse distribution of the seismicity, and a better correspondence to the fault system.

As it can be clearly noted in Fig. 5 (line b), the cumulative distribution of p_i over the time spanned by the catalog, normalized to the total number of events, is much closer to a linear trend than the starting cumulative number of events. This gives a good indication of the ability of our algorithm to remove the contribution of the triggered seismicity, without really deleting any event from the catalogue.

4. Constraining the Weighting Process for the Most Uniform Total Rate

The total earthquake rate over the region is represented by the slope of the cumulative number of events versus time. Our iterative procedure results in

Table 1

Values obtained for the parameters of the epidemic model in the progressive adjustment of the background seismicity

	Initial	Iter. 2	Iter. 3	Iter. 4	Iter. 5
K (days^{p-1})	0.0637	0.0673	0.0674	0.0678	0.0698
d_0 (km)	0.504	0.502	0.499	0.498	0.478
q	1.884	1.859	1.853	1.850	1.812
c (days)	0.0154	0.0154	0.0152	0.0155	0.0154
p	1.080	1.079	1.081	1.081	1.082
α	0.563	0.544	0.546	0.546	0.549
f_r	0.0772	0.0838	0.0840	0.0817	0.0822
ln$L$	49,985.4	50,086.5	50,101.1	50,104.4	50,107.1

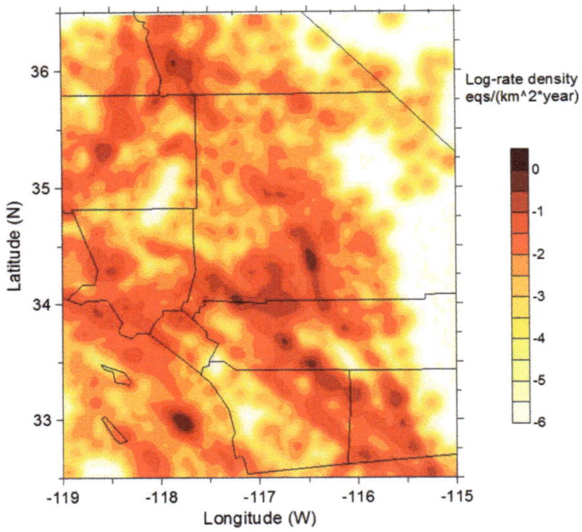

Figure 6
Smoothed distribution of the Southern California (1984–2002) seismicity obtained by the smoothing algorithm applied to the weighted catalog as described in Sect. 3 of the text and the model parameters shown in the last column of Table 1

a fairly uniform total rate of background events, yet, it is quite evident that a bump is left in the cumulative distribution. This bump shows that the effect of the aftershock sequence of the Landers (1992) earthquake has not been completely removed. We make the arbitrary assumption that an ideal weighting algorithm should achieve a uniform total rate of background seismicity. In this respect, we are interested in exploring the idea that a modification of the parameters of the ETAS model could achieve a more uniform background rate without compromising the likelihood too much.

We define the parameter D_n (the same used in the Kolmogorov–Smirnov or K–S test; see for instance Gibbons and Chakraborti, 2003) to express the mismatch between the actual cumulative distribution and the desired linear trend with the same total earthquake count:

$$D_n = \max\left|\hat{F}_n(i) - F_0(i)\right|, \tag{2}$$

where $\hat{F}_n(i)$ is the observed value of the normalized cumulative distribution at the event i, and $F_0(i)$ is the theoretical normalized linear trend, so that $\hat{F}_n(n) = F_0(n)$. The value of this parameter for the ETAS model parameters reported in the last column of Table 1 and the cumulative distribution shown in Fig. 5 (line b), is $D_n = 0.083$.

Through a set of empirical tests, we searched for the value of K in the ETAS model that minimizes D_n. Figure 7, where the results of this search are reported, shows how, increasing the K value, D_n decreases substantially, and then it increases again after having reached a minimum ($D_n = 0.033$) at $K = 0.12$ (almost twice the value obtained from the maximum likelihood best fit, see Table 1).

The new cumulative distribution, obtained using the previous set of best fit parameters, but with the substitution of $K = 0.12$, is shown in Fig. 5 (line c). Figure 8 shows the corresponding map of the smoothed spatial distribution, using the algorithm described in Sect. 3 and the model parameters with the value of K adjusted at $K = 0.12$. A comparison between the maps of Figs. 6 and 8 shows a very slight change in the smoothed seismicity distribution, though the value of the K productivity parameter has been raised by more than 70%. This can be explained considering that the ETAS model using the maximum likelihood estimate of K is already capable of suppressing nearly all the clustered activity. However, the map of Fig. 8 is characterized by a degree of smoothing slightly higher than that of Fig. 6.

We computed the log-likelihood of the same catalog used in the previous analysis, with a lower magnitude threshold equal to 3.0, using the final best fit parameters of the ETAS model reported in the last column of Table 1, but with the smoothed seismicity distribution obtained using the value of K adjusted at $K = 0.12$ as shown in Fig. 8. This test provided a

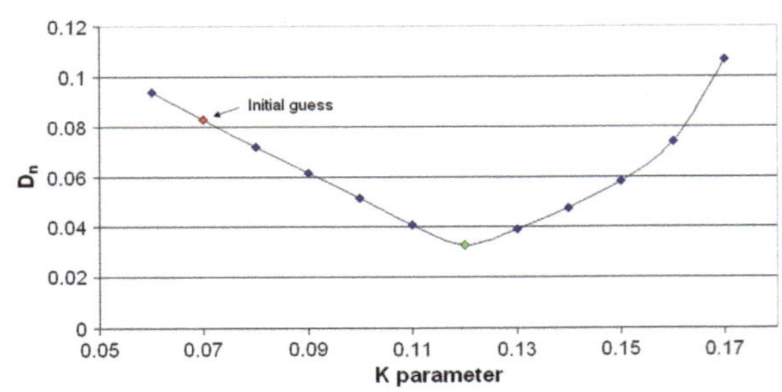

Figure 7
Plot of the D_n parameter, showing a measure of the departure of the cumulative distribution of events from a theoretical linear trend, versus the K value of the ETAS model used for giving a weight to each event in the seismic catalog

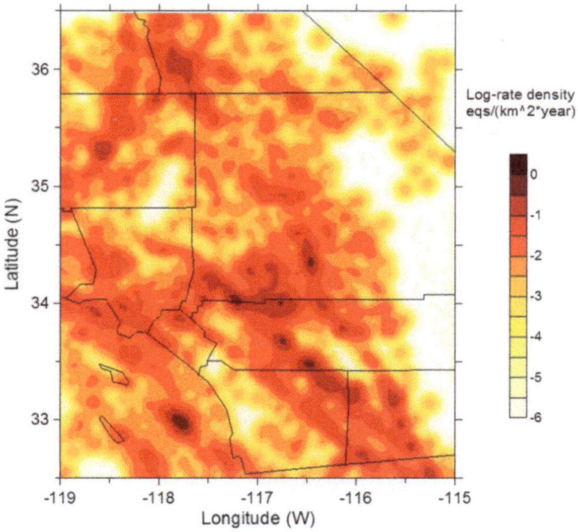

Figure 8
Smoothed distribution of the Southern California (1984–2002) seismicity obtained by the smoothing algorithm applied to the weighted catalog as described in Sect. 3 of the text and the model parameters shown in the last column of Table 1, with the substitution of the value $K = 0.12$ for this parameter

value of 50,111.4 for the log-likelihood, which is slightly better than that of the best fit obtained with the smoothed density distribution of Fig. 6, obtained from the best fit value of $K = 0.07$. Of course, we use $K = 0.12$ only for the preparation of the smoothed seismicity, because this value, if used in the ETAS model, would yield a much smaller value of the likelihood.

A visual inspection shows that the new total background rate of Fig. 5 (line c) is not yet uniform. This is also confirmed by the fact that the K–S test for a linear distribution is not passed at a 90% confidence level. A further increase of the parameter K, together the complete disappearance of the effect of the aftershock sequences on the cumulative distribution, produces a decrease of the slope just at the time of the Landers (1992) earthquake. As an example, Fig. 5 (line d) shows the plot of the cumulative distribution obtained for $K = 0.17$.

5. *Testing the Model on Independent Data*

It is generally acknowledged that testing a forecast model requires a data set which is completely independent of the one used in the learning process (see e.g., CONSOLE, 2001; CONSOLE *et al.*, 2003; HELMSTETTER *et al.*, 2007). In line with this rule, we have tested the quality of the ETAS model relative to a time-independent, spatially variable model, using a new data set for the same area of southern California, spanning the time period 2003–2007. At the same time, we have also tested the impact of the different choices for the seismicity smoothing algorithm used for obtaining the background spatial distribution. In these tests we have considered both the lower magnitude threshold $m_0 = 3.0$ used in the learning phase (including only $N_3 = 185$ target events), and an even lower magnitude threshold $m_0 = 2.0$ (including $N_2 = 1772$ target events), which allows a more robust test on a number of events about ten times larger. The main results are reported in Table 2.

Table 2 shows in a clear way the overall better performance of the models including the weighting technique for the smoothed seismicity, compared with the standard unweighted smoothing algorithm. However, if we look at the log-performance factor, obtained as the difference between the log-likelihood of the ETAS model and that of the Poisson model (and at its average, also called probability gain per earthquake), we notice that it decreases from the model using the standard unweighted smoothing algorithm to those including the weighting technique. This is clearly a consequence of the better performance of the new background distribution applied to a time-independent model. If we compare, instead, the performances of the two weighted background distributions (the one with $K = 0.07$ and the other with $K = 0.12$), we notice very little difference. The overall likelihood is improved from the former to the latter for $m_0 = 3.0$, but gets worse for $m_0 = 2.0$, The performance factor and the probability gain per earthquake obtained from these two approaches are also not significantly different.

We may conclude that the use of the procedure described in Sect. 3 is useful for obtaining a better time-independent spatial distribution of the seismicity, which significantly improves the likelihood of the earthquakes catalogs under this model. It also gives a moderate improvement to the resulting ETAS model, which includes the effect of triggered seismicity. However, the arbitrary change of the K parameter in

Table 2

Performance of the ETAS model on the seismicity of the test period (2003–2007) with different models of the time-independent background seismicity distribution

	Unweighted smoothing	Weighted smoothing with $K = 0.07$	Weighted smoothing with $K = 0.12$
$\ln L_e$ ($m_0 = 3.0$)	1,560.8	1,586.5	1,588.4
$\ln L_e - \ln L_0$ ($m_0 = 3.0$)	1,739.3	1,704.4	1,704.4
$(\ln L_e - \ln L_0)/N_3$ ($m_0 = 3.0$)	9.40	9.21	9.21
$\ln L_e$ ($m_0 = 2.0$)	20,109.0	20,149.0	20,147.5
$\ln L_e - \ln L_0$ ($m_0 = 2.0$)	18,622.0	18,071.4	18,022.9
$(\ln L_e - \ln L_0)/N_2$ ($m_0 = 2.0$)	10.51	10.20	10.17

L_0—Likelihood of the time-independent, spatially variable Poisson model

L_e—Likelihood of the time-dependent ETAS model

the smoothing algorithm, even producing a more straight cumulative number of events in time, does not significantly affect the spatial seismicity model, and does not represent an advantage for the performance of the ETAS model.

6. *Spatial Distribution of Seismic Moment*

The information about the geographical distribution of the earthquake rate λ_0 may have relevant consequences for the seismic hazard assessment in a given area. For instance, assuming that all tectonic stress is fully released by earthquakes, and that the magnitude distribution is a truncated Gutenberg–Richter distribution, it is possible to obtain the expected seismic moment rate $\dot{M}_0$ from the number of earthquakes observed per unit space and time λ_0 (Ward, 1994; Catalli *et al.*, 2008):

$$\dot{M}_0 \cong \lambda_0 M_0^* S \frac{b}{1.5 - b}\left[10^{(1.5-b)(m_{\max}-m_0)} - 1\right] \quad (3)$$

where:

b is the parameter of the Gutenberg–Richter magnitude distribution,
m_0 is the lower magnitude threshold of the catalog
$m_{\max}$ is the assumed maximum possible magnitude
M_0^* is the seismic moment of an earthquake of magnitude m_0, and S is the area of the seismogenic zone

Taking into account that only a fraction of the tectonic moment rate is released by earthquakes, Eq. 3 must be retained as a constraint putting a lower limit on the estimate of the moment rate. The practical estimate of the moment rate $\dot{M}_0$ is conditioned by the constraint that the seismogenic volume used for the application of Eq. 3 should be small enough as to allow the assumption of a spatially uniform value of $\dot{M}_0$. At the same time the number of observed earthquakes should be large enough to justify the assumption of stationarity for the seismic process. These constraints are obviously in conflict within each other, so that a reasonable compromise should be chosen.

A map of the moment rate can be obtained through Eq. 3, after having multiplied the seismic moment of each event by a weight p_i equal to its probability of independence. A value of the maximum magnitude $m_{\max} = 8.0$ (Bird and Kagan, 2004) has been assumed in this application. As for the rate density, the moment rate was then normalized by a factor equal to the ratio between the total number of events in the catalog and the sum of p_i over all these events. This was done in order to preserve the total seismic moment released by the earthquakes in the period of time covered by the catalog. Figure 9 shows the map obtained for the area considered in our study, using the spatial distribution of the occurrence rate shown in Fig. 6 ($K = 0.07$). This map is suitable for a comparison with analogous maps obtained from geodetic observations.

7. *Discussion and Conclusions*

Following the method proposed by Zhuang *et al.* (2002), and adopting an iterative procedure as

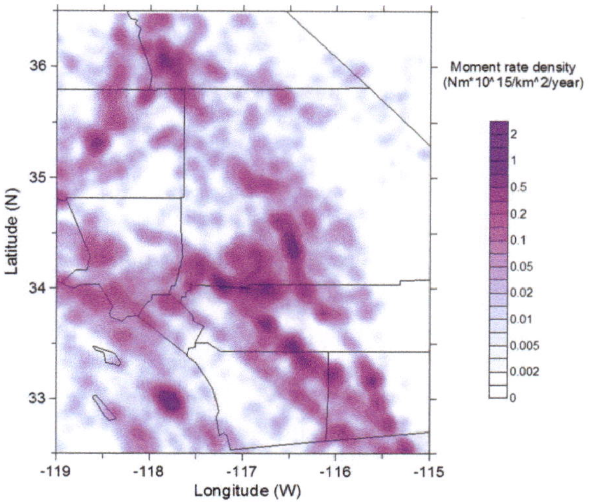

Figure 9
Map of the seismic moment rate obtained by Eq. 3 from the earthquakes observed in Southern California in the time period 1984–2002, using the value $K = 0.07$ for the productivity parameter

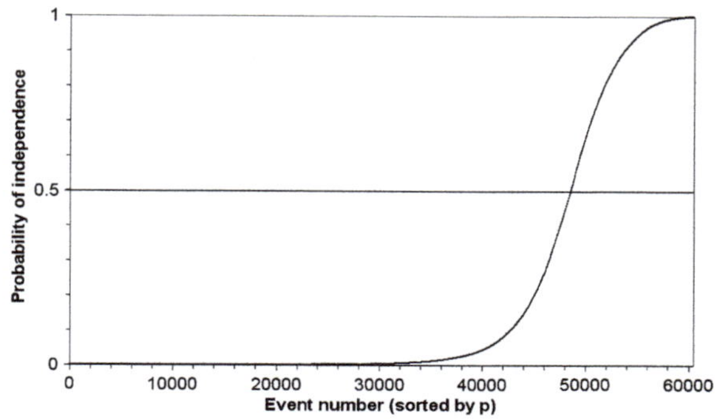

Figure 10
Distribution of the probability of independence for the 60,480 events of the Southern California earthquake catalog

suggested by Marsan and Longliné (2008), our method finds the optimal set of parameters for the ETAS model, taking also into account their effect on the best spatial distribution. The results reported in Table 1 show that the parameters related to the spatial distribution of the seismicity are most affected by the iterative procedure, while the temporal parameters are more robust.

Moreover, we have shown also that the maximum likelihood set of parameters does not guarantee a constant background rate. A more linear trend can be achieved by an appropriate change of the productivity parameter K of the ETAS model.

The method developed in this study was chosen to produce an earthquake catalog that fulfils the property of stationarity for the seismic process without the need of removing events from the catalog. Each event is assigned a weight proportional to the probability of independence according to the ETAS model with suitable values of the parameters. The probability is a number that can assume any value between 0 and 1. Figure 10 shows the distribution of the probability of independence for the 60,480 events of the SCEDC catalog analyzed in this study. It can be noted that about 50% of the events are characterized by a probability of independence close to zero. Note that among these events even earthquakes traditionally defined as main shocks could be included, if these main shocks had been preceded by significant seismic activity commonly defined as foreshocks. For instance, the main shock of the Landers (M = 7.3) 1992 sequence has been assigned a probability of independence $p = 0.00007$ only, because of the influence of some previous foreshocks. The event that most probably triggered the main shock was a small earthquake of magnitude 2.3, which occurred about 20 h before the main shock, with an epicenter shifted by 0.5 km to the west. This event had a probability of independence $p = 0.38$. These results clearly contradict the usual geophysicist's intuition that the main shocks should be independent events. Note also that the initial M 2.3 earthquake (definable as a foreshock in a strict sense) triggered 2.6 aftershocks in its first generation (in probabilistic sense), while the magnitude 7.3 main shock triggered 211.9 aftershocks in its first generation. The total number of M $\geq$ 2.0 aftershocks in the whole sequence was larger than 15,000. This means that the ETAS model assigns most of the events to next generations. This is consistent with the low value found for the productivity parameter α by the maximum likelihood best fit ($\alpha = 0.55$). In our ETAS model the numerous low magnitude events trigger a total number of events larger than the number of events triggered by the few big ones. This result, as for other ETAS results, is difficult to interpret in geophysical terms, because this model is based on a statistical, rather than a geophysical parameterization.

This analysis shows that for a very large fraction of earthquakes the probability of independence does not justify a clear separation between independent and triggered events. About 20% of earthquakes in the analyzed catalog exhibit a probability of independence larger than 0.5, but for few of them the probability is very close to 1. For comparison, MARSAN and LONGLINÉ (2008) found that the background rate of $M \geq 3$ earthquakes corresponds to 19.5% of the total rate.

The background seismicity distribution obtained by down-weighting the events of the catalog according to their probability of independence still exhibits large peaks of the spatial density. This is clearly the case for the Landers and Oceanside areas, as shown in both Figs. 6 and 8. A detailed analysis of the probability of independence of the earthquakes included in these small zones shows that these high values of the spatial density do not come from the aftershock sequences, but from a rather persistent moderate seismic activity. For instance, for a square of 10 km × 10 km centered on the epicenter of the Landers (1992) main shock (34.2°N, 116.4°W), the analysis shows that the total weight of the events included in the first day of the aftershock series (101 events) is only 0.0082. For the first 10 days (447 events) the total weight is 0.097, and extending the count to the first 100 days (941 events) we obtained a total weight equal to 0.94. Note that the weight is fairly proportional to the time duration of the considered period, and not to the number of events contained in every period. The smaller proportion of activity in the first day can be ascribed to the increased detection threshold of the first hours in the network.

In the case of the Oceanside area (33.0°N, 117.8°W), the same kind of analysis shows a seismic activity even more sparse in time, with many more events of relevant probability of independence distributed all over the observation period. Therefore, the ETAS model achieving the maximum likelihood accommodates this circumstance with a relatively high level of the time-independent component.

Any attempt to derive a background earthquake rate using a short catalog is going to encounter serious difficulties and challenges. It is clear from the density maps in this paper, as well as from many maps published by other researchers in the 2007(1) SRL special issue (FIELD, 2007), that such a background rate model is not unique. Even when we used all of the available historic data to produce the map (KAGAN *et al.*, 2007), one can still see that the rate is likely to be too low in certain places, and too high in others. Perhaps, only by using geodetic and geologic information (SHEN *et al.*, 2007; BIRD and LIU, 2007) one could obtain a map of a really long-term background rate. However, such a geodetic/geologic map would fail to correctly predict earthquakes on the 5- or 15-year time-span, since these events are likely to occur near sites of recent activity. Thus, the "background" rate needs a proper definition — what kind of forecast is being considered, is it 5-, 15-, 30-, or 50-year? KAGAN and JACKSON (1994) argue that the time horizon of a prediction should be comparable to the length of an earthquake catalog used in the forecast.

KAGAN *et al.* (2010) discuss general drawbacks of the ETAS model. Particularly, because the c value in the Omori's law parameterization is not scaled with magnitude, the model fit to earthquake patterns might be strongly biased. This is related to aftershock number deficiency after strong earthquakes (see e.g., KAGAN, 2004; HELMSTETTER *et al.*, 2006). However, KAGAN *et al.* (2010) (this issue) removed all of the close-in-time aftershocks and recalculated the model parameters, and they found that the results do not change significantly. This is consistent with the strong similarity we have noted between the background seismicity distribution obtained by a large range of values for the K parameter.

Even taking into account the above-mentioned difficulties in the search for the background rate, and meaning of the background rate itself, maps of the smoothed seismicity produced by our algorithm may represent a spatially variable, time-independent Poisson model useful as a null hypothesis against which to test short-term or medium-term time-dependent forecast models.

Acknowledgments

The authors are grateful for comments by reviewers Jiancang Zhuang and Agnes Helmstetter and by the Associate Editor David Rhoades that have been helpful in the revision of the manuscript. The UCLA authors (DDJ and YYK) appreciate partial support from the Southern California Earthquake Center (SCEC). SCEC is funded by NSF Cooperative

Agreement EAR-0529922 and the U.S. Geologic Survey (USGS) Cooperative Agreement 07HQAG0008. Publication 1257, SCEC.

Appendix

The expected occurrence rate density of earthquakes, $\lambda(x,y,t,m)$, at any time and location, is modeled as the sum of the independent, or time-invariant "spontaneous" activity and the contribution of every previous event:

$$\lambda(x,y,t,m) = f_r \cdot \lambda_0(x,y,m) + \sum_{j=1}^{N} H(t-t_j) \cdot \lambda_j(x,y,m,t), \quad (4)$$

where $\lambda_0(x,y,m)$ is the rate density of the long-term average seismicity, expressing the Gutenberg–Richter magnitude distribution as

$$\lambda_0(x,y,m) = \lambda_0(x,y)\beta e^{-\beta(m-m_0)}; \quad (5)$$

f_r is the failure rate (fraction of spontaneous events over the total number of events) of the process;
β is related to the most widely known b value by the relationship $\beta = b \ln 10$;
m_0 is a reference magnitude;
$H(t)$ is the step function;

and $\lambda_j(x,y,t,m)$ is a kernel function that depends on the magnitude of the triggering earthquake, the spatial distance from the triggering event, and the time interval between the triggering event and the time of interest. We factor this function in three terms depending, respectively, on time, space and magnitude, as:

$$\lambda_j(x,y,t,m) = K \cdot f(x-x_j, y-y_j) \cdot h(t-t_j) \cdot \beta e^{-\beta(m-m_0)}, \quad (6)$$

where K is a constant parameter, while $f(x,y)$ and $h(t)$ represent the space and time distributions, respectively.

The spatial distribution of the triggered seismicity is modeled by a function with circular symmetry around the point of coordinates (x_j,y_j). This function in polar coordinates (r,θ) can be written as:

$$f(r,\theta) = \left(\frac{d_j^2}{r^2+d_j^2}\right)^q, \quad (7)$$

where r is the distance from the point (x_j,y_j), q is a free parameter modeling the decay with distance, and d_j is the characteristic triggering distance. We assume that d_j is related to the magnitude m_j of the triggering earthquake:

$$d_j = d_0 10^{\alpha(m_j-m_0)/2}, \quad (8)$$

where d_0 is the characteristic triggering distance of an earthquake of magnitude m_0 and α is a free parameter, i.e., the distance distribution is scaled with magnitude.

For the time dependence we adopt the modified Omori law (Ogata, 1983):

$$h(t) = (t+c)^{-p}(p>1), \quad (9)$$

where c and p are characteristic parameters of the process.

The free parameters for the ETAS model actually estimated in this study are the following:

- K (productivity coefficient),
- d_0 (characteristic triggering distance),
- q (exponent of the spatial distribution of triggered events),
- α (coefficient of the exponential magnitude productivity law),
- c (time constant of the generalized Omori law) and
- p (exponent of the generalized Omori law).

The fraction of spontaneous events over the total number of events of the process, f_r, is constrained by these free parameters. The b value is estimated from the entire catalog independently from the other parameters.

We may note that neither the spatial (Eq. 7) nor the temporal (Eq. 9) kernel distributions are normalized. This might produce biases in the parameters of the model that control the productivity of triggered seismicity. In particular, this is the case for K and the spatial parameters d_0, q and α. In spite of these biases, the best fit process always converges to stable results, independently of the guess choices for the parameters' values to initiate the optimization algorithm.

References

BIRD, P. and KAGAN, Y. Y. (2004), *Plate-tectonic analysis of shallow seismicity: Apparent boundary width, beta, corner magnitude, coupled lithosphere thickness, and coupling in seven tectonic settings*, Bull. Seismol. Soc. Am. *94*(6), 2380–2399 (plus electronic supplement).

BIRD, P. and LIU, Z. (2007), *Seismic hazard inferred from tectonics: California*, Seism. Res. Lett. *78*(1), 37–48.

CATALLI, F., COCCO, M., CONSOLE, R. and CHIARALUCE, L. (2008), *Modeling seismicity rate changes during the 1997 Umbria-Marche sequence (central Italy) through a rate- and state-dependent model*, in preparation.

CONSOLE, R., *Testing earthquake forecasting hypotheses* (2001), Tectonophysics, *338*, 261–268.

CONSOLE, R., MURRU, M., and LOMBARDI, A. M. (2003), *Refining earthquake clustering models*, J. Geophys. Res. *108*, 2468, doi: 10.1029/2002JB002130.

CONSOLE, R. and MURRU, M. (2001), *A simple and testable model for earthquake clustering*, J. Geophys. Res. *106*, 8699–8711.

CONSOLE, R., MURRU, M., CATALLI, F., and FALCONE, G. (2007), *Real time forecasts through an earthquake clustering model constrained by the rate-and-state constitutive law: Comparison with a purely stochastic ETAS model*, Seism. Res. Lett. *78*, 49–56.

FIELD, E. H. (2007), *Overview of the Working Group for the Development of Regional Earthquake Likelihood Models (RELM)*, Seism. Res. Lett. *78*(1), 7–16.

FRANKEL, A. (1995), *Mapping seismic hazard in the central and eastern United States*, Seism. Res. Lett. *66*, 8–21.

GARDNER, J. K. and KNOPOFF, L. (1974), *Is the sequence of earthquakes in California, with aftershocks removed, Poissonian?*, Bull. Seism. Soc. Am. *64*, 1363–1367.

GIBBONS, J. D. and CHAKRABORTI, S. *Non-parametric Statistical Inference*, 4th ed. (New York: Marcel Dekker 2003), 645 pp.

GUTENBERG, B. and RICHTER, C. (1944), *Frequency of earthquakes in California*, Bull. Seism. Soc. Am. *34*, 185–188.

HELMSTETTER, A. and SORNETTE, D. (2002), *Subcritical and supercritical regimes in epidemic models of earthquake aftershocks*, J. Geophys. Res. *107*, 2237, doi:10.1029/2001JB001580.

HELMSTETTER, A. and SORNETTE, D. (2003), *Importance of direct and indirect triggered seismicity in the ETAS model of seismicity*, Geophys. Res. Lett. *30*, 1576, doi:10.1029/2003GL017670.

HELMSTETTER, A., KAGAN, Y. Y., and JACKSON, D. D. (2006), *Comparison of short-term and time-independent earthquake forecast models for southern California*, Bull. Seism. Soc. Am. *96*(1), 90–106.

HELMSTETTER, A., KAGAN, Y. Y., and JACKSON, D. D. (2007), *High-resolution time-independent grid-based forecast for* $M \geq 5$ *earthquakes in California*, Seism. Res. Lett. *78*(1), 78–86.

JACKSON, D. D. and KAGAN, Y. Y. (1999), *Testable earthquake forecasts for 1999*, Seism. Res. Lett. *70*(4), 393–403.

KAGAN, Y. Y. (1991), *Likelihood analysis of earthquake catalogues*, Geophys. J. Int. *106*(1), 135–148.

KAGAN, Y. Y. (2002), *Seismic moment distribution revisited: II. Moment conservation principle*, Geophys. J. Int. *149*(3), 731–754.

KAGAN, Y. Y. (2004), *Short-term properties of earthquake catalogs and models of earthquake source*, Bull. Seism. Soc. Am. *94*(4), 1207–1228.

KAGAN, Y. Y., BIRD, P., and JACKSON, D. D. (2010), *Earthquake Patterns in Diverse Tectonic Zones of the Globe*, Pure Appl. Geophys. (this issue).

KAGAN, Y. Y. and JACKSON, D. D. (1994), *Long-term probabilistic forecasting of earthquakes*, J. Geophys. Res. *99*, 13,685–13,700.

KAGAN, Y. Y. and JACKSON, D. D. (2000), *Probabilistic forecasting of earthquakes, (Leon Knopoff's Festschrift)*, Geophys. J. Int. *143*, 438–453.

KAGAN, Y. Y., JACKSON, D. D., and RONG, Y. F. (2007), *A testable five-year forecast of moderate and large earthquakes in southern California based on smoothed seismicity*, Seism. Res. Lett. *78*(1), 94–98.

KAGAN, Y. Y. and KNOPOFF, L. (1976), *Statistical search for non-random features of the seismicity of strong earthquakes*, Phys. Earth Planet. Inter. *12*(4), 291–318.

MARSAN, D. and LONGLINÉ, O. (2008), *Extending earthquakes' reach through cascading*, Science *319*, 1076–1079.

OGATA, Y. (1983), *Estimation of the parameters in the modified Omori formula for aftershock frequencies by the maximum likelihood procedure*, J. Phys. Earth *31*, 115–124.

OGATA, Y. (1998), *Space-time point-process models for earthquake occurrences*, Ann. Inst. Statist. Math. *50*(2), 379–402.

OGATA, Y. (1999), *Seismicity analysis through point-process modeling: A review*, Pure Appl. Geophys. *155*, 471–507.

REASENBERG, P. (1985), *Second-order moment of Central California seismicity*, J. Geophys. Res. *90*, 5479–5495.

SHEN, Z. K., JACKSON, D. D., and KAGAN, Y. Y. (2007), *Implications of geodetic strain rate for future earthquakes, with a five-year forecast of M5 earthquakes in southern California*, Seism. Res. Lett. *78*(1), 116–120.

WARD, S. N. (1994), *A multidisciplinary approach to seismic hazard in California*, Bull. Seism. Soc. Am. *84*, 1293–1309.

ZHUANG, J. and OGATA, Y. (2006), *Properties of the probability distribution associated with the largest event in an earthquake cluster and their implications to foreshocks*, Phys. Rev. E, doi: 10.1103/PhysRevE.73.046134.

ZHUANG, J., OGATA, Y., and VERE-JONES, D. (2002), *Stochastic declustering of space-time earthquake occurrences*, J. Am. Stat. Ass. *97*, 458, 369–380.

ZHUANG, J., OGATA, Y., and VERE-JONES, D. (2004), *Analyzing earthquake clustering features by using stochastic reconstruction*, J. Geophys. Res. *109* (B5), B05301, doi:10.1029/2003JB002879.

ZHUANG, J., CHRISTOPHERSEN, A., SAVAGE, M. K., VERE-JONES, D., OGATA, Y., and JACKSON, D. D. (2008), *Differences between spontaneous and triggered earthquakes: Their influences on foreshock probabilities*, J. Geophys. Res. *113*, B11302, doi: 10.1029/2008JB005579.

(Received July 20, 2008, revised January 30, 2009, accepted March 5, 2009, Published online March 10, 2010)

Pure Appl. Geophys. 167 (2010), 831–843

DOI 10.1007/s00024-010-0066-4

Pure and Applied Geophysics

Seismicity Models of Moderate Earthquakes in Kanto, Japan Utilizing Multiple Predictive Parameters

Masajiro Imoto[1] and David A. Rhoades[2]

Abstract—We construct a single hazard function from multiple predictive parameters independently developed for moderate earthquakes in Kanto, Japan, during a learning period from 1990 to 1999, and applied to a testing period from 2000 to 2005. Here, we consider as predictive parameters the a and b values of the Gutenberg–Richter relation, the v value (change in b value), and the Every Earthquake a Precursor According to Scale (EEPAS) model rate. To study the correlations among the parameters, we prepare two groups of space–time coordinate sets for assessment, namely the background and conditional groups selected from the learning period. The background group contains ten thousand sets of coordinates randomly selected from the space–time volume of our study. The conditional group contains 33 sets of space–time coordinates corresponding to the epicenters of the target earthquakes ($M \geq 5.0$) just before their times of occurrence. Each parameter for the background group is transformed so that its distribution conforms to the standard Normal function. The mean and variance of the conditional distribution is then estimated after applying the same transformation to the conditional group. Using the means and variances of b values, v values and EEPAS rates and the correlation matrices in the background and conditional distributions, we construct a combined hazard function following the procedure developed for normally distributed parameters. The information gain per event (*IGpe*) of the new hazard function is 0.26 and 0.3 units larger than that of the EEPAS rate for the learning and testing period, respectively. The R-test confirms the statistical significance of the difference in the *IGpe* value for the testing period.

Key words: Seismicity model, multiple parameters, b value, EEPAS, Japan.

1. Introduction

Utsu (1977, 1982), Rhoades and Evison (1979), and others (Aki, 1981; Hamada, 1983; Grandori *et al.*, 1988) have formulated earthquake-probability calculations based on multiple precursory phenomena. These calculations assume each precursory phenomenon occurs independently of the others, conditional on the occurrence (or non-occurrence) of a following earthquake. Under this assumption (the conditional independence assumption), an approximate formula for the combined probability gain has been derived as a product of the probability gains for individual observations and the probability of secular seismicity (Aki, 1981). If multiple precursory phenomena were to occur independently of each other, the ratio of the number of earthquakes preceded by every type of precursory phenomena to the total number of target earthquakes, which is called reliability (Reasenberg and Matthews, 1988), would become progressively smaller as additional precursory phenomena are included. In Utsu and Aki's formulation, physical observations are classified as either anomalies or normal values and the probability is calculated for each class. The threshold at which an observation is deemed to be anomalous is another point of issue.

Imoto (2007) clarified the problem by expressing the probability gain as a quantity of information. For normally distributed observations (or predictive parameters), he derived a formula for the information gain in the general case where multidimensional observations are correlated in both the conditional and background distributions, extending earlier results for multidimensional observations correlated

[1] National Research Institute for Earth Science and Disaster Prevention, Tennodai 3-1, Tsukaba-shi, Ibaraki-ken 305-0006, Japan. E-mail: imoto@bosai.go.jp

[2] Institute of Geological and Nuclear Sciences, 1 Fairway Drive, Avalon, Lower Hutt 5010, P.O. Box 30-368, Lower Hutt 5040, New Zealand.

only in the conditional distribution (Imoto, 2006a). Applying this procedure to three parameters (the Gutenberg–Richter a and b values and the v value, which measures changes in b value) observed in Kanto, central Japan, Imoto (2008) constructed a seismicity model (referred to as the ABT model) in Kanto, Japan.

The "Every Earthquake a Precursor According to Scale" (EEPAS) model (Rhoades and Evison, 2004) is a method of forecasting earthquakes based on the notion that the phenomenon of the precursory scale increase (Ψ) occurs at all scales in the seismogenic process. The hazard rate of future earthquakes is estimated directly from past earthquakes. The phenomenon has been retrospectively recognized to occur as a long-term precursor of some tens of major earthquakes in California and northern Mexico, Japan, Greece and NW Turkey, and New Zealand and Italy (Evison and Rhoades, 2002, 2004). Central Japan is of great interest for earthquake hazard mitigation. The catalog produced from the network operated in the Kanto area of central Japan by the National Research Institute for Earth Science and Disaster Prevention (NIED) has a high degree of homogeneity and completeness to much lower magnitudes from 1979 onwards than the Japan Meteorological Agency catalogue. By applying EEPAS to depth-restricted subsets of the NIED catalog of the Kanto area, for magnitudes above 4.75, the EEPAS model is found to be more informative than a quasi-static baseline model based on proximity to past earthquakes (PPE), and considerably more informative than the stationary uniform Poisson model (Rhoades and Evison, 2006).

In the present study, we will attempt to combine the EEPAS model with the elements of the ABT model to formulate a superior model, following the procedure proposed by Imoto (2007). As stated above, the ABT model is constructed from three parameters, the a, b, and v values. The EEPAS model calculates hazard rates in a more complex manner and cannot be decomposed into a combination of several parameters. We propose to overcome this problem by regarding the EEPAS hazard rate itself as a precursory parameter (surrogate precursor). We can then build up a composite model combining the a, b, and v values and the EEPAS rate.

2. Method

Consider a vector-valued random variable, $\boldsymbol{\Theta}$, defined on a space–time study domain $\mathbf{X}$. The observed value $\boldsymbol{\theta}$ of $\boldsymbol{\Theta}$ at the space–time coordinates $\mathbf{x}$ represents the values of the predictive parameters at $\mathbf{x}$. Let $\mathbf{E}$ be the subset of $\mathbf{X}$ comprising the space–time coordinates of targeted earthquakes. The distribution of $\boldsymbol{\Theta}$ over the whole space–time domain $\mathbf{X}$ is called the background distribution. This distribution can be approximated by sampling $\boldsymbol{\theta}$ at values of $\mathbf{x}$ distributed randomly in $\mathbf{X}$. Its probability density (the "background" density) is denoted by $g(\boldsymbol{\theta})$. The distribution of $\boldsymbol{\Theta}$ over $\mathbf{E}$ is called the conditional distribution. This distribution can be approximated by sampling $\boldsymbol{\theta}$ at values of $\mathbf{x}$ distributed randomly in $\mathbf{E}$. It represents the conditional distribution of $\boldsymbol{\theta}$ given that at least one earthquake occurs. Its probability density (the "conditional" density) is denoted by $f(\boldsymbol{\theta})$.

The hazard function $h(\mathbf{x}|\boldsymbol{\theta})$ at the space–time coordinates $\mathbf{x}$ conditioned on a value of $\boldsymbol{\theta}$ is such that the expected number of earthquakes in a space–time volume $d\mathbf{x}$ is $h(\mathbf{x}|\boldsymbol{\theta})d\mathbf{x}$ (Daley and Vere-Jones, 2003, p. 229). It follows that

$$h(\mathbf{x}|\boldsymbol{\theta}) = \frac{f(\boldsymbol{\theta})m_0}{g(\boldsymbol{\theta})V_0}, \tag{1}$$

where m_0 is the number of target earthquakes and V_0 is the volume of the space–time domain $\mathbf{X}$.

Taking the Poisson model as the standard, the information gain per event (*IGpe*: Daley and Vere-Jones, 2003, p. 276; Imoto, 2004) for a large number of target events is given by

$$IGpe = \int_R f(\boldsymbol{\theta}) \ln \frac{f(\boldsymbol{\theta})}{g(\boldsymbol{\theta})} d\boldsymbol{\theta}, \tag{2}$$

where the integral is performed over the whole range R of $\boldsymbol{\theta}$. Assuming that $f(\boldsymbol{\theta})$ and $g(\boldsymbol{\theta})$ are multivariate normal distributions, Imoto (2007) derived an analytical form to estimate the *IGpe* value. In this case, $f(\boldsymbol{\theta})$ and $g(\boldsymbol{\theta})$ are defined by the mean and covariance matrix of the parameter vector $\boldsymbol{\theta} = (\theta_1, \theta_2, \ldots, \theta_n)$ in the background and conditional distributions. Let the covariance matrices be $\mathbf{B}$ for the background and $\mathbf{C}$ for the conditional distribution. The joint densities $g(\boldsymbol{\theta})$ and $f(\boldsymbol{\theta})$, are then given as follows:

$$g(\boldsymbol{\theta}) = \frac{1}{\sqrt{2\pi}^n \sqrt{\det(\mathbf{B})}} \exp\left\{ -\frac{1}{2} \boldsymbol{\theta}^t \mathbf{B}^{-1} \boldsymbol{\theta} \right\}, \quad (3)$$

and

$$f(\boldsymbol{\theta}) = \frac{1}{\sqrt{2\pi}^n \sqrt{\det(\mathbf{C})}} \exp\left\{ -\frac{1}{2} (\boldsymbol{\theta} - \boldsymbol{\mu})^t \mathbf{C}^{-1} (\boldsymbol{\theta} - \boldsymbol{\mu}) \right\}, \quad (4)$$

where the superscript -1 refers to the inverse of a matrix, and $\boldsymbol{\mu}$ is the vector of mean values in the conditional distribution. Without loss of generality, the means and variances of $(\theta_1, \theta_2, \ldots \theta_n)$ in the background distribution have been fixed at 0 and 1, respectively. The covariance matrices **B** and **C** can be expressed as

$$\mathbf{B} = \begin{pmatrix} 1 & & \gamma_{ij} & \\ & 1 & & \\ & & \ddots & \\ & \gamma_{ij} & & 1 \end{pmatrix}, \quad (5)$$

and

$$\mathbf{C} = \begin{pmatrix} \sigma_1^2 & & \rho_{ij}\sigma_i\sigma_j & \\ & \sigma_2^2 & & \\ & & \ddots & \\ & \rho_{ij}\sigma_i\sigma_j & & \sigma_n^2 \end{pmatrix}, \quad (6)$$

where γ_{ij} is the correlation coefficient between θ_i and θ_j in the background distribution, and ρ_{ij} is that in the conditional distribution, and σ_i^2, is the variance of the conditional distribution of θ_i.

For a single parameter θ_1, the *IGpe* value for θ_1, $IGpe(\theta_1)$ can be represented as follows:

$$IGpe(\theta_1) = \frac{1}{2} \ln \frac{1}{\sigma_1^2} + \frac{\sigma_1^2 + \mu_1^2 - 1}{2}, \quad (7)$$

where μ_1 is the mean of $f(\theta_1)$. For the general case, the *IGpe* value can be estimated with the formula of Imoto (2007).

In the present study, each parameter is transformed in such a way that its background distribution is that of a standard normal random variable. A normal distribution is then fitted to the conditional distribution of the transformed parameters. The hazard function is built up for the new parameter system, θ', after estimating the means and variances of the conditional distributions and the correlation matrices for the background and conditional distributions of the transformed parameters. We then obtain the hazard function of the combined model by substituting Eqs. (3) and (4) into Eq. (1). This function estimates the hazard rate at any coordinates of interest conditioned on the parameter values observed there.

3. Model Components

We attempt to build a seismicity model for moderate and large earthquakes (target events with $M \geq 5.0$) in Kanto, central Japan, based on three hazard functions referred to as $h_{ab}(x| a,b)$, $h_v(x| v)$ and $h_{eep}(x)$. The first hazard function depends on the *a* and *b* values in the Gutenberg–Richter relation (GR relation) (Imoto, 2006b, 2008), and the second on the *v* value, a parameter representing temporal change in the *b* value (Imoto, 2003), which is obtained from the difference between two mean event sizes, the long-term and short-term means. The third hazard function is the EEPAS model, optimized to the seismicity in Kanto, Japan (Rhoades and Evison, 2006).

Hazard functions $h_{ab}(x| a,b)$ and $h_v(x| v)$ are represented as

$$h_{ab}(x|a,b) = a_1 \cdot 10^{a - b_1(b-b_2)^2} \quad (8)$$

and

$$h_v(x|v) = c_0 B(v|c_1, c_2), \quad (9)$$

where *B* denotes the Beta function, which is given as

$$B(v) = v^{c_1 - 1}(1-v)^{c_2 - 1} \Big/ \int_0^1 x^{c_1 - 1}(1-x)^{c_2 - 1} \mathrm{d}x, \quad (10)$$

and a_1, b_1, b_2, c_0, c_1 and c_2 are model parameters that have been optimized by the maximum likelihood method in a point-process analysis.

In the EEPAS model, the hazard function $h_{eep}(x)$ of earthquake occurrence is defined for any time *t*, magnitude *m* and location (*x*,*y*), where *m* exceeds a threshold magnitude m_c, and (*x*,*y*) is a point in the region of surveillance *R*. Each earthquake (t_i, m_i, x_i, y_i) contributes a transient increment $\lambda_i(t,m,x,y)$ to the future rate density in its vicinity, given by

$$\lambda_i(t, m, x, y) = f_{1i}(t) g_{1i}(m) h_{1i}(x, y), \quad (11)$$

where f_{1i} is the density of the probability distributions for time, g_{1i} is that for magnitude, and h_{1i} is that for location. These densities take the forms

$$g_{1i}(m) = \frac{1}{\sigma_M\sqrt{2\pi}} \exp\left[-\frac{1}{2}\left(\frac{m - a_M - b_M m_i}{\sigma_M}\right)^2\right], \quad (12)$$

$$f_{1i}(t) = \frac{H(t - t_i)}{(t - t_i)\ln(10)\sigma_T\sqrt{2\pi}} \times \exp\left[-\frac{1}{2}\left(\frac{\log_{10}(t - t_i) - a_T - b_T m_i}{\sigma_T}\right)^2\right], \quad (13)$$

and

$$h_{1i}(x, y) = \frac{1}{2\pi\sigma_A^2 10^{b_A M_i}} \exp\left[-\frac{(x - xi)^2 + (y - y_i)^2}{2\sigma_A^2 10^{b_A M_i}}\right], \quad (14)$$

where a_M, b_M, σ_M, a_T, b_T, σ_T, σ_A and b_A are parameters, and $H(s) = 1$ if $s > 0$ and 0 otherwise.

The total rate density is obtained by summing over all past occurrences, including earthquakes outside R, that could affect the rate density within R:

$$\lambda(t, m, x, y) = \mu\lambda_0(t, m, x, y) + \sum_{t_i \geq t_0, m_i \geq m_0} \eta(m_i)\lambda_i(t, m, x, y), \quad (15)$$

where μ is a parameter, λ_0 is a baseline rate density depending on the PPE (Jackson and Kagan, 1999), t_0 is the time of the beginning of the catalog, and η is a normalizing function

$$\eta(m) = b_M(1 - \mu)\exp\left[-\beta\left(a_M + (b_M - 1)m + \frac{\sigma_M^2}{2}\right)\right], \quad (16)$$

that makes the long-run average rate density under the EEPAS model, as a function of magnitude, i.e., averaged over time and location, conform to the Gutenberg–Richter frequency-magnitude relation. Here, $\beta = b \ln(10)$, which can be estimated at any time by the average over past earthquakes.

4. Data Analysis

In this study we use the catalog of earthquakes for 1980 to 2005 located by the Kanto–Tokai network operated by the National Research Institute for Earth Science and Disaster Prevention (NIED) (Hamada *et al.*, 1985). In 2006 this seismological network was merged into the national network (Hi-net) (Obara *et al.*, 2005) and the data processing system was terminated on that occasion. Since then the catalogue of earthquakes in this region has been produced using a different data processing system. This restructuring of the network has probably disrupted the homogeneity of the catalogue, and thus our testing period is terminated at the end of 2005.

Epicenters of earthquakes used in the study ($M \geq 2.0$) are plotted in Fig. 1a, where a solid box indicates the area under examination. The magnitude–frequency relation of the earthquakes in the study domain, plotted in Fig. 1b, shows a linear relationship between magnitude and the cumulative number of earthquakes, down to magnitude 2.0. Therefore, it is assumed that earthquakes down to this size are homogeneously located in this space–time domain.

Taking into account the limited duration of the catalog and the need for stable estimation, a time window of 3650 days is selected for the a and b values. In order to detect short-term changes in b value, a shorter time window of 960 days is selected for the v value. A spherical volume with radius 20 km and a cutoff magnitude of 2.0 are commonly used for the three parameters to simplify conditions (Imoto and Yamamoto, 2006).

We do not distinguish the magnitudes of target earthquakes. We consider the hazard function for all earthquakes larger than 5.0. Therefore, the integral form of the rate density (15)

$$h_{eep}(t, x, y) = \int_{m=5.0}^{\infty} \lambda(t, m, x, y)dm \quad (17)$$

is used.

In constructing a model, we estimated the a, b, and v values and h_{eep} rates at 10,000 space–time coordinate sets and at 33 hypocenters of target events just before the times of occurrence of earthquakes within the study volume ($200 \times 200 \times 90$ km^3) from January 1990 to December 1999 (the learning period). Here, the a and b values are considered separately since $h_{ab}(\mathbf{x}| a,b)$ is the product of the

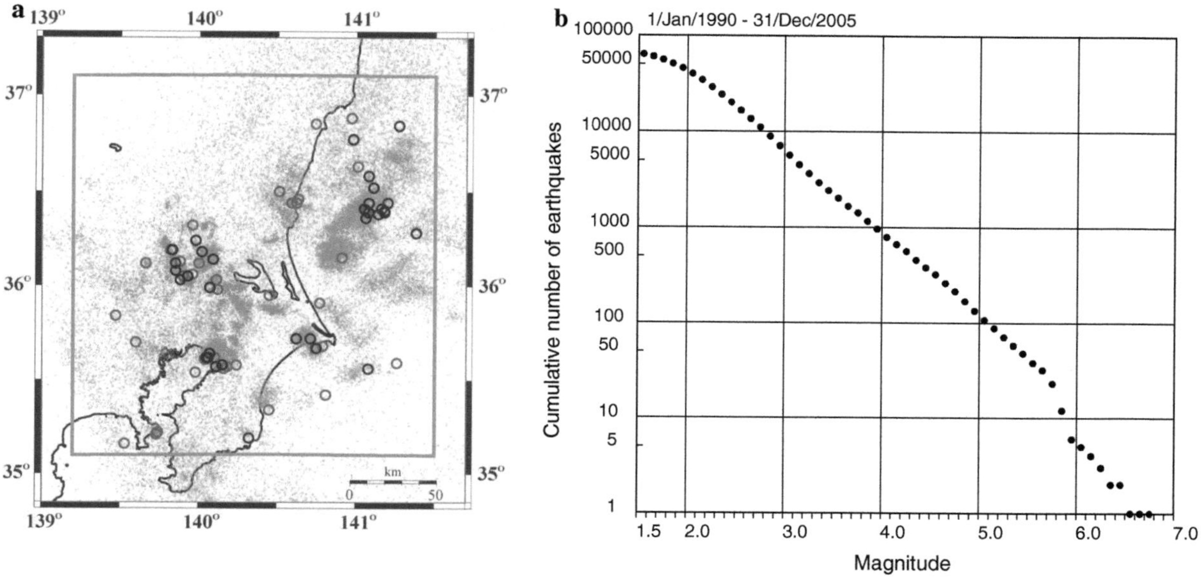

Figure 1
a Map of micro-earthquakes with magnitude 2.0 or larger from Jan. 1, 1990 to Dec. 31, 2005. *Circle symbols* (*gray*: learning period, and *black*: testing period) indicate the location of the main shocks. The *box* encloses the study region. **b** The Gutenberg–Richter magnitude–frequency relation of the earthquakes in the study region

a value term and the *b* value term. The 10,000 coordinate sets are randomly selected from regularly spaced grid points at 2-km-spacing and 10-day intervals in the above space–time domain.

We obtained empirical conditional and background distributions from the estimates at the 33 hypocenters and 10,000 coordinate sets, respectively. Figures 2a–d illustrate the empirical background (solid line) and conditional (gray line) distributions for the *a* value, *b* value, *v* value and h_{eep} rate.

The hazard rate contributed by the *b* value in (8) is not monotonic and takes its maximum at $b = b_2$. The same is true for the *v* value rate in (9). Considering these characteristics, the *b* and *v* value are transformed into normally distributed parameters. We assumed that the background distributions of the *a*, *b*, and *v* values and h_{eep} rate are respectively represented by an exponential function (dashed line in Fig. 2a), a normal function (dashed line in Fig. 2b), a Beta function (dashed line in Fig. 2c; Dougherty, 1990), and a Weibull function (dashed line in Fig. 2d; Dougherty, 1990), which are fitted by the maximum-likelihood method. Accordingly, the *a*, *b*, and *v* values and the h_{eep} rate are transformed into the a', b', v' and h_{eep}' values through their probability distribution functions.

Figures 3a–d illustrate the background distribution in black and the conditional distribution in gray for a', b', v' and h_{eep}'. Normal distributions fitted to the conditional and background distributions are indicated with dashed lines in gray and black. These figures indicate that the background distributions for all four transformed parameters and the conditional distributions for b' and v' are satisfactorily approximated by normal distributions. For the conditional distributions for a' and h_{eep}' the normal approximation is not a close fit, but is deemed adequate for only 33 data points. The parameters of the approximating normal distributions are summarized in Table 1. The last column in Table 1 indicates *IGpe* for each parameter, calculated using equation (7), where both distributions are assumed to be normally distributed.

5. *Results and Testing*

Table 2 indicates the correlation matrices, lower left for the background distribution and upper right

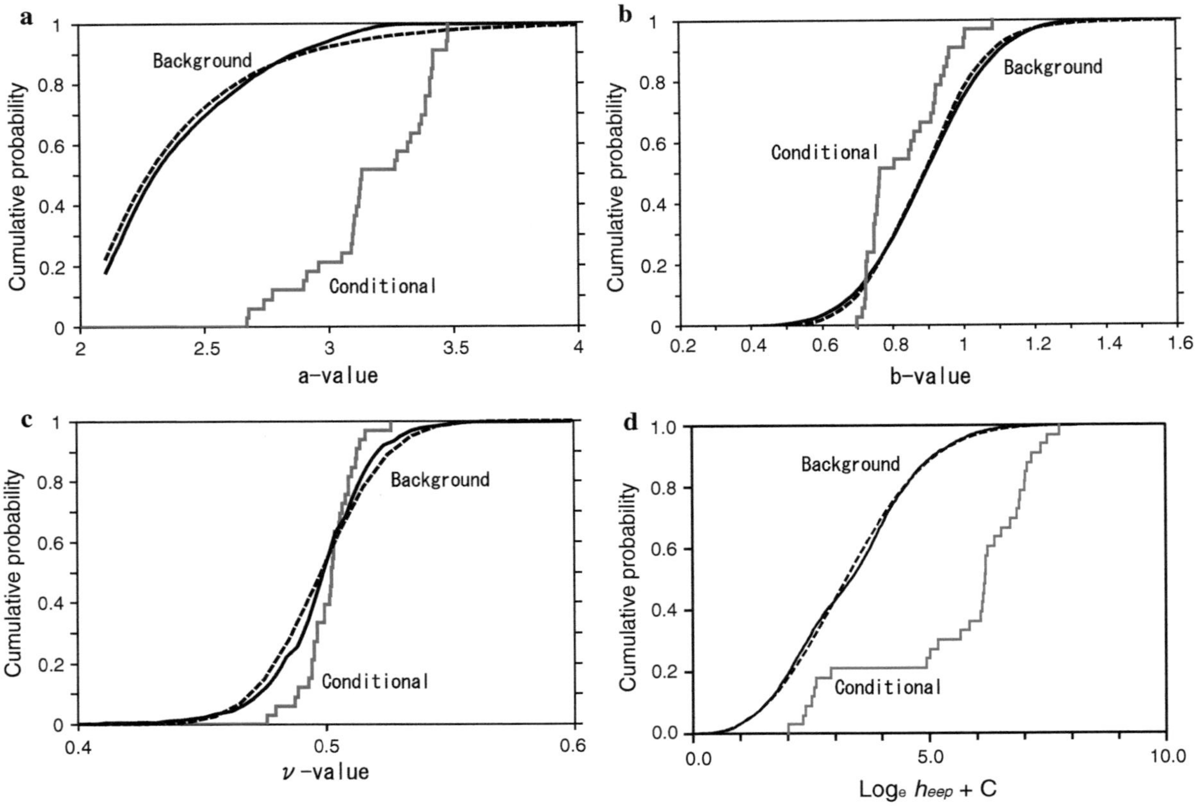

Figure 2
Cumulative background and conditional distributions for the parameters. **a** Empirical background distribution of the a value is plotted as a *black solid curve*, and conditional distribution as a *gray stepped line*. An exponential distribution is fitted to the background distribution (*dashed curve*). **b** Same as **a** but for the b value. A normal distribution is fitted to the background distribution (*dashed curve*). **c** Same as **a** but for ν value. A Beta function is fitted to the background distribution (*dashed curve*). **d** Same as **a** but for the h_{eep} rate. A Weibull function is fitted to the background distribution (*dashed curve*)

for the conditional distribution. Strong correlations of 0.63 and 0.81 between the a' value and h_{eep}' are observed in the background and conditional distribution, respectively, probably because the baseline model of EEPAS is PPE, which is a sum of contributions from previous earthquakes. Roughly speaking, both the a value and h_{eep} strongly depend on the number of earthquakes around the point of assessment. Therefore, we exclude the a value from the candidate components of a combined model and attempt to build a model based on the three parameters: b value, ν value and h_{eep} rate.

We can calculate the *IGpe* value expected from the combined model using the procedure given in the previous section. However in order to confirm the validity of *IGpe* estimates, we calculate the probability gain at each target event directly from the hazard function given in Eq. (1) for each parameter. Substituting Eqs. (3) and (4) with values from Tables 1 and 2 into Eq. (1), we obtain the hazard function of the combined (Ebt) model. Using this Ebt hazard function, we performed a retrospective analysis of the data from 1990 to 1999 (the learning period) and a forward analysis of the data from 2000 to 2005 (the testing period). The results obtained are listed in the second and third rows in Table 3. The *IGpe* values in the first row are calculated by Eqs. (2) and (7) and those in the second and third rows are estimated as the logarithm of the geometrical mean of probability gains at the times of target events. Table 4 summarizes the hazard rates obtained from the ABT, EEPAS and Ebt models. A comparison of each *IGpe*

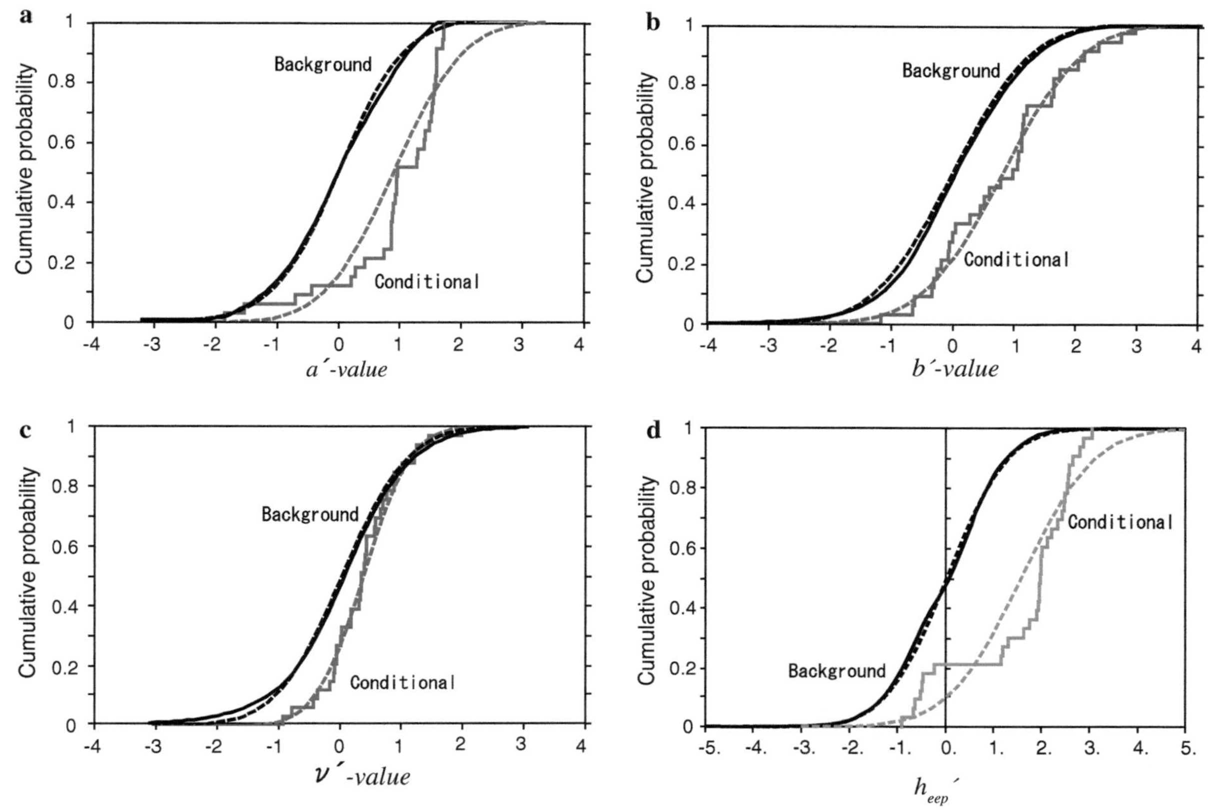

Figure 3
Cumulative background and conditional distributions for the transformed parameters. **a** Empirical background distribution for the transformed *a* value is plotted as a *black solid curve*, and the conditional distribution, as a *gray stepped line*. Their fitted normal distributions are plotted as *dashed lines* in black and gray. **b** Same as **a** but for the *b* value. **c** Same as **a** but for *v* value. **d** Same as **a** but for the h_{eep} rate

Table 1

Parameters of normal distributions for the a, b, and v values and the EEPAS rate, and their IGpe values

	Background		Conditional		*IGpe*
	Av.	Std.	Av.	Std.	
a'	0	1	0.959	1.059	0.46
b'	0	1	0.699	0.978	0.24
v'	0	1	0.302	0.644	0.19
h_{eep}'	0	1	1.565	1.200	1.26

Table 2

Correlation matrices

	a'	b'	v'	h_{eep}'
a'	*	−0.215	−0.402	0.813
b'	0.039	*	−0.022	−0.081
v'	0.088	0.077	*	−0.297
h_{eep}'	0.633	0.211	0.067	*

Lower left: Estimated from background distribution. *Upper right*: Estimated from conditional distribution

value in the second row to the corresponding value in the first row indicates that the model is successfully built up. The *IGpe* value of the Ebt model in the retrospective analysis is 0.26 units larger than that of EEPAS. Figure 4 shows comparisons of the hazard rates under the Ebt and EEPAS models for a few selected examples.

Results obtained from the testing period data indicate slight differences between the expectations and observations. The *IGpe* value of the Ebt model is 0.30 units larger than that of EEPAS, which is a little larger than that of the retrospective analysis in the learning period but a little smaller than that of the initial assessment. It should be noted that the *IGpe* values of EEPAS and the Ebt model in the testing

Table 3

Summary of the IGpe values

	Events	b'	ν'	h_{eep}'	Ebt
Initial assessment		0.24	0.19	1.26	1.62
Learning period	33	0.25	0.20	1.16	1.42
Testing period	32	0.20	0.33	1.77	2.07

Those expected from the formula of IMOTO (2007) are listed in the first row (initial assessment). Those calculated as average values of probability gains for each target event are listed in the second row for the learning period and in the third row for the testing period. The *IGpe* value for the Ebt model is given in the last column

period are 0.5 and 0.45 units, respectively, larger than those of the initial assessment. The seismicity in the testing period is about 1.5 times greater than in the learning period. This increased seismicity increases *IGpe* by 0.4 units for hazard rates based on a number of earthquakes. Therefore, we consider the differences between the *IGpe* values of both the EEPAS and Ebt models in the learning period and those in the testing period to be due to activation of seismicity in the testing period.

To confirm the results in the testing period, we perform the *N*-, *L*- and *R*-tests for EEPAS and the Ebt model (KAGAN and JACKSON, 1995). In the present study we do not generate simulated random sequences, as in a widely used procedure (SCHORLEMMER *et al.*, 2007), but calculate means and variances of test statistics analytically. The derivations are given in the Appendix.

Figures 5a, b compare the cumulative numbers of target earthquakes, from January 1, 2000, up to a date at a point of the abscissa, with those expected from sequences conforming, respectively, to the Ebt and EEPAS models. This is the *N*-test. The upper and lower error ranges are indicated with dark (±1 standard deviation) and light (±2 standard deviations) shading. These figures indicate that each model is self-consistent from the viewpoint of earthquake productivity.

Figures 6a, b compare the likelihoods of the Ebt and EEPAS models calculated from the observations with those calculated for sequences conforming to the proposed models. This is the *L*-test. The upper and the lower error ranges are shaded as in the *N*-test. These figures indicate that neither model is rejected by the *L*-test. It is not surprising that the results of the

Table 4

List of main shocks and hazard rates obtained from the ABT, EEPAS and Ebt models

Origin time	Location			Mag.	Rate			
Date	Lat.	Long.	Depth		h_{abt}	h_{eep}	h_{ebt}	
02/12/90	36.40	141.06	37.5	5.4	2.88	6.07	5.51	
05/03/90	36.46	140.63	49.9	5.5	4.00	5.15	7.97	
06/01/90	35.67	140.74	49.0	5.8	3.16	4.51	6.32	
07/04/90	36.03	140.11	67.0	5.1	2.82	7.30	5.72	
08/23/90	35.34	140.45	45.9	5.5	1.47	112.64	112.64	
10/06/90	36.44	140.62	50.8	5.2	3.31	7.23	10.21	
06/25/91	36.63	141.01	41.5	5.3	1.35	3.18	2.02	
11/19/91	35.61	140.04	76.2	5.1	3.99	19.95	21.20	
02/02/92	35.22	139.74	85.3	6.1	6.25	0.11	1.43	
04/10/92	35.70	139.60	83.8	5.2	0.70	0.23	0.27	
04/14/92	36.19	139.82	55.9	5.3	3.23	6.64	5.73	
05/11/92	36.50	140.51	56.4	5.8	1.84	1.52	1.15	
05/20/92	35.21	139.73	86.0	5.2	0.40	0.15	0.67	
05/29/92	35.68	140.79	50.9	5.2	5.17	2.65	5.25	
08/27/92	36.06	139.96	44.9	5.0	3.73	19.45	30.82	
05/21/93	36.05	139.90	48.0	5.8	3.60	20.17	32.27	
09/18/93	36.15	140.91	37.0	5.2	2.77	2.09	1.33	
12/13/93	36.39	141.07	37.9	5.0	5.62	6.62	6.26	
01/01/95	35.62	140.10	73.5	5.0	9.06	16.32	27.86	
01/07/95	36.32	139.96	63.9	5.7	1.49	1.46	1.47	
03/23/95	36.12	140.00	50.4	5.2	4.89	3.58	3.39	
07/03/95	35.16	139.53	112.0	5.6	0.52	0.10	0.10	*
09/11/96	35.59	141.26	55.5	6.7	0.01	0.13	0.13	*
10/12/96	36.12	139.66	91.4	5.3	0.63	0.12	0.41	
12/21/96	36.13	139.88	52.1	5.8	3.77	7.34	8.22	
03/23/97	35.98	140.12	70.4	5.6	3.21	5.13	5.52	
07/09/97	35.56	140.16	74.3	5.3	8.51	9.80	17.89	
08/09/97	35.84	139.47	65.9	5.0	0.50	0.14	0.25	
09/08/97	35.54	139.98	107.7	5.4	0.26	0.18	0.18	*
01/14/98	35.58	140.24	73.1	5.2	6.48	4.82	7.40	
04/09/98	36.85	140.74	81.3	5.0	0.39	0.17	0.08	
04/09/98	36.88	140.97	85.9	5.7	3.13	0.19	0.74	
06/14/98	35.42	140.81	41.5	5.5	0.46	0.39	0.39	*
08/29/98	35.61	140.05	63.6	5.4	9.98	15.74	27.33	
03/26/99	36.44	140.59	53.2	5.2	3.70	21.08	37.19	
07/15/99	35.95	140.45	47.9	5.0	1.10	0.71	0.59	
09/13/99	35.57	140.18	72.6	5.2	6.20	11.47	7.99	
12/04/99	35.91	140.77	98.1	5.2	0.02	0.05	0.05	*
06/03/00	35.67	140.75	45.5	5.8	3.23	4.78	7.58	
07/21/00	36.52	141.11	29.6	6.3	5.37	2.28	2.54	
07/20/01	36.19	139.83	53.2	5.1	4.48	7.57	10.31	
02/12/02	36.58	141.08	32.6	5.9	5.95	2.91	3.04	
06/14/02	36.24	139.98	53.5	5.2	3.98	2.20	3.45	
10/21/02	36.38	141.14	32.9	5.1	4.52	9.63	10.12	
01/09/03	36.41	141.16	45.6	5.2	6.81	3.25	5.06	
01/21/03	36.36	141.06	44.8	5.0	5.95	7.55	10.67	
03/13/03	36.08	139.85	45.3	5.4	3.55	5.25	7.96	
04/25/03	36.40	141.18	43.9	5.0	5.10	4.86	5.02	
06/16/03	36.84	141.27	77.5	5.4	0.06	1.16	0.83	
09/20/03	35.19	140.32	68.4	5.7	0.23	0.27	0.27	*
10/15/03	35.62	140.05	71.5	5.5	16.64	16.04	16.04	
11/15/03	36.44	141.20	42.1	5.5	6.71	6.10	10.25	

Table 4 continued

Origin time	Location			Mag.	Rate		
Date	Lat.	Long.	Depth		h_{abt}	h_{eep}	h_{ebt}
04/04/04	36.39	141.18	45.3	5.5	7.31	12.34	22.24
08/06/04	35.62	140.06	70.8	5.1	10.50	31.76	31.76
10/06/04	35.99	140.07	62.9	6.0	3.30	6.25	8.24
10/17/04	36.28	141.38	32.5	5.6	0.62	1.07	0.51
01/01/05	36.77	140.98	88.9	5.6	1.67	0.11	0.32
02/08/05	36.14	140.09	66.2	5.3	2.58	10.01	13.00
02/16/05	36.03	139.88	43.3	5.8	4.33	2.66	5.00
04/11/05	35.72	140.62	52.6	6.4	3.68	4.71	7.49
05/19/05	35.56	141.08	30.5	5.0	3.42	1.86	3.70
06/09/05	35.64	140.07	67.9	5.0	3.82	35.34	35.34
06/20/05	35.72	140.71	50.2	5.7	2.73	18.27	16.12
06/24/05	36.44	141.08	42.5	5.1	7.79	15.47	26.87
07/11/05	36.41	141.05	44.5	5.1	7.43	15.90	28.34
07/23/05	35.58	140.15	69.3	6.2	3.87	8.09	12.11
07/28/05	36.12	139.85	49.0	5.4	5.93	12.21	24.05
08/07/05	35.57	140.11	68.8	5.3	3.55	8.13	12.59
10/16/05	36.05	139.93	46.8	5.4	4.38	10.34	13.29
10/19/05	36.39	141.08	45.4	5.5	7.55	33.49	33.49
12/28/05	36.18	140.02	51.6	5.3	5.16	4.28	7.59

Each rate is given in a unit of 5×10^{-7} event/cell. An asterisk in the last column indicates that the EEPAS rate is adopted as the Ebt rate in the absence of sufficient samples for calculating the b value and/or v value

L-test are similar to those of the *N*-test because, when the rates are very low, the number of target earthquakes included in the test sequence has a strong influence on the log-likelihood value.

Figure 7a compares the difference (Δ ln L) between the log-likelihood of the Ebt and EEPAS models over the testing period with that expected under the Ebt model. A black line indicates the Δ ln L value of the observed sequence of earthquakes, and a white line indicates the expected value from sequences conforming to the Ebt model. The upper and lower error bands for sequences conforming to the Ebt model are indicated with dark (± 1 standard deviation) and light (± 2 standard deviations) shading. The observed Δ ln L values stay mostly within the ± 1 standard deviation band. On this result, the null hypothesis that earthquake occurrences conform to the Ebt model cannot be rejected.

Figure 7b compares the observed Δ ln L with that expected under the EEPAS model. The observed Δ ln L of 5.2 in late 2005 is located outside the ± 2 standard deviation band. This suggests that the Δ ln L of 5.2 is not consistent with the EEPAS model, but is considerably more consistent with the Ebt model. The null hypothesis that earthquake occurrences conform to the EEPAS model can be rejected with about 95% confidence, in favor of the Ebt model.

6. *Discussion and Conclusions*

A previous study (Imoto, 2008) developed a seismicity model (ABT) of moderate and large earthquakes in Kanto, Japan, based on three parameters, the a, b, and v values. According to that study, the information gain per event (*IGpe*) values of each parameter and the ABT model are 0.47, 0.22, 0.18 and 0.98. It should be noted that the *IGpe* value of the ABT model, 0.98, is slightly larger than that obtained by a simple summation of the three components. This difference has been interpreted as the effect of correlations among parameters. Combining the b value, v value, and EEPAS rate, which have individual *IGpe* values of 0.24, 0.19 and 1.26, respectively, the Ebt model attains an estimated *IGpe* value of 1.62, which is slightly less than the simple summation of the three components. This result is satisfactory, since the *IGpe* of the combined model depends on many factors such as the means, variances and correlation matrices of the component parameters. Table 3 indicates appreciable differences in the *IGpe* of the EEPAS model (h_{eep}': the second to the last column) and the Ebt model (the last column) between the initial assessment (the first row) and the learning period (the second row). The *IGpe* value for the EEPAS model is 0.1 lower in the learning period than in the initial assessment, and that of the Ebt model is 0.2 lower. In a simulation study (Imoto, 2008), the errors of *IGpe* estimation ranged from 0.1 to 0.3, where the value of 0.3 appeared in the estimation of the combined model. Therefore, we consider the losses in *IGpe* values between the initial assessment and the learning period are mainly due to errors of *IGpe* estimation. Another factor possibly contributing to the losses is that the parameter distributions may not be exactly normal, as they are assumed to be for estimation purposes. In spite of these losses, the *IGpe* value of the Ebt model is still larger by 0.26 than that of the EEPAS model.

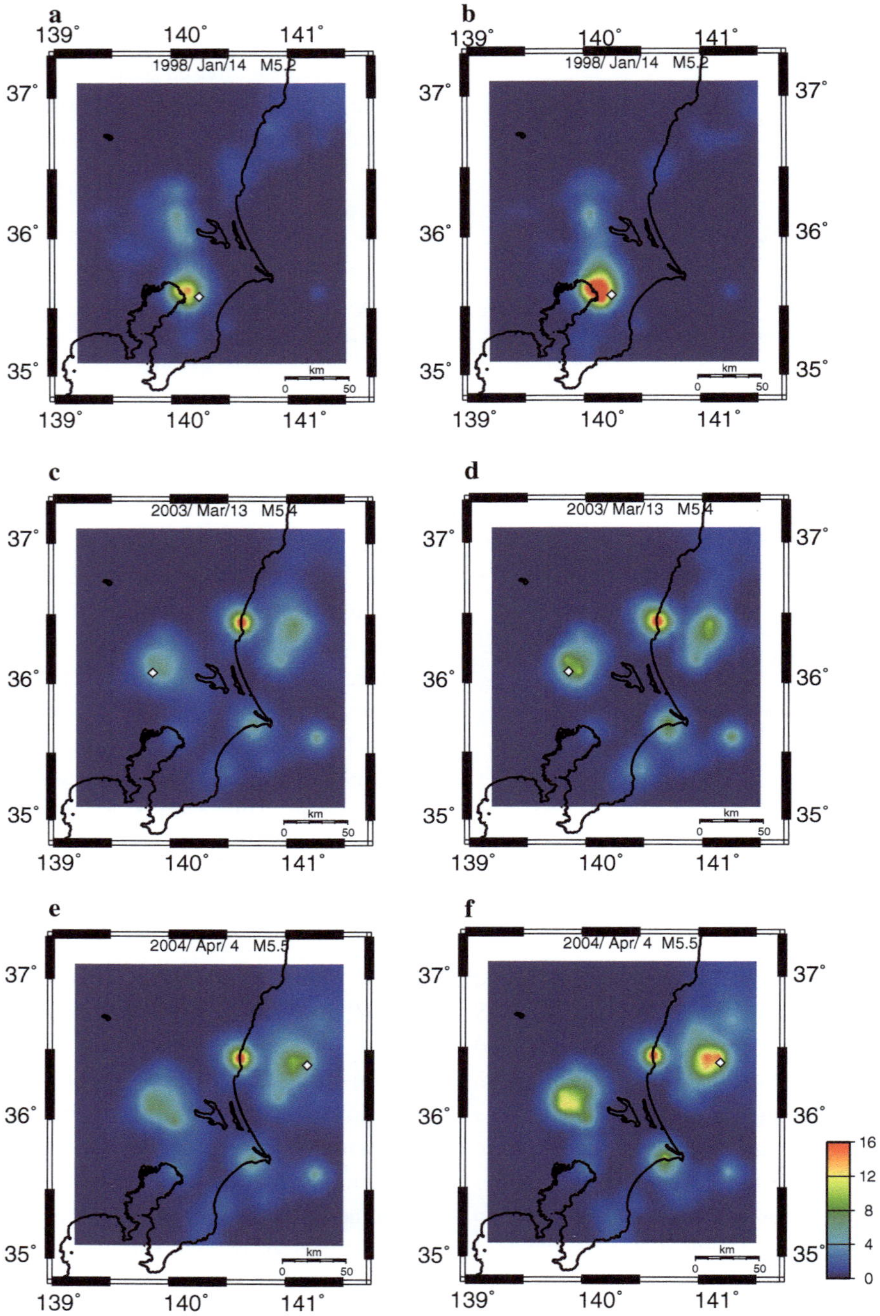

Figure 4
Comparisons of hazard rate under the Ebt model to that under the EEPAS model. Hazard rates are expressed in a unit of 5×10^{-7} event/cell. **a**, **c** and **e** pertain to the EEPAS model and **b**, **d** and **e** to the Ebt model. **a**, **b**—14 Jan. 1998, **c**, **d**—13 Mar. 2003, **e**, **f**—4 April, 2004

In summary, we have developed an Ebt model of moderate and large earthquakes based on three parameters, the b value of the Gutenberg–Richter magnitude frequency relation, its temporal change, and the EEPAS rate. Applying the formula of Imoto (2007) to the data observed by the Kanto–Tokai

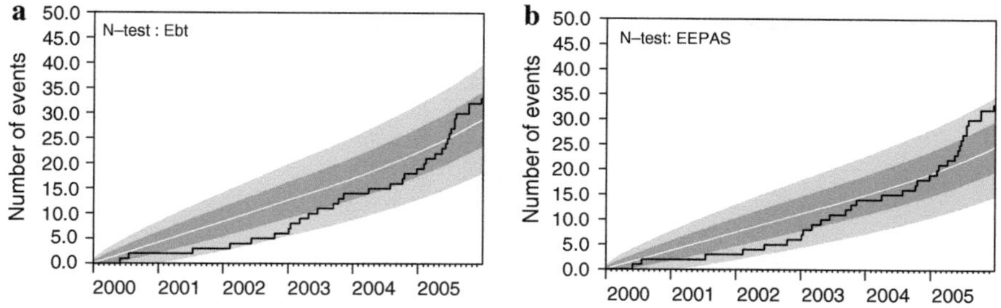

Figure 5
N-test for Jan. 2000 up to the respective date. The *solid line* indicates the cumulative number of target earthquakes. The *white line* indicates its expected value under the model during testing. Dark (*light*) shadow zones indicate ranges less than one standard deviation (two standard deviations) from the mean. **a** Ebt model tested. **b** EEPAS model tested

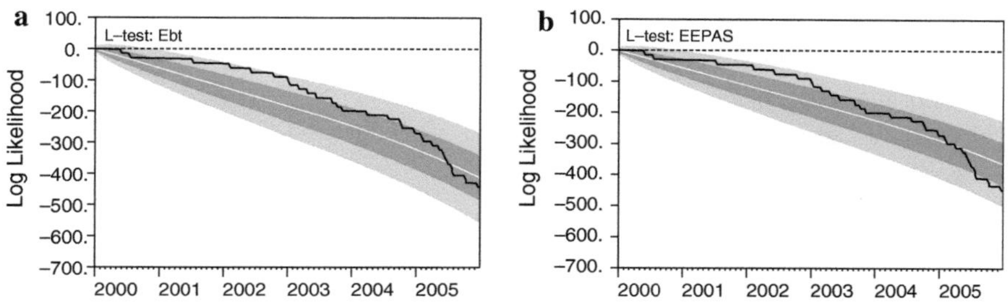

Figure 6
Same as Fig. 5 but for *L*-test. **a** Ebt model tested. **b** EEPAS model tested

network (NIED), we estimated that the *IGpe* value of the model becomes about 0.1–0.2 units greater than that of the EEPAS model. These results demonstrate that the concept of surrogate precursors enables us to develop a single model, utilizing multiple parameters, which is superior to a model based on any of the individual parameters. The model developed here proved superior to the individual models in retrospective tests on data from both the learning and testing periods. However, it remains to be seen how it would perform in a truly prospective test.

Acknowledgments

The authors thank Y. Y. Kagan and two anonymous reviewers for their discussions and comments on this manuscript. The second author was supported by the Foundation for Research, Science and Technology under contract CO5X0402, and by the GNS Science Capability Fund.

Appendix

The proposed model could be tested with *N*-, *L*-, and *R*-statistics (Kagan and Jackson, 1995). In these tests, statistics obtained from the observed data are compared with a distribution of statistics, which could be empirically obtained with a large number of randomly generated sequences. Where the observed statistics are not consistent with statistics computed assuming a given model to be correct, the null hypothesis that the real earthquake sequences conform to the model may be rejected. In the present study, we will perform these three tests analytically, not with random catalogs.

If an observed score is outside a certain range (for example, ± 2 standard deviations from the expected value), the model should be rejected at the respective level of significance (5%). However, a higher likelihood score, even if outside the $+2$ standard deviations, is regarded as a better model in the *L*-test (Schorlemmer *et al.*, 2007).

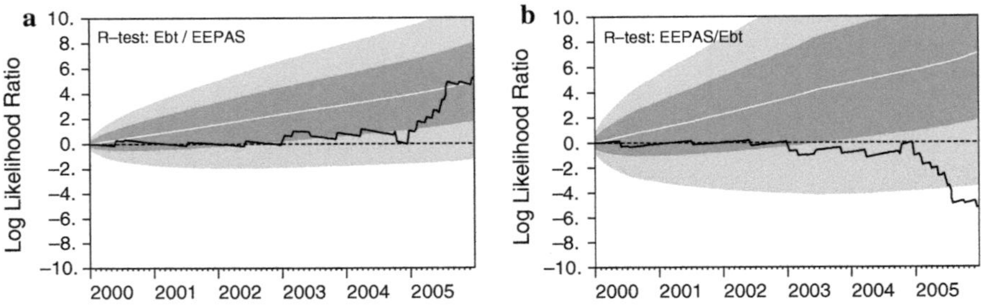

Figure 7
a *R*-test of the Ebt model against the EEPAS model, where a sequence of target earthquakes conforming to the Ebt model is assumed. See Fig. 5 for notation. **b** Same as **a**, except a sequence of target earthquakes conforming to the EEPAS model is assumed

The *N*-test checks the consistency of the observed number with the earthquake productivity of the proposed model. Assuming the study space–time domain is divided into K cells, an earthquake in each cell follows the Poisson process. The Poisson rate in the *j*-th cell is noted as λ_j, the expected number of event, and its variance $\mathrm{Var}[n_j] = \lambda_j$. The total expected number, E[*n*], is given by

$$\mathrm{E}[n] = \sum_{j=1}^{K} \lambda_j, \tag{A1}$$

where an earthquake occurrence in one cell is assumed to be independent from that in another cell. The variance of the total number is given by

$$\mathrm{Var}[n] = \sum_{j=1}^{K} \mathrm{Var}[n_j]. \tag{A2}$$

The *L*-test determines the consistency of the observed likelihood with that expected from sequences conforming to the model. The log likelihood for the *j*-th cell is given by

$$l_j = Y_j \,\mathrm{Ln}\,\lambda_j + (1 - Y_j)\mathrm{Ln}(1 - \lambda_j), \tag{A3}$$

where Y_j is 1 when an earthquake occurs and 0 otherwise, and Ln refers to the natural logarithm. It is assumed that λ_j is far less than unity and that at most one earthquake occurs in one cell. The expected value of the log-likelihood, $\mathrm{E}[l_j]$ is expressed by

$$\mathrm{E}[l_j] = \lambda_j \,\mathrm{Ln}\,\lambda_j + (1 - \lambda_j)\mathrm{Ln}(1 - \lambda_j). \tag{A4}$$

The variance is given by

$$\begin{aligned}\mathrm{Var}[l_j] &= \sum_i \left\{ (l_j - \mathrm{E}[l_j])^2 \mathrm{f}_j(Y_i) \right\} \\ &= \left\{ \mathrm{Ln}\,\lambda_j - \mathrm{Ln}(1 - \lambda_j) \right\}^2 \sum_i (Yi - \lambda_j)^2 \mathrm{f}_j(Y_i), \\ &= \left\{ \mathrm{Ln}\,\lambda_j - \mathrm{Ln}(1 - \lambda_j) \right\}^2 \mathrm{Var}[n_j].\end{aligned} \tag{A5}$$

When an earthquake in one cell is independent from that in another cell, the mean E[*l*] and variance Var[*l*] of the log-likelihood for the whole domain are given by

$$\mathrm{E}[l] = \sum_{j=1}^{K} \left[\lambda_j \,\mathrm{Ln}\,\lambda_j + (1 - \lambda_j)\mathrm{Ln}(1 - \lambda_j) \right], \tag{A6}$$

and

$$\mathrm{Var}[l] = \sum_{j=1}^{K} \left[\left\{ \mathrm{Ln}\,\lambda_j - \mathrm{Ln}(1 - \lambda_j) \right\}^2 \lambda_j \right]. \tag{A7}$$

The *R*-test examines whether a difference between the likelihoods of two models is significant. The expected value of the likelihood ratio, $\mathrm{E}[l_j^{12}]$ is expressed by

$$\begin{aligned}\mathrm{E}\left[l_j^{12}\right] &= \lambda_j^1 \mathrm{Ln}\left(\lambda_j^1 / \lambda_j^2\right) \\ &\quad + \left(1 - \lambda_j^1\right)\mathrm{Ln}\left(\left(1 - \lambda_j^1\right) / \left(1 - \lambda_j^2\right)\right),\end{aligned} \tag{A8}$$

where superscripts 1 and 2 refer to the first and second models. It is assumed that earthquake sequences conform to the first model. The variance is given by

$$\begin{aligned}\mathrm{Var}\left[l_j^{12}\right] &= \sum_i \left\{ \left(l_j^{12} - E\left[l_j^{12}\right] \right)^2 f_j(Y_i^1) \right\}, \\ &= \left\{ \mathrm{Ln}\left(\lambda_j^1/\lambda_j^2\right) - \mathrm{Ln}\left(\left(1-\lambda_j^1\right)/\left(1-\lambda_j^2\right)\right) \right\}^2 \\ &\quad \times \sum_i \left(Y_i^1 - \lambda_j^1\right)^2 \mathrm{f}(Y_i^1), \\ &= \left\{ \mathrm{Ln}\left(\lambda_j^1/\lambda_j^2\right) - \mathrm{Ln}\left(\left(1-\lambda_j^1\right)/\left(1-\lambda_j^2\right)\right) \right\}^2 \\ &\quad \times \mathrm{Var}\left[n_j^1\right]. \end{aligned} \tag{A9}$$

The mean $\mathrm{E}[l^{12}]$ and variance $\mathrm{Var}[l^{12}]$ of the log-likelihood ratio for the whole domain are given by

$$\mathrm{E}\left[l^{12}\right] = \sum_{j=1}^{K} E\left[l_j^{12}\right], \tag{A10}$$

and

$$\mathrm{Var}[l] = \sum_{j=1}^{K} \mathrm{Var}\left[l_j^{12}\right]. \tag{A11}$$

References

Aki, K., *A probabilistic synthesis of precursory phenomena.* In Earthquake Prediction, (eds., D. W. Simpson and P. G. Richards) (AGU, 1981) pp. 566–574.

Daley, D. J. and Vere-Jones, D., *An introduction to the theory of point processes*, vol. 1, *Elementary theory and methods*, Second edition (Springer, New York 2003).

Dougherty, E. R., *Probability and statistics for the engineering, computing and physical sciences* (Prentice-Hall, Inc., Englewood Cliffs, 1990).

Evison, F. F. and Rhoades, D. A. (2002), *Precursory scale increase and long-term seismogenesis in California and northern Mexico*, Ann. Geophys. *454*, 479–495.

Evison, F. F. and Rhoades, D. A. (2004), *Demarcation and scaling of long-term seismogenesis*, Pure Appl. Geophys. *161*, 21–45.

Grandori, G., Guagenti, E., and Perotti, F. (1988), *Alarm systems based on a pair of short-term earthquake precursors*, Bull. Seism. Soc. Am. *78*, 1538–1549.

Hamada, K. (1983), *A probability model for earthquake prediction*, Earthquake Prediction Res. *2*, 227–234.

Hamada, K., Ohtake, M., Okada, Y., Matsumura, S., and Sato, H. (1985), *A high quality digital network for microearthquake and ground tilt observations in the Kanto-Tokai area*, Japan, Earthq. Predict. Res. *3*, 447–469.

Imoto, M. (2003), *A testable model of earthquake probability based on changes in mean event size*, J. Geophys. Res. *108*, ESE 7.1-12 No. B2, 2082, doi: 10.1029/2002JB001774.

Imoto, M. (2004), *Probability gains expected for renewal process models*, Earth Planets Space *56*, 563–571.

Imoto, M. (2006a), *Earthquake probability based on multidisciplinary observations with correlations*, Earth Planets Space *57*, 1447–1454.

Imoto, M. (2006b), *Statistical models based on the Gutenberg-Richter a and b values for estimating probabilities of moderate earthquakes in Kanto, Japan.* In Proceed. The 4th International Workshop on Statistical Seismology, January 9–13, 2006, *ISM Report on Research and Education*, ISM, Tokyo, Japan, *23*, 116–119.

Imoto, M. (2007), *Information gain of a model based on multidisciplinary observations with correlations*, J. Geophys. Res. *112*, B05306, doi: 10.1029/2006JB004662.

Imoto, M. (2008), *Performance of a seismicity model based on three parameters for earthquakes ($M \geq 5.0$) in Kanto, central Japan*, Ann. Geophys. *51*, 727–736.

Imoto, M. and Yamamoto, N. (2006), *Verification test of the mean event size model for moderate earthquakes in the Kanto region, central Japan*, Tectonophys. *417*, 131–140.

Jackson, D. D. and Kagan, Y. Y. (1999), *Testable earthquake forecasts for 1999.* Seismol. Res. Lett. *70*, 393–403.

Kagan, Y. Y. and Jackson, D. D. (1995), *New seismic gap hypothesis: Five years after.* J. Geophy. Res. *100*, 3943–3960.

Obara, K., Kasahara, K., Hori, S., and Okada, Y. (2005), *A densely distributed high-sensitivity seismograph network in Japan: Hi-net by National Research Institute for Earth Science and Disaster Prevention*, Rev. Sci. Instrum. *76*, 021301.

Reasenberg, P. and Matthews, M. (1988), *Precursory seismic quiescence: A preliminary assessment of the hypothesis*, Pure Appl. Geophys. *126*, 373–406.

Rhoades, D. A. and Evison, F. F. (1979), *Long-range earthquake forecasting based on a single predictor*, Geophys. J. R. Astr. Soc. *59*, 43–56.

Rhoades, D. A. and Evison, F. F. (2004), *Long-range earthquake forecasting with every earthquake a precursor according to scale*, Pure Appl. Geophys. *161*, 47–72.

Rhoades, D. A. and Evison, F. F. (2006), *The EEPAS forecasting model and the probability of moderate-to-large earthquakes in central Japan*, Tectonophys. *417*, 119–130.

Schorlemmer, D., M. Gerstenberger, M. C., Wiemer, S., Jackson, D. D., and Rhoades, D. A. (2007), *Earthquake likelihood model testing*, Seismol. Res. Lett. *78*, 17–29.

Utsu, T. (1977), *Probabilities in earthquake prediction*, Zisin II *30*, 179–185 (in Japanese).

Utsu, T. (1982), *Probabilities in earthquake prediction (the second paper)*, Bull. Earthq. Res. Inst. *57*, 499–524 (in Japanese).

(Received July 31, 2008, revised February 20, 2009, accepted February 20, 2009, Published online March 10, 2010)

Pure Appl. Geophys. 167 (2010), 845–853
© 2010 Birkhäuser / Springer Basel AG
DOI 10.1007/s00024-010-0078-0

Pure and Applied Geophysics

Recurrence of Large Earthquakes: Bayesian Inference from Catalogs in the Presence of Magnitude Uncertainties

G. Zöller,[1] S. Hainzl,[2] and M. Holschneider[1]

Abstract—We present a Bayesian method that allows continuous updating the aperiodicity of the recurrence time distribution of large earthquakes based on a catalog with magnitudes above a completeness threshold. The approach uses a recently proposed renewal model for seismicity and allows the inclusion of magnitude uncertainties in a straightforward manner. Errors accounting for grouped magnitudes and random errors are studied and discussed. The results indicate that a stable and realistic value of the aperiodicity can be predicted in an early state of seismicity evolution, even though only a small number of large earthquakes has occurred to date. Furthermore, we demonstrate that magnitude uncertainties can drastically influence the results and can therefore not be neglected. We show how to correct for the bias caused by magnitude errors. For the region of Parkfield we find that the aperiodicity, or the coefficient of variation, is clearly higher than in studies which are solely based on the large earthquakes.

Key words: Earthquake dynamics, Bayesian analysis, earthquake recurrence, uncertainties.

1. Introduction

The extraction of information from earthquake catalogs is a puzzle. Especially if one focuses on the largest events, the data become sparse and thus statistical properties of only the largest events will become increasingly unstable. The emergence of enormous uncertainties makes the situation even worse. In order to overcome this problem, numerical models are useful for both the understanding of the physics of earthquakes and the test of the feasibility of statistical methods. In particular, if a model resembles main features of observed seismicity, it may become capable for purposes of data assimilation and finally for forecasting the next big event.

The question of earthquake recurrence is one of the key problems in seismic hazard assessment. Even in fault regions that are governed by overall periodic recurrence of large events, considerable deviations from the periodicity are observed, e.g., on the Parkfield segment of the San Andreas fault, where the most recent *M* 6 earthquake in 2004 emanated with about 16 years delay from the expected time of occurrence. From observations alone it is impossible to judge whether such fluctuations are the rule or the exception. There are mainly two approaches to deal with this situation: the first one aims at modeling the evolution of stress and seismicity with physical models. However, the more realistic these models are, the more complicated they become and the more adjustable parameters are involved. The second approach uses statistical methods, particularly the fitting of preselected distributions to a small number of data. Although some of the distributions may have a physical background, like the Weibull distribution (Weibull, 1951), the applicability to the earthquake problem remains questionable. In a recent work, Zöller *et al.* (2008) have shown that the Brownian passage time distribution covers a broad range of scenarios as a consequence of the central limit theorem. Furthermore, it has been demonstrated, how stress changes from small and intermediate earthquakes can cause the advanced or delayed occurrence of the next large event. Imposing empirical relations of earthquake statistics like the Gutenberg–Richter law (Gutenberg and Richter, 1956) for the frequency-size

[1] Institute of Mathematics and Interdisciplinary Center for the Dynamics of Complex Systems, University of Potsdam, Am Neuen Palais 10, 14469 Potsdam, Germany. E-mail: zoeller@uni-potsdam.de; hols@math.uni-potsdam.de

[2] Deutsches GeoForschungsZentrum, Potsdam, Telegrafenberg, 14473 Potsdam, Germany. E-mail: hainzl@gfz-potsdam.de

distribution makes possible robust estimates of the parameters of the Brownian passage time distribution and makes the fitting of distributions to a small number of large earthquakes unnecessary. The analysis of a long catalog of synthetic earthquakes shows that the method leads to reasonable estimates of the coefficient of variation describing the degree of periodicity of large earthquakes. However, in practical applications, earthquake recordings are added in time and parameters have to be recalculated continuously. For this purpose, we derive a Bayesian approach in order to update the coefficient of variation based on the model of ZÖLLER *et al.* (2008). Finally, we discuss the influence of emerging uncertainties on the results. Therefore, we consider the true magnitude to be a random variable drawn from a given probability distribution.

2. Method

The methodology is decomposed into three parts: first, the fractal oscillator; second, the calculation and maintenance of the aperiodicity and third, the imposed magnitude uncertainties.

2.1. The Fractal Oscillator Model

We present a brief review of the method to estimate the aperiodicity from an earthquake catalog. A comprehensive description is provided in ZÖLLER *et al.* (2008).

The backbone of the method is the "fractal oscillator" (FO), a modification of the Brownian relaxation oscillator (BRO) proposed by MATTHEWS *et al.* (2002) and extended by ZÖLLER and HAINZL (2007), which serves as a model for recurrent large earthquakes in a fault region. Here we deal with stochastic models for seismicity on long time scales. Instead of simulating the whole earthquake process deterministically on all scales, we focus on the recurrence of the largest earthquakes and plug all small-scale processes into a stochastic component. This includes the occurrence of aftershocks and background seismicity, aseismic stress release and other effects. In ZÖLLER *et al.* (2008), we have shown that the evolution of the loading and unloading characteristics in the fault region can be effectively the same as a stochastic variable with drift. In this study, we follow this idea and assume that a strictly periodic recurrence of large earthquakes according to Reid's elastic rebound theory is corrected by the loading and unloading processes between two large events. Although the stochastic modeling of the subscale processes is certainly oversimplified with respect to the details of the underlying physics, we claim that the evolution of seismic release is reasonable in a statistical sense, e.g., on long time scales. Therefore, stochastic models serve as a powerful tool when dealing with problems related to seismic hazard.

Both models, the fractal oscillator and the Brownian relaxation oscillator, introduce a load state variable τ representing seismic moment release. The temporal evolution of τ is governed by two processes: first, a linear increase modeling the tectonic plate motion; second, random fluctuations accounting for small-scale processes, especially small earthquakes. While the BRO imposes Gaussian random fluctuations for the latter component, the FO uses doubly-truncated fractal fluctuations mimicking Gutenberg–Richter type seismicity (GUTENBERG and RICHTER, 1956). A large earthquake occurs, as soon as τ reaches a predefined static threshold τ_s. Such a threshold dynamics is widely accepted in earthquake modeling, see e.g., BEN-ZION and RICE (1993) and ZÖLLER *et al.* (2004), where earthquakes of all sizes are initiated, as soon as a threshold (given by the static friction) is reached. In this study, the failure criterion only applies for the largest earthquakes, while the smaller ones are associated with the stochastic fluctuations. Thus, from the modeling perspective, our model incorporates two types of earthquakes: first a "large" earthquake, with a size depending mainly on the dimension of the fault zone, and second, intermediate and small earthquakes following Gutenberg–Richter statistics, where the magnitudes can be calculated from the seismic release (stochastic component).

Mathematically, the calculation of the time to the next large event is a first passage time problem of Gaussian or fractal fluctuations with constant drift (tectonic loading). For the BRO, the solution of the first passage time problem is the well-known

Brownian passage time distribution (for a detailed description see REDNER, 2001). However, assuming a large number of fluctuations during a seismic cycle in the FO, the sum of the fluctuations will also converge to Gaussian noise because of the central limit theorem. Consequently, the recurrence time distribution of this model is also the Brownian passage time distribution,

$$f(t;\mu_T,\alpha) = \sqrt{\frac{\mu_T^3}{2\pi\alpha^2 t^3}}\exp\left\{-\frac{(t-\mu_T)^2}{2\mu_T\alpha^2 t}\right\}, \quad (1)$$

where μ_T is the mean recurrence time of large earthquakes, and the aperiodicity α is equal to the coefficient of variation

$$c_V = \alpha = \frac{\sigma_T}{\mu_T} \quad (2)$$

with the standard deviation σ_T of the recurrence times.

The recurrence time distribution for a specific fault zone may be obtained by fitting Eq. 2 to the large earthquakes from the catalog. However, the results are expected to be unstable because of the small number of large earthquakes. As an example, we refer to the Parkfield segment, which is probably the best monitored seismically active region in the world: The Parkfield catalog includes only seven M 6 events leading to six recurrence times.

In their FO model, ZÖLLER *et al.* (2008) have shown that c_V for large earthquakes (magnitude m_{main}) can be related to the Richter b value in the limit case of an infinite number of intermediate and small Gutenberg–Richter distributed earthquakes with $m \in [-\infty; m_{\text{main}}]$ within a seismic cycle:

$$c_V^0 = \sqrt{\frac{b}{3-b}}. \quad (3)$$

However, deviations from the scaling behavior for high magnitudes are observed frequently in natural seismicity. Apart from statistical fluctuations, the observation of characteristic earthquake behavior is probably related to physical conditions in a specific fault region. To account for this case, a corrected value is given by

$$c_V = c_V^0 \cdot \sqrt{\frac{N_{\text{exp}}}{N_{\text{real}}}} \quad (4)$$

for $N_{\text{real}} \geq N_{\text{exp}}$. Here c_V^0 can be calculated from a maximum likelihood estimation of the b value in the scaling regime $[m_0; m_1]$ using Eq. 3. The value N_{exp} is the number of *expected* large earthquakes from the Gutenberg–Richter law $\log N_{\text{exp}} = a - bm_{\text{main}}$ in $[m_0; m_1]$, while N_{real} is the real number of large earthquakes in the catalog. Thus the relation for c_V includes both, information on the large earthquakes and on the intermediate and small earthquakes.

2.2. Bayesian Inference

Bayes' theorem provides an elegant way to update a given a priori probability distribution $p_0(x)$ of a random variable x, if new observations $X_1, \ldots, X_n$ of x become available (see e.g., ZÖLLER *et al.*, 2007). The conditional a posteriori distribution $p_1(x|X_1, \ldots, X_n)$ is then given according to Bayes' theorem by

$$p_1(x|X_1,\ldots,X_n) = \frac{L(X_1,\ldots,X_n|x)}{\int_0^\infty L(X_1,\ldots,X_n|x)p_0(x)\mathrm{d}x} \cdot p_0(x), \quad (5)$$

where L denotes the likelihood function. Further observations $X_{n+1}, \ldots, X_{n+m}$ of x can be accounted for by using $p_1(x|X_1, \ldots, X_n)$ as a new prior and calculating the new posterior $p_2(x|X_1, \ldots, X_{n+m})$ with Eq. 5.

In many practical applications, the main difficulty is to find the likelihood function. However, in this study the calculation of the likelihood function becomes straightforward, if the Gutenberg–Richter law is considered to be true. With the magnitude m as the random variable (x in Eq. 5), the Gutenberg–Richter density is

$$p_\beta(m) = \beta e^{-\beta(m-m_0)} \quad (6)$$

with the abbreviation $\beta = b\ln(10)$ and the minimum magnitude m_0. Here, the Gutenberg–Richter law is only truncated for small magnitudes accounting for limited observational capacities. The additional truncation in the FO model for large magnitudes corresponds to the separation into small and large earthquakes. The likelihood function for n earthquakes with magnitudes $m_1, \ldots, m_n$ is thus

$$L(m_1,\ldots,m_n|\beta) = \prod_i p_\beta(m_i), \quad (7)$$

and can be written as

$$L(m_1,\ldots,m_n|\beta) = \beta^n \exp\left(-n\frac{\beta}{\hat{\beta}}\right) \tag{8}$$

with and $\hat{\beta} = (\langle m\rangle - m_0)^{-1}$, where $\langle m\rangle$ is the mean magnitude in the catalog. Maximizing $L(m_1,\ldots,m_n|\beta)$ with respect to β leads to the formula of AKI (1965),

$$\beta_{ML} = \hat{\beta} = \frac{1}{\langle m\rangle - m_0}. \tag{9}$$

Equations 5 and 7 allow continuous updating of the Richter b value with a growing earthquake catalog. Finally, the probability density function of the b value has to be transformed to those of the coefficient of variation (Eq. 4). Since $c_V^0 = f(b) = \sqrt{b/(3-b)}$ is a monotonic function, the probability density of $c_V^0(b)$ is given by

$$p(c_V^0) = \left|\frac{1}{f'(f^{-1}(c_V^0))}\right| \cdot p(f^{-1}(c_V^0)) \tag{10}$$

leading to

$$p(c_V^0) = \frac{6c_V^0}{\left(1+(c_V^0)^2\right)^2} \cdot p(b(c_V^0)). \tag{11}$$

In order to mimic a realistic situation, we consider c_V as a function of time t by using only catalog information for times $t' \leq t$ from the past.

2.3. Magnitude Uncertainties

Information in earthquake catalogs is usually accomplished by uncertainties. In a first approach these uncertainties may be neglected. However, magnitude errors due to rounding (RHOADES, 1996), systematic errors, and random errors (TINTI and MULARGIA, 1985) can significantly influence the calculation of seismological parameters, e.g., the b value. As a simple illustration, a uniformly distributed magnitude error of $\Delta m = 0.05$ for all earthquakes in a given catalog with n events causes a possible increase of the "true" catalog size by a factor up to 12% ($10^{b\Delta m}$), assuming the Gutenberg–Richter law with $b = 1$. Here we consider errors drawn from a uniform distribution of random numbers in a finite interval centered around the true magnitude. The interval length 0.1, 0.01 etc. account for rounding errors or grouped magnitudes.

In detail, we assume that the probability to observe an earthquake with magnitude m, given that the *true* magnitude is m', is a box function with half-width δ around m':

$$f(m-m') = \begin{cases} \frac{1}{2\delta} & \text{for } |m-m'| < \delta \\ 0 & \text{otherwise.} \end{cases} \tag{12}$$

Despite its simplicity, a box function is a reasonable choice because of the finite width. In contrast, Gaussian distributed uncertainties (TINTI and MULARGIA, 1985), allow for the occurrence of unrealisticly high values. Assuming that the true magnitudes follow a Gutenberg–Richter law, the probability density for the *observed* magnitudes is then given by the convolution of function f with the Gutenberg–Richter density $p_\beta(m)$:

$$\begin{aligned} P_\beta^\delta(m) &= (p_\beta * f)(m) = \int_{-\infty}^{\infty} p(m')f(m-m')dm' \\ &= \frac{\beta e^{-\beta\delta}}{2\delta} \int_{m-\delta}^{m+\delta} e^{-\beta(m'-m_0)}dm'. \end{aligned} \tag{13}$$

It is important to note that the factor $e^{-\beta\delta}$ accounts for the normalization, because the magnitude can take values between $m_0 - \delta$ and ∞ (instead of $m \in [m_0; \infty]$ in the case without uncertainties). This approach assumes a Gutenberg–Richter model, where earthquakes with $m < m_0 - \delta$ are impossible, which is, at least in first order, reasonable. However, in a more refined analysis, the absence of events with $m < m_0 - \delta$ in a catalog will be expressed by a measurement function, while the Gutenberg–Richter model will be valid for all magnitudes. This study is left for future work.

Calculating the integral in Eq. 13 leads to

$$\begin{aligned} P_\beta^\delta(m) &= \frac{\sinh(\beta\delta)}{\delta} e^{-\beta(m-m_0+\delta)} \\ &= \text{sinhc}(\beta\delta)e^{-\beta\delta} \cdot p_\beta(m). \end{aligned} \tag{14}$$

The likelihood function is obtained by replacing the distribution p in Eq. 7:

$$L(m_1,\ldots,m_n|\beta) = \prod_i P_\beta^\delta(m_i). \tag{15}$$

In analogy with Eq. 8, the likelihood can be written as

$$L(m_1,\ldots,m_n|\beta) = \left(\frac{\sinh(\delta\beta)}{\delta}\right)^n e^{-n\beta(\langle m\rangle - m_0+\delta)}. \tag{16}$$

Maximizing $L(m_1,\ldots,m_n|\beta)$ with respect to β leads to

$$\frac{\delta}{\tanh(\delta\beta_{ML})} = \langle m\rangle - m_0 + \delta. \tag{17}$$

A similar formula has been suggested by UTSU (1966) without correcting the minimum magnitude by subtracting δ. This problem has been reported by MARZOCCHI and SANDRI (2003); they suggest the corrected formula for the maximum likelihood estimation $\beta_{ML} = (\langle m\rangle - m_0 + \delta)^{-1}$, which is different from Eq. 17 and has also been used by Utsu himself (UTSU, 1999, 2002). However, for $\delta\beta \ll 1$, the tangens hyperbolicus can be approximated by the linear term of its Taylor expansion:

$$\tanh(\delta\beta_{ML}) \approx \delta\beta_{ML} = \frac{\delta}{\langle m\rangle - m_0 + \delta}; \tag{18}$$

in this approximation, the formula of MARZOCCHI and SANDRI (2003) is in agreement with Eq. 17. For most practical purposes, however, especially for grouped magnitudes with $\delta = 0.05$ or $\delta = 0.005$ and realistic b values ($b \approx 1$), the approximation in Eq. 18 is sufficient, e.g., for an uncorrected value $b_0 = (\langle m\rangle - m_0)^{-1} = 1$ and magnitudes rounded to one decimal place ($\delta = 0.05$), the change of the corrected b value arising from the approximation in Eq. 18 is only $\Delta b_{ML} = 0.003$.

We claim that an (uncorrected) maximum likelihood estimation $\beta_0 = (\langle m\rangle - m_0)^{-1}$ calculated from a catalog with magnitude errors according to Eq. 12 is biased; the corrected value based on Eq. 17 is

$$\beta_{ML} = \frac{1}{\delta}\tanh^{-1}\left(\frac{\delta\beta_0}{1+\beta_0\delta}\right) \approx \frac{\beta_0}{1+\beta_0\delta}. \tag{19}$$

3. Applications

The methodology derived in the previous section is now applied to two data sets: First, an observational earthquake catalog from the Parkfield segment of the San Andreas fault, and second, a simulated earthquake catalog covering about 40,000 years of seismicity evolution. The latter data are particularly useful to study the bias-correction (Eq. 19), because the "true" magnitudes are known.

3.1. The b Value in the Presence of Magnitude Uncertainty

As a first application, we consider an example for the influence of magnitude uncertainty on a maximum likelihood estimation of the Richter b value. The uncertainty is parameterized as in Eq. 12. In catalogs where magnitudes are given with one decimal place, the choice $\delta = 0.05$ in Eq. 12 is an estimate for the rounding error. Apart from rounding errors other sources of uncertainty such as measurement errors, uncertainties in the equations for magnitude determination and other random errors occur. In this section we plug all these errors into a higher value of δ in Eq. 12.

The calculation is carried out for the Parkfield segment of the San Andreas fault in California, where seven M 6 earthquakes occurred between 1857 and 2004 (LANGBEIN *et al.,* 2005). We use the ANSS composite catalog in the time interval between 1987 to present, including 1,041 earthquakes with magnitudes $2.0 \le M \le 5.96$ along the Parkfield segment. As shown in ZÖLLER *et al.* (2008), the cumulative frequency-size distribution suggests the completeness of the catalog in this range of magnitudes. A maximum likelihood fit of the Gutenberg–Richter law provides the b value without accounting for magnitude uncertainties $b = 0.89$.

Figure 1 shows the logarithm of the likelihood function (Eq. 15) depending on $\beta = b\ln(10)$. The solid line is the case without uncertainties ($\delta = 0$), where Eq. 15 reduces to Eq. 7. While magnitudes are usually rounded to one decimal place, the ANSS catalog provides magnitudes with two decimal places; the dashed line refers to this situation by using magnitude errors with $\delta = 0.005$. The dotted line mimics rounded magnitudes with one decimal place ($\delta = 0.05$), and the dash-dotted line the presence of additional uncertainties ($\delta = 0.1$). The curves indicate that b is strongly affected by magnitude uncertainties. Surprisingly, the estimate of b changes by approximately 9% because of rounding errors to one decimal place. The reason for this effect is the decreased lower magnitude threshold m_0 (Eq. 13). The results for the maximum likelihood estimation b_{max} and their errors are provided in the caption of Fig. 1.

3.2. *Simulated Seismicity: Evolution of the Coefficient of Variation with Time*

Ben-Zion and Rice (1993) and Zöller *et al.* (2004, 2005) have presented a numerical model of a 2D fault embedded in a 3D elastic half-space that resembles various characteristics of Parkfield seismicity. We consider a realization of the model

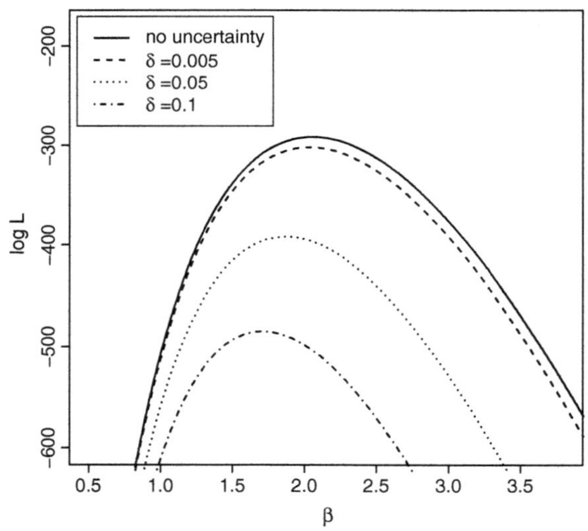

Figure 1
Logarithm of the likelihood function (Eqs. 7 and 15) in dependence of $\beta = b \ln(10)$ calculated for Parkfield seismicity. The parameter δ refers to the magnitude uncertainty (see Eq. 12). The maximum likelihood estimations of the Richter b value are $b_{max} = 0.89 \pm 0.06$ for $\delta = 0$ (*solid line*), $b_{max} = 0.89 \pm 0.06$ for $\delta = 0.005$ (*dashed line*), $b_{max} = 0.81 \pm 0.11$ for $\delta = 0.05$ (*dotted line*), and $b_{max} = 0.75 \pm 0.10$ for $\delta = 0.10$ (*dash-dotted line*). The errors are defined as the half of the interval of b where L has decreased to $0.1 \cdot L(\beta_{max})$.

consisting mainly of constant tectonic loading and co-seismic stress transfer on a smooth fault in accordance with dislocation theory. The corresponding earthquake catalog covers about 40,000 years and includes 536,697 earthquakes with magnitude $M \geq 4.05$. Following Zöller *et al.* (2008). we define large earthquakes by the characteristic earthquake peak ($m \geq 6.75$) leading to 1,279 events. The frequency-size distribution is considered in a scaling regime up to magnitudes $m = 6.3$. This data set is now used to demonstrate the Bayesian approach for updating the estimate of the coefficient of variation c_V. The calculation can be compared with the evolution of c_V calculated directly from the large events. Because this "true" value is unstable for a small number of events, we start the calculations with a learning period of 25 large events (1,000 years) and then continue with time steps of 5 years, using only past data. The coefficient of variation as function of time is shown as a thin dashed line in Fig. 2. Despite the considerable fluctuations, the value becomes more stable due to the increasing number of large events with time. The thick curve denotes the value of c_V calculated from the FO model (Eq. 4) with error bars appearing as the thickness of the line. Here, the error is defined as the half-width of decay of the likelihood function to 10% of the maximum value. It is emphasized that the prediction of the aperiodicity is solely based on small past earthquakes ($4.05 \leq m \leq 6.3$) entering in the a value (in N_{exp}) and the b value in Eq. 4. As an important result we observe that a stable value of c_V close to the "true" value is reached in an early state of the catalog. Recalling that the FO model is based on simple but realistic assumptions and that the large number of small earthquakes stabilizes the calculations results in a feasible estimation of the aperiodicity, even if the number of large earthquakes is small.

Now we introduce magnitude uncertainties according to Eq. 12 with $\delta = 0.05$. Since the true magnitudes in the synthetic data are known, the effect of rounding errors and the feasibility of the bias correction can be studied easily. Therefore, we round the magnitudes to one decimal place and introduce the threshold $m_{min} = 4.1$. Then we calculate the biased estimate of c_V ($\delta = 0$) as function of time (Fig. 3, curve 1). This curve is compared with the true magnitudes above $4.1 - \delta = 4.05$ (Fig. 3, curve 2). Finally the biased values of curve 1 are corrected using Eq. 19 leading to curve 3. Figure 3 demonstrates that only the rounding of the magnitudes increases the maximum likelihood estimate of c_v by a factor of two. Additional instrumental errors will have further impact on the estimates. However, the bias can be corrected efficiently, if the assumption of the Gutenberg–Richter law is fulfilled to some degree. In the case of perfect agreement, curves 2 and 3 should be identical.

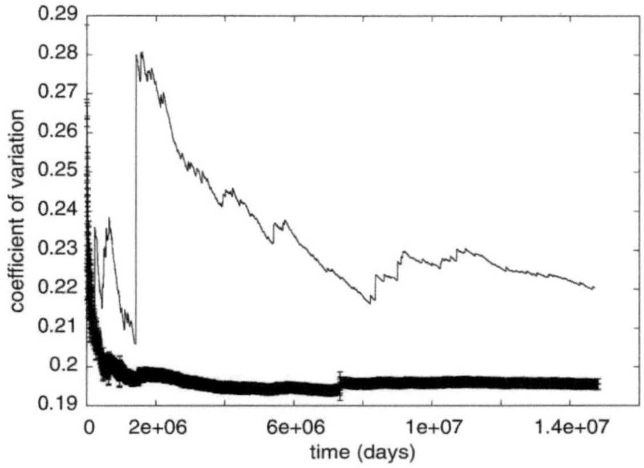

Figure 2
Thin dashed line Coefficient of variation c_V as function of time t for earthquakes with $M \geq 6.75$ in the simulated catalog; only data from the past ($t' < t$) are used. *Thick line* prediction of c_V from the FO model updated with time using the Bayesian approach without magnitude uncertainty ($\delta = 0$). The curves are calculated with a time step of 5 years. The thickness of the line indicates the error as described in the text.

3.3. *Revisiting Parkfield*

The methods developed in the previous sections can now be applied to the Parkfield catalog. This data record, ranging from 1987 to present, includes only one M 6 event, namely the 2004 earthquake. In the magnitude range $m \in [2.0; 5.0]$, the frequency-size distribution follows a power law and includes no characteristic earthquake peak. Therefore we use the basic approach presented in ZÖLLER *et al.* (2008), in which the coefficient of variation is given by Eq. 3; the term $\sqrt{N_{\text{exp}}/N_{\text{real}}}$ in Eq. 4 only corrects for deviations from the scaling.

The yearly update of the aperiodicity seems to saturate in a value between 0.6 and 0.65 (see circles with error bars in Fig. 4). This value predicted by the fractal oscillator model is higher than values found in other studies, e.g., $c_V = 0.44$ in ELLSWORTH *et al.* (1999), or the $c_V = 0.38$ based on the empirical moments of the seven M 6 earthquakes between 1857 and 2004. Although the real value of c_V remains unknown, we argue that the fractal oscillator model includes more information than just the occurrence

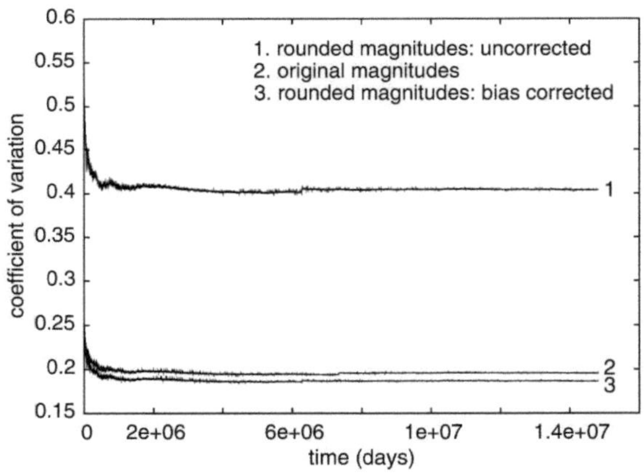

Figure 3
Coefficient of variation calculated from synthetic earthquake catalogs. *Middle curve* c_V from the original catalog (identical with thick line in Fig. 2); *top curve* magnitudes rounded to one decimal place (error-corrupted); *bottom curve* bias correction of curve 1 (Eqs. 19 and 4).

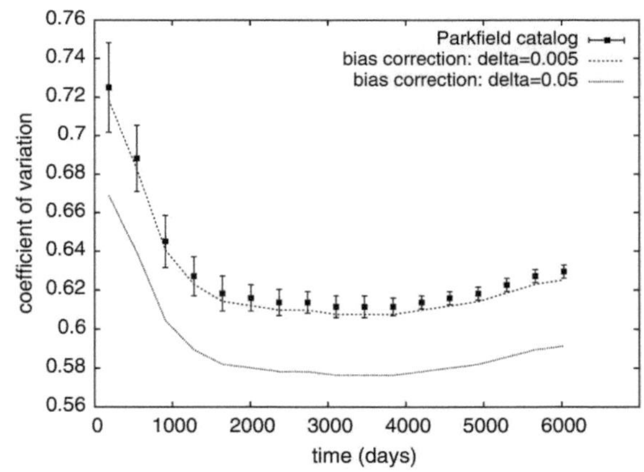

Figure 4
Coefficient of variation calculated for Parkfield seismicity. The circles with *error bars* are based on the ANSS catalog providing magnitudes with two decimal places. *Error bars* are calculated as in Fig. 3. The corresponding bias-corrected values ($\delta = 0.005$ and $\delta = 0.05$) are given by the *lines*.

time of the largest earthquakes, and is thus probably more realistic.

The solid line in Fig. 4 shows the bias-corrected values of c_V calculated from Eq. 19 with $\delta = 0.005$ accounting for the fact that the ANSS catalog provides magnitudes with two decimal places and the dashed line corresponds to $\delta = 0.05$, accounting for additional error sources. Due to this relatively high precision, the bias resulting from the magnitude binning is negligible.

4. Discussion

In this study we have employed the fractal oscillator (FO) model recently proposed by Zöller *et al.* (2008) in order to gain information pertaining to the recurrence time distribution of large earthquakes, especially the aperiodicity or the coefficient of variation. The empirical calculation of this quantity is usually unstable because of the small number of large events. However, the FO model estimates the occurrence time of the next large event from stress changes due to smaller earthquakes and is thus more robust.

The use of Bayes' theorem allows continuous updating the estimation of the aperiodicity, if new information (new earthquakes) becomes available. Using a synthetic earthquake catalog from a numerical fault model, the approach is demonstrated. The results indicate that a stable value of c_V is reached in an early state of seismicity evolution. The calculated value slightly underestimates the empirical value; however, the model prediction provides a reasonably close estimate becoming better with time.

Second, we have shown that magnitude uncertainties usually significantly influence the results. Even the *b* value changes significantly in the presence of relatively small uncertainties, e.g., caused by grouped magnitudes (rounding errors) or observational errors. Here we have studied magnitude uncertainties following a uniform distribution in a finite interval around the true magnitude. In the presence of these errors, the *b* value as well as c_V are biased systematically. We claim that estimations of the Gutenberg–Richter (GR) law (or parameters that are based on GR law) have to be corrected, if observational data are used. Using synthetic seismicity, it is demonstrated that the aperiodicity is biased significantly, if only the effect of magnitude binning is taken into account. However, the bias can be corrected almost perfectly if the rounding error is incorporated explicitely in the likelihood method. Future work will deal with other types of aleatoric uncertainties; especially the treatment of probability densities with infinite long tails is not straightforward due to the required correction of the minimum magnitude in the Gutenberg–Richter model.

In sum, the presented method has been determined to be powerful to extract information from sparse and noisy data. The bridge between renewal models and observational data is build by a Bayesian approach. Instead of dealing with a small number of large earthquakes, more information is gained, because more observational data are taken into account, leading to an increase of robustness and reliability of the results.

Acknowledgements

We thank Editor Matt Gerstenberger and two anonymous reviewers for useful comments that helped enhance the manuscript. G. Z. and S. H. acknowledge support from the German Research Society (DFG HA4789/2-1).

References

Aki, K. (1965), *Maximum likelihood estimation of b in the formula* $\log N = a - bM$ *and its confidence limits*, Bull. Earthquake Res. Inst. Tokyo Univ. *43*, 237–239.

Ben-Zion, Y. and Rice, J. R. (1993), *Earthquake failure sequences along a cellular fault zone in a three-dimensional elastic solid containing asperity and nonasperity regions*, J. Geophys. Res. *98* 14109–14131.

Ellsworth, W., Matthews, M., Nadeau, R., Nishenko, S., Reasenberg, P., and Simpson, R. (1999), *A physically based earthquake recurrence model for estimation of long-term earthquake probabilities*, U.S. Geol. Surv. Open-File Rept. pp. 99–522.

Gutenberg, B. and Richter, C. (1956) *Earthquake magnitude, intensity, energy and acceleration*, Bull. Seismol. Soc. Am. *46*, 105–145.

Langbein, J., Borcherdt, R., Dreger, D., Fletcher, J., Hardebeck, J., Hellweg, M., Ji, C., Johnston, M., Murray, J., Nadeau, R., Rymer, M., and Treiman, J. A. (2005), *Preliminary Report on the*

28 September 2004, M 6.0 Parkfield, Seismol. Res. Lett. *76*, 1–17.

MARZOCCHI, W. and SANDRI, L. (2003), *A review and new insights on the estimation of the b-value and its uncertainty, earthquake*, Ann. Geophys. *46*, 1271–1282.

MATTHEWS, M., ELLSWORTH, W., and REASENBERG, P. (2002), *A Brownian model for recurrent earthquakes*, Bull. Seismol. Soc. Am. *92*, 2233–2250.

REDNER, S., *A Guide to First Passage-time Processes* (Cambridge Univ. Press 2001).

RHOADES, D. A. (1996), *Estimation of the Gutenberg–Richter relation allowing for individual earthquake magnitude uncertainties*, Tectonophys. *258*, 71–83.

TINTI, S. and MULARGIA, F. (1985), *Effects of magnitude uncertainty on estimating the parameters of the Gutenberg–Richter frequency-magnitude law, earthquake*, Bull. Seismol. Soc. Am. *75*, 1681–1697.

UTSU, T. (1966), *A statistical significance test of the difference in b-value between two earthquake groups*, J. Phys. Earth *14*, 34–40.

UTSU, T. (1999), *Representation and analysis of the earthquake size distribution: a historical review and some new approaches*, Pure Appl. Geophys. *155*, 509–535.

UTSU, T., *Statistical features of seismology*. In International Handbook of Earthquake and Engineering Seismology Part A (ed. W. H. K. Lee *et al.*) (Academic Press 2002) pp. 719–732.

WEIBULL, W. (1951), *A statistical distribution of wide applicability*, J. Appl. Mech. *18*, 293–297.

ZÖLLER, G. and HAINZL, S. (2007), *Recurrence time distributions of large earthquakes in a stochastic model for coupled fault systems: the role of fault interaction*, Bull. Seismol. Soc. Am. *97*, 1679–1687, doi:10.1785/0120060262.

ZÖLLER, G., HOLSCHNEIDER, M., and BEN-ZION Y. (2004), *Quasi-static and quasi-dynamic modeling of earthquake failure at intermediate scales*, Pure Appl. Geophys. *161*, 2103–2118, doi: 10.1007/s00024-004-2551-0.

ZÖLLER, G., HAINZL, S., HOLSCHNEIDER, M., and BEN-ZION, Y. (2005), *Aftershocks resulting from creeping sections in a heterogeneous fault*, Geophys. Res. Lett. *32*, L03308, doi:10.1029/2004GL021871.

ZÖLLER, G., BEN-ZION, Y., HOLSCHNEIDER, M., and HAINZL, S. (2007), *Estimating recurrence times and seismic hazard of large earthquakes on an individual fault*, Geophys. J. Int. *170*, 1300–1310, doi:10.1111/j.1365-246X.2007.03480.x.

ZÖLLER, G., HAINZL, S., and HOLSCHNEIDER, M. (2008), *Recurrent large earthquakes in a fault region: What can be inferred from small and intermediate events*? Bull. Seismol. Soc. Am. *98*, 2641–2651, doi:10.1785/0120080146.

(Received August 7, 2008, revised June 18, 2009, accepted June 19, 2009, Published online February 25, 2010)

MIX
Papier aus verantwortungsvollen Quellen
Paper from responsible sources
FSC® C105338

If you have any concerns about our products,
you can contact us on
ProductSafety@springernature.com

In case Publisher is established outside the EU,
the EU authorized representative is:
Springer Nature Customer Service Center GmbH
Europaplatz 3, 69115 Heidelberg, Germany

Printed by Libri Plureos GmbH
in Hamburg, Germany